MULTIDOMAIN OPERATIONS

For policymakers, strategists, and senior military leaders, especially on this side of the Atlantic, an indispensable read to understand the roots, ambitions, limits, dead ends and perspectives of Multi Domain Operations. A necessary wake up call to prevent MDO from being excessive, misinterpreted, hollowed out and disconnected from what is achievable in time and is relevant to dominate the future battle space. An enormous amount of food for thought!

Lieutenant General ret. Alfons Mais,
21st Chief of German Army / 2020-2025

The character of warfare has been evolving rapidly and there is much that we can learn from current conflicts. This excellent book reminds that all wars are sui generis and therefore not all lessons are applicable to what happens next. What most observers would agree is that the Ukraine War has seen a vast leap forward in autonomy (to make up for capability limitations elsewhere) and the integration of vast quantities of data to achieve greater information advantage. This book rightly highlights the potential that the integration of the operational domains has to offer, and it also presents wise advice on how to execute. I commend it to you.

General Sir Nick Carter GCB CBE DSO ADC Gen

Multidomain Operations

The Pursuit of Battlefield Dominance

in the 21st Century

AMOS C. FOX
FRANZ-STEFAN GADY

Howgate Publishing Limited

First published in 2026 by
Howgate Publishing Limited
Station House
50 North Street
Havant
Hampshire
PO9 1QU
Email: info@howgatepublishing.com
Web: www.howgatepublishing.com

British Library Cataloguing-in-Publication Data
A catalogue record for this book is available from the British Library

ISBN 978-1-912440-80-1 (pbk)
ISBN 978-1-912440-81-8 (hbk)
ISBN 978-1-912440-82-5 (ebk – ePUB)

The views expressed in this book are those of the authors and do not necessarily reflect official policy or position.

CONTENTS

Figures and Tables   *vii*
Foreword   *viii*
Abbreviations   *xiv*

Introduction   1
    Amos C. Fox and Franz-Stefan Gady

PART 1: ORIGINS, EVOLUTION, AND MAINSTREAMING

1. The Army in Multidomain Operations: The Intellectual Journey
   of an Organization   10
   J.P. Clark

2. Multidomain Operations : A Bridge Between Industrial and
   Information Age Warfare?   24
   Jesse Skates

PART 2: PRACTICAL CONSIDERATIONS

3. Who Left Brigade Combat Teams Out of Multidomain Operations? 39
   Bill Murray

4. A Deficit of Technique: Addressing Tactical Shortcomings in
   Multidomain Operations Doctrine   61
   Bryan Quinn

5. Without a Why There Is No How: How Bureaucratic Complexity
   Detached MDO from Strategic Reality   80
   Robert Rose

PART 3: TENSIONS WITH CONTEMPORARY CONFLICT AND
FUTURE DEVELOPMENTS

6. The Muddle of Multidomain Operations: Why and How
   the Concept Should Be Reined In   101
   Heather Venable

7. Addressing the Flaws in Multidomain Operations:
   The Pursuit of Dominance in 21st Century Warfare   117
   Amos C. Fox

8. Multidomain Operations As Military Strategy   130
   Jeffrey W. Meiser

9. Attrition, Maneuver, and Lessons from Ukraine for Future
   Warfighting   148
   Michael Kofman and Franz-Stefan Gady

PART 4: INTERNATIONAL AND ALLIANCE PERSPECTIVES

10. The Expanding Role of Ground Forces in Multidomain
    Operations across East Asia and the Indo-Pacific   169
    Andre Luiz Viana Cruz de Carvalho and Sandro Texieira Moita

11. Multidomain Operations with Japanese Characteristics   190
    Rintaro Inoue and Yuka Koshino

12. Warfighting Concepts and the State of MDO Development:
    Does the Emperor Have Any Clothes?   208
    Davis Ellison and Tim Sweijs

13. Multidomain by Design: The Australian Defence Force
    and MDO   225
    Andrew Carr

14. Educating Officers for Joint and Multidomain Operations:
    Is There a Difference?   239
    Thomas Crosbie and Holger Lindhardtsen

15. More Cowbell: Is Multidomain Operations Just
    Another Concept?   260
    William F. Owen

Conclusion
    Amos C. Fox and Franz-Stefan Gady   273

Bibliography   284
Contributing Authors   311
Index   317

FIGURES

Figure 3.1 Zones on the Battlefield   41
Figure 4.1 Military Adaptation Framework   65
Figure 4.2 Range of Doctrine   71
Figure 6.1 Converging Capabilities to Generate Cross-Domain
            Synergy and Layered Options   111
Figure 6.2 Synchronization of disparate planning timelines for
            converged effects   113
Figure 7.1 Scale of Relative Dominance   120
Figure 7.2 Notional Russian Invasion of Europe   126

TABLES

Table 10.1 Role of Ground Forces Across Flash Points   173
Table 10.2 Balance of Firepower, Protection, and Mobility   188
Table 15.1 Doctrine Buzzwords by Year and Publication   271

FOREWORD

The US Army's Multidomain Operations (MDO) concept, like many meta concepts, labors under the burden of many purposes and the risks inherent to defense planning under conditions of uncertainty. Overall, MDO represents a necessary but evolutionary step forward into a complex environment. Having watched the development and subsequent iterations of this initiative, I believe there are three points about the Army's incomplete approach to future crises and wars that should be raised. These include a) its theory of success or victory; b) the lack of substantial recognition of the human and cognitive domain; and c) the resultant challenges to effective leadership and what the US military calls Mission Command.

Theory of Success

Good concepts and good strategies should have an explicit theory of success or hypothesis that explains how and why it leads to desired objectives. As detailed by Tim Sweijs's excellent chapter, many MDO adopters have laid out theories of success that are only implicit at best. Different national military institutions will adopt a theory of success or victory based on their unique culture and geographic position. I agree with Dr Sweij's contention that countries with opaque or nonexistent theories of victory risk being overly vague, which can severely limit their overall utility in force design and force development (acquisition). The aim of any concept is to offer a hypothesis about resolving a key operational challenge, and to pose options to be tested and refined via wargames and experimentation.

Early versions of MDO included language reflecting a theory of victory linked to central aspects of maneuver warfare including the generation of dilemmas and decision paralysis. Several academics and analysts, including contributors to this volume, found fault with that language. These arguments deny or downplay the importance of human will, psychological dislocation, and cognitive influence in favor of the attrition of materiel.[1]

Kinetic action can never be divorced from warfare. At the same time, the human and moral forces in war cannot be eliminated from the conduct of conflict. The allure of technology and attrition in American military history is easy to recognize. The American predisposition to offensive and kinetic operations has long been an element of American strategic and military culture.[2]

Yet, the idea of a sharp distinction between destruction by fires or maneuver, or between physical or psychological effects, reflects a limited historical grasp of conflict.[3] To be sure, maneuver today is challenged by an increasingly transparent battlefield and the proliferation of precision weapons. Yet, the interactive and reciprocal effects of combat operations should not be overlooked. Reducing an adversary's physical capability via attrition is often the prerequisite to generating the confusion or shock that induces a commander to realize that a campaign or battle has been lost. The synergy of combinations of defeat mechanisms, to include cognitive effects, is needed to create the necessary dilemma(s) or perception of futility.[4]

The latest versions of MDO do not provide clarity on this debate. But as the contribution by Sweijs and Allison shows, a causal argument as to how a new concept will actually lead to a desired result in battle is central to development efforts in terms of organizational change, desired equipment features, doctrine, and education.

---

1 For credible arguments on these points see Heather Venable, "Paralysis in Peer Conflict? The Material Versus the Mental in 100 Years of Military Thinking," *War on the Rocks*, December 1, 2020; Franz-Stefan Gady, "Manoeuvre Versus Attrition in US Military Operations," *Survival* 63, no. 4 (2021): 131-148; Michael Kofman, "A Bad Romance: US Operational Concepts Need to Ditch their Love Affair with Cognitive Paralysis and Make Peace with Attrition," Modern Warfare Institute, March 31, 2021.
2 For insights on both strategic and military culture see Peter Mansoor and Williamson Murray, eds., *The Culture of Military Organizations* (Cambridge, UK: Cambridge University Press, 2019).
3 For a brief defense of and update to Maneuver Warfare see George P. Garrett and Frank Hoffman, "Maneuver Warfare is Not Dead, But it Needs to Evolve," *Proceedings* 149, no. 11 (November 2023).
4 See my take on defeat mechanisms, Frank G. Hoffman, "Defeat Mechanisms in Modern Warfare," *Parameters* s 51, no. 4 (Winter, 2021): 49-66.

The Human Domain

One of the early comments about the nascent MDO concept was what Colonel Todd Schmidt called "the missing domain."[5] The notion of a human or cognitive domain in warfare faces a steep uphill fight for acceptance within the US military establishment. MDO does mention the need to "maximize human potential," and notes that resilient leaders and Soldiers will require training, educating, equipping, and support in order to execute Multidomain Operations. But it seems to make this peripheral rather than central to the solution. This is because the very notion of a human domain and dimension runs counter to prevailing values in the US strategic and military cultures which privilege kinetic actions in the physical realm over more unconventional approaches. As Colin Gray noted long ago, US military culture is thoroughly conventional, firepower centric, and technologically oriented.[6] Despite its own military experience, and rhetorical statements about the primacy of people over technology, US military doctrine remains generally hostile to the existence of a human or cognitive domain.[7]

There are whispers of growing awareness, however. The current Joint Warfighting Concept includes a tenet for "Expanded Maneuver," which urges the Joint Force to think "creatively about moving through space and time, including—but not limited to—maneuver through land, sea, air, space, cyber, the electromagnetic spectrum, information space, *and the cognitive realm*" (emphasis added).[8] This suggests a growing appreciation of the cognitive domain inside the Joint Staff. Further evidence can be gleaned from a speech to a special forces audience delivered by the new Joint Chiefs of Staff Chairman, General Dan Caine, in which, he observed that in order to be successful in this era, "We have to find every ounce of combat capability, *every ounce of cognitive effect* that we can both inside our forces with our allies and partners and the interagency…[9] (emphasis added).

---

5 Todd Schmidt, "The Missing Domain of War: Achieving Cognitive Overmatch on Tomorrow's Battlefield," Modern War Institute, April 7, 2020.

6 Colin Gray, "Irregular Enemies and the Essence of Strategy: Can the American Way of War Adapt?" Strategic Studies Institute, Carlisle Barracks, PA, 2014.

7 See Frank Hoffman and Michael Davies, "Joint Force 2020 and the Human Domain," *Small Wars Journal*, June 10, 2013, 1–13.

8 General Mark A. Milley, "Strategic Inflection Point: The Most Historically Significant and Fundamental Change in the Character of War Is Happening Now," *Joint Force Quarterly* 110 (3rd Quarter, 2023.

9 Quoted in Andrew White, "Joint Chiefs Chairman Caine has an 'algorithm' for US 'winning,'" *Breaking Defense*, May 8, 2025, https://breakingdefense.com/2025/05/joint-chiefs-chairman-caine-has-an-algorithm-for-us-winning/.

We lack a holistic conception of how the many forms and modes of conflict, including cyber tools and neurological science, can be applied by a cunning opponent who is prepared to orchestrate combined arms by blending military and non-military tools, or what are often labeled as kinetic and non-kinetic instruments, in unconventional ways.[10] Our principal competitor does not overlook the potential in their conception of Systems Confrontation or cognitive warfare.[11] In our major operating concepts we should endeavor to recognize how our adversaries think and how they are prepared to "fight" in both the physical and the cognitive domains. Unfortunately, one finds little of that in the Army's MDO doctrine, which orients on penetrating adversary anti-access systems. That is a legitimate operational challenge, but the Chinese also focus on penetrating the communications and decision-making processes of our leaders and operational commands. Prevailing against what the PLA calls Cognitive Domain Operations will require counter-cognitive warfare preparations in addition to serious investments in leader development, training systems, human performance enhancements, and professional education. Yet one finds only brief nods to human factors and the intellectual preparation needed to remain competitive in the coming age in MDO debates.

War, even when executed by swarms of autonomous machines in the metaverse, will remain a very human endeavor. The preparation of tomorrow's leaders for the intellectual, moral, and ethical challenges of war in the coming era mandates renewed investments in rigorous education and experiential development to hone a competitive edge that can be applied under the stress and uncertainty of multidomain combat.[12] If we only focus on materiel and technology, any efforts to enhance our combat effectiveness against a large-scale competitor will be incomplete.. Fortunately, several chapters in this volume are dedicated to leadership development and education.

---

10 I use the latter terms guardedly, see Lawrence Freedman, "The Language of War," Substack, March 3, 2023, https://samf.substack.com/p/the-language-of-war.

11 Jeffrey Engstrom, *Systems Confrontation and Systems Destruction Warfare* (Santa Monica, CA: RAND, 2018); Nathan Beauchamp-Mustafaga, "Cognitive Domain Operations: The PLA's New Holistic Concept for Influence Operations," Jamestown Foundation *China Brief*, 19, no.16, September 6, 2019.

12 On the human and moral dimensions of contemporary war, see Mick Ryan, *War Transformed: The Future of Twenty-First-Century Great Power Competition and Conflict* (Annapolis, MD: Naval Institute Press, 2022), 165–208.

Mission Command

Professionals on both sides of the Atlantic have endorsed the adoption of mission command as essential to success in the projected operating environment. Under Mission Command, leadership decisions are distributed throughout the breadth and depth of the battlespace, framed by the higher commander's intent and a shared understanding of the context. Its successful application depends not only on delegating decision-making authority to the lowest echelons feasible, but on the disciplined initiative by many subordinates.

The US Army has long promoted this theory as has the Marine Corps.[13] The UK adopted the concept in 1989, and various NATO allies have as well.[14] It is now a foundational part of Joint doctrine.[15]

Though Mission Command is now doctrine, it is not without its critics, particularly in the US Army.[16] Some claim that it has not taken root in the US Army operational culture, while others have challenged its utility.[17]

One principal drawback claimed is that mission command seeks to impose a singular, universal leadership or command style for all modes of war, in all operational contexts. These critics contend that situational understanding should be the decisive factor in guiding a commander's approach to his/her selection of command and control modes.[18] They argue that Army should embrace a continuum of styles and offer guidance for commanders to constantly assess the operational context they are operating within their experience levels and confidence, their unit's proficiency, the experience and judgment of subordinate leaders, the complexity and ambiguity of the environment, and the character of the opposition when

---

13 The Army's official doctrine is at ADP 6-0, *Mission Command: Command and Control of Army Forces,* Headquarters, USArmy, July 2019.

14 Niklas Nilsson, "Practicing mission command for future battlefield challenges: the case of the Swedish army," 20, no. 4 *Defence Studies* (2020): 436-452.

15 Initiated by the Chairman of the Joint Chiefs of Staff, General Martin Dempsey, USArmy, Mission Command White Paper, Joint Chiefs of Staff, April 3, 2012; culminating with Joint Publication 1-0. *Joint Warfighting,* Washington, DC, Joint Chiefs of Staff (2023), III-1.

16 Amos Fox, "Cutting our Feet to Fit Our Shoes, An Analysis of Mission Command in the US Army," *Military Review* (January-February 2017): 49-57.

17 Stephen J. Townsend, D. Crissman, D., J. C. Slider, and Keith Nightingale, "Reinvigorating the Army's approach to command and control," Military Review (September/October 2019): 1-9, https://www.armyupress.army.mil/Journals/Military-Review/Online-Exclusive/2019-OLE/July/Townsend-Reinvigorating/.

18 Andrew Hill and Heath Niemi, "The Trouble with Mission Command: Flexive Command and the Future of Command and Control," 86 *Joint Force Quarterly* (4th Quarter 2017): 98.

determining which approach to employ.[19] This would appear to be a very obvious solution.

Others find that MDO itself makes Mission Command problematic due to the centralizing tendency inherent to the technologies being employed.[20] The authorities to employ weapons and high-end systems may never be delegated to lower-level commanders. These commanders may not have the requisite skills to effectively command and employ assets that cross domains, and thus not be able to take the initiative in a meaningful manner.

Another concern lies in the application of Mission Command to coalition warfare or with interagency task forces. Since Mission Command is predicated on cohesion, trust, and mutual understanding built up over time, the concept could be impossible for ad hoc coalitions or interagency task force to apply. Trust, confidence, and cohesion require time, and pick up teams rarely win championships in sports or wars. Effective command and control in the 21st century require a more adaptive approach to command styles than mission command presently offers.

This anthology, carefully curated by a pair of independent and critical thinkers, explores the MDO concept thoroughly and exposes its potential limitations. The contributors offer sober and scholarly assessments, and the chapters from international partners provide an invaluable mirror to how our efforts to posture the US Army for success in tomorrow's crises look from overseas. Hopefully, this book will impel debate and progress. Most importantly, it offers numerous ideas to ensure that the emperor's proverbial wardrobe is fit for purpose.

Frank Hoffman
Fairfax Station, Virginia
December 15, 2025

---

19 Fox, "Cutting Our Shoes," 56.
20 Conrad Crane, "Mission Command and Multidomain Battle Don't Mix," *War on the Rocks*, August 23, 2017.

ABBREVIATIONS

A2AD	Anti-Access and Area-Denial
ACC	Air Combat Command (US Air Force)
ACE	Agile Combat Employment
AFC	Army Futures Command (US Army)
ALB	AirLand Battle
ARCIC	Army Capabilities and Integration Center
CDLD	Concept Development and Learning Division
CJADC2	Combined Joint All-Domain Operations
COFM	Correlation of Forces and Means
COG	Center of Gravity
COIN	Counterinsurgency
CRC	Concept-Required Capability
DA	Department of the Army (United States)
DMO	Distributed Maritime Operations
DOD	Department of Defense (United States)
DOTMLPF-P	Doctrine, Organization, Training, Materiel, Education and Leadership, Personnel, and Policy
DOTMLPF-I	Doctrine, Organization, Training, Materiel, Education and Leadership, Personnel, and Interoperability
EABO	Expeditionary Advanced Base Operations
FCS	Future Combat Systems
FORSCOM	Army Forces Command (US Army)
IHL	International Humanitarian Law
ISIS	Islamic State in Iraq and Syria
JACD	Joint and Army Concepts Division
JADC2	Joint All-Domain Operations
LOCE	Littoral Operations in a Contested Environment
LRF	Long-Range Fires
LRPF	Long-Range Precision Fires

MCCDC	Marine Corps Combat Development Command (US Marine Corps)
MDB	Multidomain Battle (also, Multi-Domain Battle)
MDO	Multidomain Operations
US	United States
UK	United Kingdom
NATO	North Atlantic Treaty Organization
OIF	Operation Iraqi Freedom
OIR	Operation Inherent Resolve
PGM	Precision Guided Munition
PRC	People's Republic of China
RMA	Revolution in Military Affairs
THAAD	Terminal High Altitude Area Defense
TRADOC	Training and Doctrine Command
UAS	Unmanned Aerial System
UAV	Unmanned Aerial Vehicle

Introduction

Amos C. Fox and Franz-Stefan Gady

In *Multidomain Operations: The Pursuit of Battlefield Dominance in the 21st Century* we seek to answer an important question – is Multidomain Operations, or MDO, the proper military doctrine for the 21st century? If so, why is that the case; and if not, what must be done to either discard the doctrine, or adapt it to make it a worthwhile idea upon which to frame 21st century armed conflict.

To address these questions, *Multidomain Operations: The Pursuit of Battlefield Dominance in the 21st Century* examines that puzzle through four lenses – MDOs origins, practical considerations from the field, analysis from academia and professional military education, and international (i.e., non-American) perspectives on the subject. In doing so, *Multidomain Operations* comes to a handful of conclusions that seek to influence those who make the real decisions that drive doctrine, organizations, training, materiel, leadership and education, personnel, facilities, policy, and interoperability (DOTMLPF-P/I).

In Part I of the book, we briefly examine the origins of MDO through the lens of two individuals that helped craft the concept and the doctrine – Dr. J.P. Clark and Lieutenant Colonel (Retired) Jesse Skates. In Chapters 1 and 2, Clark and Skates outline how the US Army framed what became MDO in relation to a Russian threat to NATO and the Baltic countries. The framing of the product of three interconnected factors. The first of those factors was Russia's annexation of Crimea and seizure of Ukraine's Donbas region in 2014 - 2015. The second factor included how to account for the emerging eminence of anti-access and area denial (A2AD) systems. The third factor included how to maximize the US Army's (and joint force) advancement in long-range fires, precision strike capabilities, and networked sensors and command-and-control (C2) systems. Together, these ideas formed what became known as Multidomain Battle (MBD). Clark and Skates highlights that changes in the philosophy guiding MDO evolved as Army leaders driving the process changed and as concept developers move out of, and into concept development positions in the Army Capabilities Integration Center (ARCIC).

Clark and Skates describe how senior leader dialogue, war-gaming, and geostrategic considerations resulted in calibrated force posture, converging capabilities, denial of *fait accompli*, and the employment of resilient formations became some of the foundational ideas with the MBD, the MDO concept, and later, MDO doctrine.

In the end, however, Clark and Skates illustrate that what sought to be a revolutionary concept, became little more than the next bureaucratic modernization concept. MDO was rushed from idea, to concept, to doctrine in record time, with insufficient research to support many of its basic arguments, outcomes, and required capabilities. In the end, MDO appears to be yet another feature in a long line of Army operating concepts (AOC) and doctrines that align with Department of Defense's five-year budget planning, or POM (Program Objective Memorandum) cycle.

Part II of *Multidomain Operations*, military practitioners from the US Army examine a range challenge and practical considerations associated with MDO. In Chapter 3, William Murray, a Lieutenant Colonel in the US Army, highlights that the US Army's prime combat formation, the brigade combat team, or BCT, is all but absent from MDO doctrine, if not in name, certainly in practice. In FM 3-0 *Operations*, the BCT's contribution to MDO is all but overlooked. FM 3-0 places a premium on continuity for BCTs, while remodeling how Army divisions, corps, and theater armies fight. Yet, at the same time, as the Army has moved forward under the banner of MDO, it has continued to cut force structure within the BCTs to free up the funds, personnel, and force structure to finish building out its plan for Multidomain Task Forces (MDTFs), drones, and other tech-centric force structure ideas. Accordingly, Murray argues that this has left the BCT nearly incapable of the most basic combat functions, much less anything that reduces duration or complexity. Thus, Murray argues, the Army (and its allies and partners) mustn't forget the BCTs as they continue to develop ways to operate, organize, and equip for 21st century wars.

In Chapter 4, Bryan Quinn, a Major in the US Army, looks at MDO from an institutional development and implementation standpoint; that is, how it transitioned from concept, to doctrine, and how materiel solutions were implemented to support it. Although approached from very different directions, Quinn's assessment is similar to Murray's. Quinn finds that though the US Army developed a theory of victory for a specific military problem, the Army has failed to develop accompanying tactical doctrine, tactics, and warfighting systems at the lowest tactical level. As a result, the US Army actually lacks the ability to implement MDO across the US Army's tactical level—from divisions to platoons—because of the tactical level's neglect.

In Chapter 5, Robert Rose, also a Major in the US Army, argues that MDO lacks a coherent theory of victory, supported by tactical clarity and contextual grounding. Instead, as the concept drifted from being focused on the threat of Russian aggression against NATO and the Baltic states to being a technology-integration springboard, MDO lost its potential gravitas. Rose refers to the success of the German and Israeli militaries successful development of threat-based doctrines prior to World War II and the Arab-Israeli War, respectively; while also illuminating France's failures to effectively innovate in the interwar period between 1918 – 1939.

Part III consists of insights provided by scholars and analysts from across professional military education (PME), academia, and think tanks. Dr Heather Venables, a professor at the US Air Force's Air Command and Staff College argues that though MDO was developed to help solve the problem of A2AD and maintaining air superiority, it has since morphed into a jargon-laden framework that is all-encompassing, and which lacks coherence and any semblance of clarity. Unlike good doctrine, which clarifies and provides explicit guidance, MDO has fueled uncertainty among the force, yet paradoxically is little more than a recasting of joint operations and combined arms warfare. Venable suggests rather that the US military, and its allies and partners, need to reform their doctrine such that it is narrowly focused on a threat and/or an operational challenge. To be sure, defeating advanced air defense systems and preserving combat power for close combat is a great place to start, instead of seeking to do everything, everywhere, with everything.

In Chapter 7, Dr Amos Fox, of Arizona State University, outlines two related shortfalls in MDO doctrine—its use of dominance and the reliance on precision guided munitions. Fox asserts that both the MDO concept and doctrine place a premium on multidomain operation's ability to generate battlefield dominance through the sophisticated application of precision-guided munitions. Yet, neither the MDO concept nor the doctrine examine how difficult it actually is to generate, much less maintain dominance. Likewise, with those considerations in mind, what the true magnitude of costs are both in terms of munitions (precision, long-range fires, and otherwise) and in actual dollars. To address this problem, Fox provides recommendations to better operationalize MDO, to include thinking in terms of, and applying a concept he refers to as *zones of proximal dominance*.

Next, in Chapter 8, Dr Jeffrey Meiser, professor in the University of Portland's Department of Political Science and Global Affairs, discusses MDO through the lens of strategy. To that end, Meiser contends that MDO is the US military's strategy for competing with and defeating near-peer competitors like Russia and China. Meiser's framework focuses on three theories to examine MDO—theory of the challenge, theory of success, and theory of failure. Meiser's use of this framework of strategic analysis finds that MDO is insufficient because of its emphasis on stand-off warfare, precision strike, and converging technologies, at the expense of a coherent balance of ends, ways, means, and risk in regard to operational challenges. To rectify these challenges, Meiser calls for reforms in MDO to incorporate what he calls *windows of convergence*, which closely matches Fox's zones of proximal dominance.

Military analysts Franz-Stefan Gady and Michael Kofman address additional challenges with MDO in Chapter 9. Citing their field reports and lessons learn from their battlefield observations in Ukraine, Gady and Kofman outline how MDO insufficiently addresses war's attritional character. To be sure, MDO's faith in stand-off warfare, long-range strike, and perceived efficacy of precision strike is misaligned with empirical findings from large-scale combat operations in Ukraine. Gady and Kofman note how MDO's basic tenets, nor US and Western preference for maneuver warfare exist only on a fleeting period of warfare, and the battlefield quickly becomes consumed with positional and attritional warfighting. As a result, MDO should be revisited to incorporate many of the findings from Ukraine's battlefield to better account for threat- and challenge-oriented realities of warfare.

The book concludes with Part IV, which is a collection of insights, reflections and recommendations from individuals from across US allied and partner states. Part IV begins with a chapter by André Luiz Viana Cruz de Carvalho and Sandro Teixeira Moita from Brazil. Carvalho and Moita look at MDO through the lens of defeating A2AD and countering China aggression in the South China Sea and other regional flashpoints in the Indo-Pacific. The authors indicate that success in this region depends on the smart application of MDTFs, while ensuring the balance between firepower, protection, and mobility throughout a vast and dispersed environment dominated by water and disparate island chains.

In chapter 11, Yuka Koshino and Rintaro Inoue discuss MDO through the lens of Japanese regional strategy. Koshino and Inoue examine cross-domain operations (CDO), Japan's spin on MDO philosophy and practice, and how the Japanese concept's incorporation of unmanned systems, AI-enabled command-and-control networks, and long-range precision fires remains underdeveloped. However, to counterbalance this lack of technology infrastructure, they posit that a focus on dislocation and disruption during the competition and crisis phases, or what they stylize as temporal warfare, can help Japan compete against regional in the Indo-Pacific.

Davis Ellison and Tim Sweijs, from the Hague Center for Strategic Studies, argue in Chapter 12 that MDO lacks clarity, coherence, technological feasibility, and a good theory of success. Ellison and Sweijs highlight how MDO is just the post-9/11 period's network-centric 2.0, coming with many of the same foibles are its predecessors to include Rapid Dominance (or Shock and Awe) and Effects-Based Operations (EBO). To improve upon concept and doctrine development, they provide six fundamentals. First, good concepts and doctrines must provide clarity. Second, they must fit the actor for whom it supports. Third, concepts and doctrines must match technological development with the time in which the concept and doctrine will exist. Fourth, good concepts and doctrines must be informed by threats and real operational challenges. Fifth, they must have an appropriate theory of success. And finally, good concepts and doctrines must be informed by risk. Accordingly, Ellison and Sweijs find MDO wanting in nearly each of these six fundamentals. To be sure, they find that MDO over-emphasizes the US and its allies and partners' ability to maintain connectivity in an increasingly interconnected battlespace, as well as the ability to operate at seemingly dizzying speeds that will leave adversaries stupefied by their operational and tactical prowess. Ellison and Sweijs contend that MDO is quicker encroaching on becoming a rhetorical façade and concept and doctrine of theoretical proportion. They conclude by calling for the US and its allies and partners to return to concepts and doctrines bound by their six fundamentals of sound military thought and tried and true fundamentals of joint operations.

In Chapter 13, the Australian National University's Dr Andrew Carr discusses how Australia has approached MDO, and how that approach fits with the US's approach to Indo-Pacific threats. Carr notes how MDO is helping the Australian Defense Force (ADF) address its own regional challenges, to include Chinese deterrence and gray-zone threats. Moreover, he also argues that Australia serves as an important ally with the US via MDO due to its strategic location in the Indo-Pacific, helping serve as a logistics bastion within the region. In addition, in the event that China seeks to annex Taiwan, Australia would play a critical MDO counterpunch against such a threat. In highlighting these points, Carr records a handful of alliance challenges that remain glaring, to include the need for the US, Australia, and other allies and partners to improve interoperability and intelligence sharing. Nevertheless, Carr makes it clear that Australia is a critical component in a globally networked MDO confederation against Indo-Pacific threats emanating from China.

In Chapter 14, professors Thomas Crosbie and Holger Lindhardtsen, from the Royal Danish Defense College, examine MDO, and joint-all domain operations (JADO) to see what, if anything, is different from the existing concept of joint operations. Those take this approach to understand how the introduction of MDO impacts PME. They find that very little differs between MDO, JADO, and joint operations; the differences being a new (and perhaps embellished) lexicon, and the further integration of emerging domains, information and communication systems, and warfighting technology. However, this requires PME to shift from thinking and teaching in terms of service contributions to MDO and joint operations, but rather to prepare all service members to think with a cross-domain mindset and be capable of applying multidomain capabilities, regardless of their own branch of service.

We conclude the book with William F. Owen's incisive commentary in Chapter 15. Owen posits that MDO is little more than joint operations and combined arms warfare with a set of new taglines, buzzwords, and an even more confusing and inarticulate way of applying combat power on the battlefield. Furthermore, he contends that MDO is little more than the heightened application of long-range strike, at the expense of close combat forces, or as he puts it, MDO is just "more cowbell". What's more, Owen comes to similar findings as Amos Fox, Franz-Stefan Gady, and Michael Kofman. To be sure, he describes the doctrine as a strike-centric approach to warfare that further erodes the importance and relevance that maneuver warfare plays in 21st century wars and accentuates the ever-increasing relevance of attritional warfare.

The goal is that the range of topics and guest authors forces the US and its allies and partners, to include institutions like NATO, to take pause and truly examine MDO. Collectively, this book represents a pause of its own; one done by a community of scholars, military officers, and analysts who are concerned about the direction MDO is taking Western militaries and in how concepts and doctrine is developed. This book, a point of reflection, yields four essential findings.

First, MDO—the concept and associated doctrine—were rushed through the development process, and thus not fully ready for implementation when the US Army formally adopted it as doctrine. Second, amongst some practitioners in the field, MDO provides insufficient detail and consideration to the tactical edge of warfighting doctrine. Third, the US military, NATO, and other partners have adopted MDO as a premature conclusion to the challenges found along the continuum of conflict without consideration to the question of military strategy, operations, and tactical warfighting. Put another way, MDO is a solution looking for a question. Thus, the concept (in general) and the doctrine (in particular) are ripe for a revisit in order to amend the forgone conclusions baked into MDO. Fourth, the concept is not ideally suited to the US's international partners. This is for a variety of reasons, to include MDO being culturally incompatible, financially prohibitive, and not much different from the existing doctrine of joint operations. Lastly, MDO is a technology-focused concept and doctrine, instead of threat-focused. As a result, the development and application of technology takes priority over the need for the US and its allies and partners to possess a realistic operating concept and doctrine fit to address germane, and not imagined, military challenges.

As you will find, *Multidomain Operations* provides a set of reasonable and balanced recommendations that address policy, strategy, tactical considerations, and PME. First, the US and its allies and partners mustn't sell off their close combat forces for the sake of technological investment and the unproven strategy of stand-off warfare.

Second, crafting military concepts and doctrine that exceed the relative simplicity of the confines of a single state is challenging. This is especially difficult when the allies and coalition partners are vastly different in terms of their physical size, industrial base, economic power, and cultural. Thus, concepts and doctrine at the policy level (i.e., of alliances and coalitions) must appropriately account for those variables and not just limit partners on the lower ends of power to the position of "plug-in" partners.

Third, MDO must be reconsidered with deference to the competitive nature of strategy, appreciating that other states and non-state actors will not play the part of willing participant on future battlefields. Likewise, MDO must lessen its faith in the unproven belief that precision fires and long-range precision strike will unlock the path to quick and less destructive wars. Moreover, MDO must account for the fact that precision fires, long-range precision strike, and force structure modifications such as the Multi-Domain Task Force (or MDTF) will lessen the requirement for rugged and robust land forces in future 21st century conflicts. Likewise, maneuver warfare will likely maintain waning relevance on battlefields dominated by the application of precision strike and long-range fires, and consequently positional and attritional warfare will become increasingly important ways of warfare in the 21st century. Thus, MDO must be updated to account for the long, deadly, and destructive nature of war and the continuous need for large land forces to consolidate battlefield gains in pursuit of a state's strategic objectives.

Fourth, MDO must be revised to provide the soldiers on the ground with a tangible applied doctrine that they can use to help them solve tactical problems. Lastly, MDO must be something that makes sense in PME. That is, the concept and the doctrine must be logically sound and pass the tests of verifiable causality so that instructors charged with teaching the concept, as well as the students in classrooms across the globe, have faith in the ideas in which they are engaged.

The hope is that *Multidomain Operations: The Pursuit of Battlefield Dominance in the 21st Century* provides an invaluable contribution to contemporary war studies. The book fills a significant gap in the field of war studies by taking a hard look at MDO and providing relevant feedback to those who make decisions along the spectrum of DOTMLPF-P/I.

PART 1

ORIGINS, EVOLUTION, AND MAINSTREAMING

1

The Army in Multidomain Operations:
The Intellectual Journey of an Organization

J.P. Clark[1]

The US Army's 2018 operating concept, *The Army in Multi-Domain Operations, 2028,* was—as all such documents are—the product of a specific organizational and operational context.[2] In the 2010s, just as the army was attempting to anticipate what challenges it would face beyond Iraq and Afghanistan, the Russian seizure of Crimea and invasion of eastern Ukraine in 2014 provided an amenable focal point for organizational attention. That threat also provided a useful vehicle for discussing the Army's contribution to the larger joint effort, as the major problem then consuming much of the US Air Force and US Navy concepts work was how to overcome various forms of anti-access/area denial systems.

Although *Multidomain Operations* (MDO) is certainly representative of the Army of the late 2010s, its final form was hardly foreordained. It might easily have been assembled in a quite different way but for the contingency of various decisions and events from its initial framing to various steps along the way. This brief account of that journey serves two purposes. For those who

---

1 Disclaimer: The views expressed are those of the author and do not necessarily reflect the official policy or position of the Department of the Army, Department of Defense, or the U.S. Government.
2 U.S. Army Training and Doctrine Command (TRADOC), *The U.S. Army in Multi-Domain Operations, 2028,* TRADOC Pamphlet 525-3-1 (6 December 2018).

seek to better understand MDO the backstory provides additional context to illuminate the ideas as well as to explain its weaknesses and blind spots. There is also utility in shedding some light on how the military produces futures concepts. A trend of the last decade or so has been for senior leaders to increasingly tout concepts, such as the *Joint Warfighting Concept*, as essential to future success. How these documents are produced, however, remains a mystery to most of the joint force.

Origins

The direct lineage of MDO begins with a query in 2015 from the Army Staff to Lieutenant General H.R. McMaster, then serving as the director of the US Army Capabilities and Integration Center (ARCIC). The Pentagon wanted his thoughts on how the Army could defeat a Russian invasion of the Baltics in the year 2030.[3] At the time, ARCIC was the Army's principal entity responsible for coordinating future force development, so setting the problem 15 years into the future was not as unusual as it might seem.[4]

The directed focus on Russia was also not a surprise. Since the seizure of Crimea and eastern Ukraine the year prior, the Russian threat loomed large for the entire US national security community. At about the same time as the e-mail to McMaster, the RAND Corporation was wargaming a Russian attack on NATO's Baltic members that when published the following year drew widespread attention; Fareed Zakaria even mentioned it on his weekly CNN commentary.[5] Within the Army, there were a number of other efforts also looking at this problem, mostly conducted by US Army Training and Doctrine Command (TRADOC)—ARCIC's parent organization. Throughout the mid- and late-2010s, various organizations within TRADOC conducted a number of wargames, studies, and seminars examining Russia as the primary, if not always the exclusive, threat. One of these concurrent efforts was a series of wargames conducted by the Maneuver Center of Excellence at Fort Benning, Georgia under then-Major General Eric Wesley. Those wargames provide a significant amount of analysis that was passed on to ARCIC. They also helped shape the thinking of Wesley, who would eventually serve as the next director of ARCIC after McMaster. Another influential effort was

---

3 Charles Hornick, e-mail to author, 4 December 2024.
4 With the creation of Army Futures Command, ARCIC was reorganized into the Futures and Concept Center.
5 David Shlapak and Michael W. Johnson, *Reinforcing Deterrence on NATO's Eastern Flank: Wargaming the Defense of the Baltics* (Santa Monica, CA: RAND Corporation, 2016).

the *Russian New Generation Warfare* study, an effort initiated by Army senior leaders at about the same time as the e-mail to McMaster.[6] The final report was classified, but the unclassified summary warned that the US Army's ability to close with and destroy the enemy was at risk.[7] Although the scope of MDO would end up expanding over time, origins are important; the initial framing for any conceptual project sets the path. Certainly, for *Multidomain Operations*, the early focal points of Russia as the threat and the closing with the enemy as the operational problem are evident in the final document.

Concerns about what came to be known as "stand-off" within MDO were hardly limited to the Army. For several years, the US Air Force and US Navy had become increasingly preoccupied with the challenge of so-called anti-access/area denial (A2AD) systems explicitly designed to limit, if not completely neutralize, the air and naval forces that have long been the strength of the US military. In 2016, Chief of Naval Operations John Richardson went so far as to decree that his service should "scale down the mention of A2AD," a comment that was widely interpreted as a ban of the term.[8] Rather than an ostrich-like denial of the problem as some assumed, Richardson's comment was actually a call to action for the Navy to seek ways to challenge long-range anti-ship missiles. The Air Force was equally concerned by its equivalent, layered integrated air defense systems that included long-range surface-to-air missiles like the Russian S-400.

Within this context, McMaster saw the opportunity to address both service and joint concerns; indeed, they were actually the same. Just as fish may fail to appreciate water, so too, many soldiers often overlook how deeply enmeshed decades of air supremacy had become within US Army force structure and doctrine. The Air Force's A2AD problem was also the Army's. Fortunately, it was beginning to develop at that time new long-range missiles that could be used against an array of targets, including enemy surface-to-air missile launchers or radars. If so, that might create a window (or "bubble") of opportunity that could be exploited by air and ground forces. McMaster directed Lieutenant Colonel Charlie Hornick—chief of the Joint Concepts

---

6   Wilson C. Blythe, Jr, interview, 3 December 2024, Carlisle, Pennsylvania.

7   Peter Jones, Ricky Waddell, Wilson C. Blythe, Jr., and Thomas Pappas, *Unclassified Summary of the U.S. Army Training and Doctrine Command Russian New Generation Warfare Study* (Ft. Eustis, VA: TRADOC, 2016).

8   Christopher P. Cavas, "CNO Bans 'A2AD' as Jargon," *Defense News*, 3 October 2016, https://www.defensenews.com/naval/2016/10/04/cno-bans-a2ad-as-jargon/; B.J. Armstrong, "The Shadow of Air-Sea Battle and the Sinking of A2AD," *War on the Rocks*, 5 October 2016, https://warontherocks.com/2016/10/the-shadow-of-air-sea-battle-and-the-sinking-of-a2ad/.

Branch but also one of McMaster's "go-to" assistants for special projects—to develop the idea in a short white paper. Hornick worked on the project for several months, completing the final version in early 2016.[9]

Multidomain Battle

In early 2016, the US Army and Marine Corps agreed to cooperatively develop a new concept. Hornick's white paper served as a useful starting point for what developed into the *Multi-Domain Battle* concept, because as a naval service—as noted earlier—A2AD was also a matter of concern.[10] In the end, Marine Corps leaders opted not to adopt *Multi-Domain Battle*, and so it was published solely as an Army document. That decision, however, came late in the development. In the meantime, the Marine Corps Combat Development Command team, led by Colonel (Retired) Mike Raimondo, made many useful contributions to both the form and content of the concept.

Within ARCIC, the concept was primarily developed by the Army Concepts Branch, a team of mainly civilian concept writers led by Lieutenant Colonel Ed Werkheiser.[11] Werkheiser often served as a sounding board for Hornick, so there was no loss of continuity from the white paper to the concept. There was, however, turnover in the leadership above Werkheiser. In the summer of 2016, Brigadier General Mark Odom arrived to lead the Concepts Development and Learning Directorate (CDLD), while Colonel Mike Runey took over the Joint and Army Concepts Division (JACD). The next February, McMaster would hastily depart ARCIC, when he was selected by President Donald Trump to serve as the National Security Advisor. His position would remain vacant for over a year, leaving the deputy, Major General Robert "Bo" Dyess, to cover two portfolios for an extended period. That left Odom and Runey considerable leeway as the principal leaders conducting direct oversight of the development of both *Multi-Domain Battle* and *Multidomain Operations*. Indeed, they were intimately involved throughout through specific direction on content, providing multiple edits of every page, and sometimes even drafting sections themselves. Odom and Runey were effective members of the writing team as well as being

---

9   Hornick e-mail.

10 U.S. Army TRADOC, *Multi-Domain Battle: Evolution of Combined Arms for the 21st Century, 2025-2040* (December 2017).

11 At various points, other members of JACD—particularly Chris Stolz, Mark Smith, and Suzy Thurmond—contributed in some way to *Multi-Domain Battle*. Mike Redman worked nearly full time on the project.

the supervisors. Also in the summer of 2016, Hornick departed for his next assignment, and I replaced him as the Joint Concepts Branch chief with JACD.

Officially, *Multi-Domain Battle* was not officially tied to any specific threat. The Marine Corps would never have agreed to a Russia-only concept.[12] Yet at least within ARCIC, Russia remained the undoubted focus. There were occasional discussions about how the ideas would work in other regions and against other adversaries; Werkheiser had served as a planner for North Korean contingencies and so brought that perspective. Nonetheless, the general description of the threat in the concept indicates that the major focus was a European battlefield. Aside from the origin of Hornick's white paper, this was also driven by all of the analysis of that problem occurring at the same time; members of the writing team often attended these events, the lessons of which were then fed into the document. Any rigorous concept must be grounded in such analysis rather than being plucked out of the air by a few people in a room with a whiteboard. McMaster was one of the strongest proponents of threat-based concepts to prevent the excesses of ambitious capabilities-based thinking during the 1990s and 2000s that led to the Army's Future Combat System (FCS) debacle.[13]

The focus on Russia had two important implications for the scope of the concept. The first was that the adversary described possessed a broad range of capabilities, a mixture of irregular warfare, conventional forces, and nuclear weapons. The second was the concept had an equally broad temporal scope, reflected in its three conditions of *competition, conflict*, and *return to competition*.

The inclusion of actions outside of conventional conflict elicited strong reactions, both for and against. On one side were those who felt that making the concept too broad watered it down. This group wanted a concept limited to conflict. The counterargument was that adversaries like Russia (but also China, Iran, and North Korea) wanted to achieve strategic objectives outside of armed conflict, and so the US Army must have some sense of what it would do to counter those efforts.

Both arguments have some merit, even if they were often expressed in unhelpful maximalist ways. The fiercest partisans suggested that if the concept was not written as they wished, some fundamental rule of

---

12 Edwin Werkheiser, e-mail to author 31 December 2024. Werkheiser also provided a number of comments that significantly improved this chapter.

13 For an extended argument about the need for threat-based concepts, Frederick W. Kagan, *Finding the Target: The Transformation of American Military Policy* (New York: Encounter Books, 2006).

the universe would be broken. In truth, concepts are merely tools to help military leaders make their organizations better. Anything that achieves that objective is valid. In the case of the Army in 2016, because Russian actions in Crimea and eastern Ukraine were one of the impetuses for the concept, it was inevitable that actions outside of conflict were going to be addressed somehow. Of course, it would have been possible to create two 40-page concepts (or some other of the countless variations) rather than a single 80-page concept. In practical terms, the partnership with the Marine Corps limited the team from pursuing any adventuresome alternatives. Senior leaders had agreed to a concept, not a series. And because JACD and Raimondo's MCCDC team were already collaborating as co-sponsors (along with U.S. Special Operations Command) for the *Joint Concept for Integrated Campaigning*, it was easy to agree on a modified use of the terms *competition* and *conflict* from that concept.[14]

The three components of the solution—calibrate force posture, converge capabilities, and employ resilient formations—were logical byproducts of the focus on Russia in competition and conflict. The first of these, *calibrate force posture*, illustrates some of the complexities of threat-based concepts. Within the concept community, it is widely accepted that concepts are guides to force development and not future war plans (or even future doctrine). The typical products of concepts are concept-required capabilities (or, CRCs), which are often related to equipment or organizations.

The key is that concepts are meant to describe some new, better way of fighting, so that the service can then build the capabilities that allow that new way of fighting. Yet as the RAND Baltic wargame study made clear, a significant part of the problem with Russia was not that the United States or even NATO lacked appropriate capabilities. Instead, the combination of small allies on Russia's border, limited high-readiness forces in Europe, and a perceived lack of political will within the alliance might tempt Russia to attempt to achieve a *fait accompli*. Even if the Russian gamble proved misguided, it would still be a costly, perhaps pyrrhic, victory for NATO.

Therefore, the RAND researchers emphasized the need for deterrence by denial.[15] As this scenario was both a focus of a lot of the analytical work going on as well as one of the key concerns of the senior leaders demanding a concept, this created a conundrum for the writing team. Was *Multi-Domain Battle* a solution for the Baltic problem, a broader concept for Russia, or an

---

14 U.S. Joint Staff, *Joint Concept for Integrated Campaigning* (28 March 2018).
15 Shlapak and Johnson.

even more expansive vision for fighting that only happened to be grounded primarily in the Russian threat? If the problem was simply to deter an attack by denial, then no new concept was necessary at all. Instead, reverting to a defense not unlike that used in defense of South Korea—lots of mines, concrete positions, and more forces—was more than sufficient. If the political will was lacking to translate the far greater economic and military resources of the alliance into a viable defense, that was not something that could be addressed by US Army leaders. The one thing that was partly—although not entirely— within their purview, was the credibility of military forces. Hence, the component of *calibrate force posture* recognized the importance for deterrence of having sufficient forward forces. Because that is a policy decision rather than a capability, some would argue that it should not be in a concept, but the writing team also felt it necessary to make the connection that the viability of the rest of the concept would depend on having sufficient forces for execution.

Among the components of the solution, *converge capabilities* drew the most attention. In contrast to *calibrate force posture*, which, as just discussed, was something for policy makers, convergence was a descriptor of action at the unit level, and so easier to envision. The idea of bringing together multiple domains in what were now called "windows of advantage" was also the most direct intellectual heir to Hornick's white paper. Many commentators have observed that this idea was not novel, but simply an extension of the traditional idea of combined arms. When this observation is meant as criticism, it demonstrates a lack of understanding about what concepts are supposed to do. Just as concepts are not future war plans, neither are they military theory. In this case, the concept's sub-title— *Evolution of Combined Arms for the 21st Century, 2025-2040*—clearly indicates the belief that this was just an evolution. The task of the concept was to envision how new capabilities like fifth-generation fighters, offensive space weapons, and cyberspace could be added to the existing combined arms approach.

A more justified line of criticism against *Multi-Domain Battle* was that it did not provide great enough fidelity in how the updated combined arms would bring together that mixture of old and new capabilities. With the benefit of hindsight, it is apparent that the writing team was still in transition from the more tactical, sometimes platform-based thinking of the original white paper, and the eventual operational vantage. This evident in the convoluted definition of convergence. *Multi-Domain Battle* described convergence as bringing capabilities together from across "domains, functions, and environments"—a clunky descriptor necessary to include information operations and the electromagnetic spectrum, neither of which is designated by the Department of Defense as a domain—to create windows of advantage that in turn enabled a position of advantage. That formulation inserted a potentially unnecessary step in the entire process; convergence was meant to produce a window that in turn enabled maneuver. But what if convergence actually achieved what was necessary and did not require some additional step?

The final component of the solution, *employ resilient formations*, was the other side of the coin from convergence. We want to converge capabilities against the enemy, but recognizing that the enemy is capable and sophisticated, we need to have resilience for when they converge capabilities against us. This seemingly commonplace observation has profound implications for force development, because it implies that the same unit organization must be able to operate within two radically different contexts. The first is when the entire US joint force is effectively working together—convergence. The second is when the same unit is physically and virtually isolated and must rely on its own resources.

Unfortunately, due to the intervention of a senior leader, the emphasis in *Multi-Domain Operations* was on the *formation* element rather than on *resilience*. This could be regarded as one of the most significant missed opportunities of the multidomain journey, because with the constrained resources of the present military, getting the balance right is the most pressing issue in force development. Convergence allows units to be stripped down to the barest essentials, because there are additional capabilities that will be coming from higher echelons or other components. Resilience requires redundancies, whether in duplicating capabilities, such as fires, or in having a capability in greater quantities, such as what is necessary for having multiple command, communications, or logistics nodes. Erring too far in either direction means either having fewer units overall because there is too much capability packed in any given organization or having "glass-jaw" units that cannot withstand the onslaught of a capable foe. One actual example of this zero-sum dilemma from the time was the requirement proposed by the Maneuver Center of Excellence for "semi-independent maneuver"; that idea was for brigade combat teams to be able to operate "semi-independently" for up to seven days.[16] While that would certainly provide resilience, as one of the logisticians

---

16 U.S. Army Maneuver Center of Excellent, *Army Functional Concept for Movement and Maneuver*, TRADOC Pamphlet 525-3-6 (February 2017).

working on concepts calculated, the amount of supplies for the brigade combat team to carry was enormous.[17] Achieving the semi-independent maneuver ambition would require many more soldiers and trucks, who in turn would need to be supplied, as well as create a large signature that could be more readily identified and targeted by the enemy. The utility of concepts is in identifying such tensions and providing a framework for someone "to do the math," and figure out the right trade-off. How isolated and for how long did we expect each echelon to be? The calculation is even harder because there is no single answer, but it varies with context. The ability of Russia to isolate US forces operating in the fluid context of the early stages of a Baltic invasion would be very different than North Korea in the relatively congested front lines of a conflict on the peninsula. When *Multi-Domain Battle* was published, the writing team was still just figuring out what some of these key questions were.

Multidomain Operations

Chief of Staff of the US Army Mark Milley's mandate to publish *Multi-Domain Battle* by the end of 2017 meant that many of the ideas would be such untested hypotheses rather than the detailed insights necessary to guide spending money on new weapons systems or creating new organizations. Odom and Runey had already determined that more work was necessary, so even before the publication of *Multi-Domain Battle* the team was thinking of "MDB 1.5." That name indicated the desire to have a follow-on Army-only concept by September 2018, that could also, hopefully, lead to a multi-service or joint "MDB 2.0" at some point in 2019.

The trajectory of what ended up becoming MDO was the incidental byproduct of what, to the best of my knowledge, was a meeting in June 2017 between the commanders of TRADOC and the US Air Force Air Combat Command (ACC), Generals Dave Perkins and Mike "Mobile" Holmes. The two were brought together by geography rather than function; the two headquarters, located just miles apart, were part of Joint Base Langley-Eustis. In function, ACC is much closer to US Army Forces Command. The two commanders had routine courtesy visits, and at one of these they agreed to co-sponsor a series of joint tabletop exercises. I do not know what was intended at the time for the output. At least on the Air Force side, it would

---

17 Dominick Edwards, "Logistics Support to Semi-Independent Operations," *Army Sustainment* 50, no. 1 (Jan/Feb 2018): 40-42.

not have been a concept, because that function was held within Headquarters Air Force. Even within TRADOC, if the intention was to tie the tabletops to concepts, that was not clearly communicated to ARCIC. The initial task of fleshing out the exercise schedule fell to the Joint Integration Division, a small office, that ran the tri-service general officer steering committee that dealt with topics such as how much air support would be allocated to army training events like combat training center rotations. That division then reached out laterally to JACD for assistance in framing the content of the tabletops.

The unusual pairing created some challenges. The ACC A3 (operations officer), led by then-Major General Andrew J. Toth, was configured to oversee active operations, not force development. Thus, Toth brought in as a contractor Brigadier General (Retired) Michael A. Longoria, one of the pioneers of air-ground integration during the early days of the Global War on Terror. Longoria was a great partner, but his routine joke about being a one-man show was not far off the mark. The power of two four-star commanders, however, did unlock significant outside resources. The RAND Corporation was contracted to facilitate the wargames, an excellent solution as it had the benefit of extensive use for both services. Additionally, although on a daily basis, Longoria was largely alone for the duration of what ended up being three tabletop exercises, Toth attended in person and the Air Force sent some of its best experts, often at the colonel-level, in the various functions of air, space, cyberspace, logistics, and battle management. The Army matched this with a team led by Odom that included experts from across the various functional centers of excellence. As one of the "grey beard" senior mentors noted at the outset of the first tabletop exercise, "It has been twenty years since there were this many army and air force colonels in the same room talking about how to fight in the future." These wargames might have been the product of chance rather than design; regardless, they provided the opportunity to further refine the ideas from *Multi-Domain Battle* and produce a more detailed concept.

The operational framework laid out in *Multi-Domain Battle* provided a useful basis for organizing the tabletop exercises. The initial impetus for developing that framework was a way in which different functional communities would often talk past each other when discussing multidomain operations. The problem was that there were many different kinds of multidomain operations, but everyone assumed that their specific "flavor" was the universal reality.

This failure to understand different contexts was most prevalent in discussions about the strike-centric *deep fires area* and the ground-maneuver-centric *close* and *deep maneuver areas*. The ethos of mission command is so deeply ingrained in ground forces that the necessity to exert centralized control exemplified by the 72-hour "air tasking order" is dismissed as wrong-headedness. This view, however, misses that as to strike some of the most well-defended targets often requires synchronizing aircraft coming from several different locations, some potentially many hours' flight time away, along with limited space and cyberspace assets, which have their own lead times for planning and employment. Conversely, those focused on deep strike combining exquisite combinations of aircraft, missile, space, or cyberspace capabilities through elaborate joint processes, often do not appreciate that those deliberate systems will break down under the scale and pace of battle in areas where large, modern militaries meet in a dynamic, unpredictable clash. The *close* and *deep maneuver areas* were meant to capture that dynamic. The further distinction between those two areas was based on the realization that the amount of support from other domains would depend on the joint commander's priorities. When ground maneuver was blessed with having the full weight of the joint force behind it, ground headquarters would need the ability to integrate exquisite capabilities like offensive space-based systems into maneuver. That was the deep maneuver area. But ground forces also had to prepare for the far more common situation of fighting largely without those scarce joint resources; in the close area, the tools available would be largely limited to organic combination of maneuver, ground fires, army aviation, and electronic warfare.

Finally, the distinction between the *strategic, operational,* and *tactical support areas* acknowledged that while forces would be contested "fort-to-foxhole," there were important distinctions to keep in mind about the nature and intensity of that contestation. Essentially, each of these areas demanded its own branch of MDO. The basic plan for the tabletop exercises was that the first one, held at Carlisle Barracks, Pennsylvania, explored the close and deep maneuver areas and was more tactically focused. The second tabletop exercise, held in Santa Monica, California, was a theater-level exercise and therefore included everything from the operational support area to the deep fires area. The third event, held in Suffolk, Virginia, explored command and control.

Much of the content in MDO resulted from what was learned while testing the *Multi-Domian Battle* ideas within the tabletop exercises. In broad terms, the games demonstrated that it would simply not be possible to conduct a DESERT SHIELD/STORM-like campaign of sequential air-centric preparation followed by ground-centric decisive actions. Or, if expressed in the terms of the operational framework, it was no longer possible to move sequentially from preparation in the deep fires area to exploitation in the deep maneuver area to finishing in the close area. That is a great way to fight because it minimizes risk and allows US comparative strengths to accumulate over time, but a top-tier adversary would simply not provide the time and space for the US-led coalition to dictate the pace of a campaign to fit our strengths.

Thus, friendly forces would have to wage simultaneous efforts to gain advantage in all three areas (while protecting the support areas). The first two tabletop exercises highlighted some of the specific challenges within each of these separate "fights." One of the major takeaways was the difficulty of achieving a decisive advantage over the adversary's key systems, such as the most capable air defense assets, intermediate-range missiles, and long-range multiple rocket launchers. Unsurprisingly with a smart, adaptive enemy, once we started making progress in targeting these keystone capabilities, the enemy would take various actions to preserve the rest. This created a dilemma for the friendly commander: should they expend even more resources and delay other operations looking for this remainder, or proceed with other actions at higher risk than we would want to accept? Within the document, this trade-off is reflected in the repeated references to *stimulating* enemy systems. The other major takeaway from the first two tabletop exercises was the extent to which the domains had begun to overlap. For any given enemy system, there were multiple combinations of capabilities that could be used against it. The fuller description of convergence than that which had been used in *Multi-Domain Battle* reflected this appreciation. For instance, a system like the Russian SS-26 ground-based missile was a simultaneous threat to the ground commander's reserve, the air commander's forward staging bases, and the joint commander's headquarters. To counter these systems, it would likely require some combination of sensors from different domains and our own long-range ground-based fires or air strikes. The best operational solution was dictated more by the given circumstances (e.g. what sensors were best optimized for that terrain and activity, and what "shooters" were available) rather than any set answer that single-, cross-, or multi-domain solutions were always best.

This realization led to the great takeaway from the final tabletop exercise—the exercise focused on command and control—that as hard as these smaller tactical problems might be, the biggest challenge was to develop a workable system that could allocate capabilities across the theater at the pace and scale necessary. In terms of the operational framework, it might be necessary to shift the effort from the deep fires area to the tactical support area and then to the deep maneuver area within the span of days or even hours. Even with today's capabilities, we have many systems that can be used in such a flexible manner, and this will only increase over time. But the command and control structure—both the technical communications backbone and the command relationships—are lacking.[18]

Conclusion

In 2018, even as the concept was nearing its final form, leaders far above the writing team made several consequential decisions that reveal much about the role of concepts within military institutions. The first was to declare *Multidomain Operations* as the Army's new operating concept. The second, related, decision was that the concept would "account" for both China and Russia but use Russia "as the present pacing threat for technical and tactical purposes."[19]

In institutional terms, this meant that MDO was a "broad" concept in the sense that it was meant to guide all Army force development. This was likely an inevitable decision in that there is no force development process for just a portion of the service destined to fight Russia. Moreover, the consistent pressure to publish *Multi-Domain Battle* and then *Multidomain Operations* on tight timelines was not manufactured; Army senior leaders facing the inexorable programming and budgeting cycle wanted anything that might help them make better decisions. They did not want to wait for a new concept for China (and might have feared the result if that concept came up with a dramatically different sets of required capabilities).

Yet as the preceding account has demonstrated, whatever might be decreed, MDO was not broad in scope. Of course, if pressed, the writing team could sketch out a theory-based argument for how the basic ideas could be applied to a conflict with People's Liberation Army in the western Pacific, but these would not have been grounded in deep study. MDO was the product of

---

18 This conclusion was further buttressed during a follow-on study for the Joint Staff director for force development, Joe Broome, J.P. Clark, Derrick Franck, Jr., and Michael Loftus, *Command in Joint All-Domain Operations: Some Considerations*, Carlisle Scholars Program, U.S. Army War College (22 July 2020).
19 *Multi-Domain Operations*, 7.

years of development, studies, and wargames focused almost exclusively on Russia. It is also revealing about the nature of concepts how much of that work was not done specifically for the sake of the concept, but something that the writing team opportunistically harvested as additional data. My observation from several years involvement in various aspects of military concepts is the general ignorance about the amount and type of analytical support required to create a good rigorous concept. Even with all the information that fed into MDO, the writing team was only starting to understand some of the tensions of the modern battlefield. Any wargame expert will tell you that they are excellent at exposing the underlying dynamics of a situation but are not analytically sufficient to prove or validate anything. Only with the benefit of the RAND tabletop exercises supplemented by so much other work like the *Russian New Generation* study were we even able to get that far. Specific decisions about force development should be driven by detailed modeling and simulation exploring a range of different start conditions, each examined over multiple iterations. Concepts should be understood as the means to frame that research, not as a final answer in themselves.

2

Multidomain Operations
A Bridge Between Industrial and Information Age Warfare?

Jesse Skates

New Concepts?

Somewhere between 2011 and 2014, the US national security establishment came to the collective realization that a new era in global geopolitics had emerged, with the intensifying rivalry between the United States, Russia and China. The source and exact timing of the epiphany is unclear. In January 2012, the Obama administration described a "rebalance toward the Asia Pacific region."[1] Colloquially referred to as the "pivot to the Pacific," the new policy was a tacit acknowledgement of the increasing perception of China as a threat. Development of the *AirSea Battle* concept in 2012 followed the Pacific pivot.[2] The Arab Spring in 2011 and emergence of ISIS in 2013, however, paused further development of related concepts until 2014, when Russia and China forced the United States to refocus on great power competition once again.

In 2014, the People's Republic of China (PRC) completed construction of military bases on contested features in the South China Sea.[3] US defense leaders interpreted this move as the PRC's first step to controlling the

---

1 'Sustaining U.S. Global Leadership: Priorities for 21st Century Defense' (Washington, DC: Department of Defense, 2012), 2. https://www.globalsecurity.org/military/library/policy/dod/defense_guidance-201201.pdf.
2 'Air-Sea Battle', Air-Sea Battle Office, accessed Jan. 13, 2025, https://dod.defense.gov/Portals/1/Documents/pubs/ASB-ConceptImplementation-Summary-May-2013.pdf.
3 P.K. Gosh, 'Artificial Islands in the South China Sea', *The Diplomat*, Sep. 23, 2014, https://thediplomat.com/2014/09/artificial-islands-in-the-south-china-sea/.

most important sea lane in the world and a significant challenge to the free navigation and trade agreements established at the conclusion of World War II. That same year, Russia invaded Ukraine undermining the belief that interstate wars in Europe had ended.[4]

Following these events, the US military conducted a myriad of strategic analysis, systems analysis, concept development, and operations research efforts. The *Russian New Generation Warfare Study* synthesized many of the findings from the above efforts identifying four meta problems posed by modernizing Russian forces.[5] The other analytical programs also identified changes in the character of conflict. The following paragraphs outline the major concepts from this period to help contextualize the assessment of the Army's multidomain concepts conducted in subsequent sections of this chapter.

In response to Russian and Chinese aggression, the US defense establishment began work on a series of operational concepts designed for a new era of warfare. This process started with *AirSea Battle* in 2012.[6] *AirSea Battle* focused on adversary precision strike networks referred to as anti-access and area denial systems, or A2AD. As the first major concept from this period, *AirSea Battle* anchored subsequent analysis in several key ideas including: defeating A2AD, leveraging US precision-strike networks to counter adversary systems, and the five domains of war (land, sea, air, space and cyberspace).[7]

The US Air Force added to *AirSea Battle* with a service-centric Future Operating Concept in 2015. Agility—one of the main principles in the Air Force Future Operating Concept—eventually evolved into its own concept in 2022 as Agile Combat Employment (ACE).[8] Broadly, the Air Force intended to increase survivability and complicate adversary targeting efforts. Specifically, the service described a plan to divide its relatively large formations into small packages and then constantly move small groups of airframes to different

---

4 David Farrell, interview with author, January 9, 2025. Initial drafts of the Russian New Generation Warfare Study started with the phrase "Maneuver war has returned to the plains of Europe."

5 Russian New Generation Warfare, https://apps.dtic.mil/sti/trecms/pdf/AD1118626.pdf.

6 Air-Sea Battle Office, 'Air-Sea Battle'. 2. At least partially authored by Andrew Krepinevich, an alumnus of the Defense Department's Office of Net Assessment who provided an initial outline of information warfare and precision strike networks in the early 1990s.

7 Air-Sea Battle Office, 'Air-Sea Battle', i.

8 Department of the Air Force, 'Air Force Future Operating Concept' (Washington, DC: Department of the Air Force, 2015), 7. https://www.af.mil/Portals/1/images/airpower/AFFOC.pdf.

locations in unpredictable ways.[9] The combination of distribution and survivability maneuvers made the air component more resilient though not necessarily more lethal.

The Navy moved beyond *AirSea Battle* and its temporary alliance with the Air Force, producing its own Distributed Maritime Operations (DMO) concept in 2018.[10] In this concept, maritime forces dispersed to the limits of their communications networks to protect surface, subsurface, and air combatants from A2AD threats. Maritime forces communicate to concentrate firepower from distributed platforms against enemy targets, creating opportunities for maneuver while simultaneously preventing the adversary from realizing its own strategic objectives.[11]

The Department of Navy did not stop with a reimagining of fleet operations; it also sought to better integrate the Marines into maritime operations. To this end, the Marine Corps contributed two concepts, including Littoral Operations in a Contested Environment (LOCE) and Expeditionary Advanced Base Operations (EABO).[12] The Marine Corps' concepts outline a series of feints with forward deployed and mobile Marines against enemy forces as a first step. When the less mobile adversary strikes back, it overextends and exposes itself to a lethal counterattack by the quicker maritime force. General David Berger, responsible for the development of EABO as the Deputy Commandant for Combat Development and Integration, went on to create the force structure to implement these concepts with *Force Design 2030* as the Commandant of the Marine Corps.[13]

The Army provided its views on the changing character of war in two concepts: "Multi-Domain Battle: The Evolution of Combined Arms for the

---

9 Department of the Air Force, 'Air Force Doctrinal Note 1-21: Agile Combat Employment' (Washington, DC: Department of the Air Force, 2022) 6. https://www.doctrine.af.mil/Portals/61/documents/AFDN_1-21/AFDN%201-21%20ACE.pdf.

10 Kevin Eyer and Steve McJessy, 'Operationalizing Distributed Maritime Operations,' *Center for International and Maritime Security.* Mar. 5, 2019, https://cimsec.org/operationalizing-distributed-maritime-operations/.

11 Dmitry Filipoff, 'Fighting DMO, Part 1: Defining Distributed Maritime Operations the Future of Naval Warfare', *Center for International and Maritime Security,* Feb. 23, 2023, https://cimsec.org/fighting-dmo-pt-1-defining-distributed-maritime-operations-and-the-future-of-naval-warfare/.

12 U.S. Marine Corps, 'Expeditionary Advanced Base Operations (EABO) Handbook: Considerations for Force Deployment and Employment', U.S. Marine Corps Warfighting Lab, Concepts and Plans Division (2018), 22, https://www.mca-marines.org/wp-content/uploads/Expeditionary-Advanced-Base-Operations-EABO-handbook-1.1.pdf.

13 Connie Lee, 'News from EWC: Marine Corps Defining New Operating Concept', *National Defense,* Oct. 22, 2019, https://www.nationaldefensemagazine.org/articles/2019/10/22/marine-corps-works-to-define-new-operating-concept.

21st Century (MDB)" and its successor, "The U.S. Army in Multi-Domain Operations 2028 (MDO)." Both concepts, but particularly MDO, combine network-centric operations with the deep maneuver of tank columns, a hallmark of ground force operations since World War II. By striking key nodes in adversary A2AD systems with multi-domain effects, the Army gains brief windows to maneuver its tank formations and attack vulnerable enemy fires units.

Thus, MDO retained the same basic technology (i.e. tanks and artillery, albeit with better range and accuracy) and logic used by German Panzer divisions in World War II: use artillery strikes to temporarily stun the enemy and provide maneuver forces an opportunity to create widespread confusion, which prevents the adversary from mobilizing an effective counterattack. The major change was a switch in targets from logistical to long-range fires units.

According to the defense establishment, by 2017, a revolution in military affairs, or RMA, was underway. Yet, descriptions of this revolution offered by the Army as well as the other services recycled ideas from decades ago. If MDB and MDO were intended to form a bridge between Industrial Age and Information Age warfare, why recycle old ideas? The answer is not clear. However, as a result, MDB and MDO did not adequately take into account the trajectory of new weapons development that would affect the future of war, such as artificial intelligence, quantum computing, and other information technologies. MDB and MDO appeared dated—like the other concepts then in development – and promised to describe a new way of war using weapons designed for an old one.

An Assessment Framework

Enterprise-level leaders enjoy unique privileges, chief among them the ability to set the grading criteria against which they are assessed. In the 20-plus years leading up to the publication of the two multidomain concepts, enterprise-level leaders in the Department of Defense and the Department of the Army (DA) promised a change in the character of war.[14] They argued that systemic changes in technology and society—stemming from the emergence of the Information Age—guaranteed new approaches to warfare, requiring

---

14 Andrew Marhsall, 'Some Thoughts on Military Revolutions – Second Version', Department of Defense: Office of Net Assessment, Aug. 23, 1993, Gail Yoshitani, 'The Character of Warfare 2030to 2050: Technological Change, the International Systems, and the State', Chief of Staff of the Army's Strategic Studies Group, 2016. https://stacks.stanford.edu/file/druid:yx275qm3713/yx275qm3713.pdf.

novel technologies, organizations and tactics.[15] Multidomain concepts are assessed against these assurances and vision of future warfare. The following framework accounts for the above considerations and clarifies the grading criteria that DoD and DA officials gave themselves.

This framework includes five independent components. The first two components—creative destruction and navigating change—assess the multidomain concepts' ability to describe a new Army and to implement recommended changes. The remaining three components of the framework address questions confronting any concept writer or theoretician contemplating the future of their organization: What to retain? How to relate to the external environment? What to change? The goal of this chapter's assessment is to determine if and to what extent the multidomain concepts (a) describe a new army purpose-built for the new era, (b) prescribe required, systemic changes to legacy institutions and weapons, and (c) deliver envisioned solutions.

The creative destruction component of the framework stems from George Morison's description of the first component of systemic change in *The New Epoch as Developed by the Manufacture of Power*. In that work, Morison notes that, "in many ways the new epoch must open as an era of destruction... There must be great destruction, both in the physical and the intellectual world, of old buildings and old boundaries and old monuments and, furthermore, of customs and ideas, systems of thought and methods of education."[16] He elaborated, "the danger is that the destructive changes will come too fast, and the developments to take their place not fast enough."[17] An effective operational concept outlining a significant shift in the character of war must describe how an organization intends to destroy old institutions and ways of operating and replace them with something new.

The navigating change component of the framework accounts for how well leaders overcome opposition. George's nephew Elting Morison describes the challenge to change thus: "[some people] identified themselves with a settled way of life they had inherited or accepted with minor modification and thus found their satisfaction in attempting to maintain that way of life unchanged... This purely personal identification with a concept, a convention, or an attitude would appear to be a powerful barrier in the way of easily

---

15 MacGregor Knox and Williamson Murray, *The Dynamics of Military Revolution 1300-2050* (New York, NY: Cambridge University Press, 2001), 6, 12.
16 George Morison, *The New Epoch: As Developed by the Manufacture of Power* (Cambridge, MA: The Riverside Press, 1903), 128.
17 Ibid, 128-129.

acceptable change."[18] Enterprise-level leaders must confront and navigate significant pushback from a generation of middle-managers both inside and outside of their institutions that need but do not want to change. If effective, they deliver new technologies, organizations, and tactics to their institution.

The first step in navigating institutional change is to provide a compelling vision of the future that maps the course of change, enabling a group of revolutionary young leaders to pilot the organization to a new plane. Such a vision answers three essential questions. First, what components of the mission to retain? Second, how does the organization relate to its external environment? Third, based on the answers to the first two questions, what does the organization have to change in order to effectively modernize? The remainder of this chapter assesses how successful MDB and MDO have been at transforming the Army into an institution that can be effective in a new period of human history.

Creative Destruction

In *The New Epoch*, George Morison described the creative destruction common to profound technological advancement. He explained that up to the industrial epoch, "the capacity of man had always been limited to his own individual strength and that of the men and animals which he could control." He went on to say, "His capacity is no longer so limited; man has now learned to manufacture power, and with the manufacture of power a new epoch began."[19] The manufacture of physical power discussed by Morison enabled humanity to become the most physically imposing creature in the world. With mechanical strength, humanity dominated the land, sea, and skies and constructed railways, interstates, hydroelectric dams, and cities. The Information Age is an epoch of similar importance in that it allows humanity to generate cognitive power beyond what is available to the individual or the people under one's control.

Norbert Wiener—a founder of the Information Age—described the effects of the coming information revolution in 1948: "Perhaps I may clarify the historical background of the present situation if I say that the first industrial revolution, the revolution of the "dark satanic mills," was the devaluation of the human arm by the competition of machinery... The modern industrial

---

18 Elting Morison, *Men, Machines, and Modern Times*, 50th Anniversary Edition (Cambridge, MA: MIT Press, 2016), 17.
19 Morison, *The New Epoch*, 4.

revolution is similarly bound to devalue the human brain…"[20] When paired together, the industrial and informational revolutions combine to minimize the value of humans in any setting requiring power—whether physical or cognitive. The associated social and technological upheaval will be immense.

What did MDB and MDO have to say about this fundamentally new era in human existence and associated technological and social trends? How does the Army expect to address these changes? What will be destroyed? And what will replace the recently lost organizations and systems?

The answer to these questions is far from clear. For instance, General David Perkins in his opening to Multi-Domain Battle argues that, "The purpose of the Multi-Domain Battle concept is to drive change and design for the future Army."[21] In the next sentence, however, he accepts business as usual for the Army, stating "It will provide the foundation on which TRADOC [U.S. Army Training and Doctrine Command] conducts capabilities-based assessments to refine required capabilities, identify gaps, and determine potential capability and policy solutions for future forces."[22] In effect, General Perkins promises to design and create a new Army for a new time by making modifications to legacy systems.

Senior leader hedging continues into *Multidomain Operations*. In General Mark Milley's opening to MDO, he notes that "emerging technologies like artificial intelligence, hypersonics, machine learning, nanotechnology, and robotics are driving a fundamental change in the character of war."[23] Then he proceeds to explain, "Therefore, the American way of war must evolve and adapt."[24] General Milley called for evolution—a relatively minor change to tactics and weaponry—when the world is facing a revolution, which triggers new approaches to warfare with novel organizations, technologies and tactics.[25] The invention of aircraft carriers, airplanes, computers, nuclear weapons and national mobilization to support operations during World War II is an example of a revolution, one extending well beyond modest changes to legacy military systems.

---

20 Norbert Wiener, *Cybernetics: Or, Control and Communication in the Animal and the Machine*, 2nd Edition (Cambridge, MA: MIT Press, 2013), 27-28.
21 U.S. Army Training and Doctrine Command, 'Multi-Domain Battle: Evolution of Combined Arms for the 21st Century' (Fort Eustis, VA: U.S. Army Training and Doctrine Command, 2017), i.
22 Ibid.
23 U.S. Army Training and Doctrine Command, 'TRADOC Pamphlet 525-3-1: The U.S. Army in Multi-Domain Operations 2028' (Fort Eustis, VA: U.S. Army Training and Doctrine Command, 2018), 28.
24 Ibid.
25 Knox and Murray, *The Dynamics of Military Revolution 1300-2050*, 12.

Despite these limitations, General Perkins, General Milley and General Stephen Townsend, the three general officers providing opening comments for the multidomain concepts, all acknowledge the need for additional work defining the future of warfare.

Accepting the need for additional work, do the concepts at least introduce a new epoch in human existence and outline some of the potential upheaval caused by the automation of physical and mental labor? The concepts discuss the compression of decision timelines, constant monitoring of US forces by adversaries, and the ability to mobilize networks of "gray zone" actors—all changes which are underpinned by information technologies. Furthermore, convergence—the effort to combine effects from all domains against critical nodes in the adversary's long-range fires system, a tenet of both MDB and MDO—is clearly a nod to the precision strike networks and network centric warfare concepts considered essential to early information warfare thinkers such as Andrew Krepinevich. MDO explicitly describes new command-and-control layers needed to manage the cognitive load of rapidly combining multidomain capabilities to conduct data collection, analysis, and target execution tasks, but does not indicate whether man or machine carries that load, leaving space for the automation of associated staff functions.

However, both MDB and MDO avoid a broader theoretical discussion about coming changes and the associated political risk of defining internal destruction and reorganization required to better leverage intellectual and physical automation.

Rather than describing technological headwinds and ways to prepare for them, the documents focus more on the rivalry between the United States and its two most capable competitors, Russia and China. Both documents note adversary capacity to see and destroy or frustrate US forces using capabilities in all domains and the gray zone and describe the adversary's ability to affect US military mobilization, disrupting deployment and sustainment operations all the way back to the American homeland. The analysis of the dangers posed by Russian and Chinese military modernization was used to justify an immediate, but ultimately incremental, US response.

Navigating Change

Elting Morison highlights multiple considerations when navigating institutional change, which can be used to help refine analysis of the multidomain concepts. He describes the need for internal experts, which tinker with technology and concepts, finding ways to apply new technologies or optimize the use of existing capabilities.[26] In addition, he notes that institutions often reject proposed changes by these innovators and overcoming organizational inertia often requires external intervention by higher authorities.[27] So how effective were concept writers and Army leaders in innovating, describing the future, and navigating institutional pushback? They achieved some degree of success.

The authors of MDB and MDO certainly tinkered with concepts and looked for ways to employ new technologies and improve the use of existing ones. Mechanisms internally referred to as "battlefield development plans"—one each for Russia and China—aided this experimentation. These plans included three books. One that described enemy systems. Another that defined US Army capabilities then in development, most at or beyond technology readiness level of six. The last book provided "plays" or tactical engagement packages consisting of multidomain capabilities for use in battles against the adversary.

Using the battlefield development plans in experiments and wargames, concept developers pitted US military capabilities against the latest weapons fielded or in development by governments in Russia and China. The ARCIC's Joint and Army Concepts Division conducted comparative analyses assessing the ranges, reload rates, endurance, survivability and basic loads of weapons on both sides. Following wargames and other analytical events, the concept writers noted capability gaps that eroded American military dominance and suggested force structure changes or procurement efforts to mitigate potential risks.

These findings informed marginal changes to Army force structure, such as the fielding of multidomain task forces and procurement of precision strike missiles and additional theater high-altitude air defense (THAAD) interceptors. Few of the multidomain forces, mostly command-and-control units above the brigade level, survived internal Army opposition and DoD scrutiny. Despite this failure, the National Defense Authorization Act for Fiscal Year 2017 captured many of the recommendations made by multidomain concept developers and Army leaders.[28] By working with Congressional policymakers to capture recommendations in law, Army leaders and

---

26 Morison, *Men, Machines, and Modern Times*, 49-50.
27 Ibid, 52.
28 Andrew Feickert, 'The 2024 Army Force Structure Transformation Initiative,' (Washington, DC: Congressional Research Service, Aug. 2024), 3-4. https://crsreports.congress.gov/product/pdf/R/R47985.

concept writers garnered external support for their agenda. Compelled to comply with legal directives, Army institutions slowly evolved, publishing multidomain doctrine and training forces on modified tactics.

The multidomain concepts provided a narrative about incremental change, which helped the authors and Army leaders navigate institutional change through fraught bureaucratic waters. Derived from much older ideas, the concepts did not offer revolutionary insights or foundational changes. They did, however, closely follow the trajectory described by Elting Morison. The concepts came from internal innovation. Concept authors and Army leaders gained external support for implementation of proposed changes, overcoming some of the anticipated organizational pushbacks. Though institutional opposition meant that few of the command-and-control changes came to fruition, the Army modernized nonetheless: changing the unit of action from the brigade to the division, fielding multidomain task forces, procuring long-range fires munitions, and publishing multidomain doctrine.

What to Retain?

All organizations have foundational identities and missions, the destruction of which threatens their survival. For the US Army, its foundational identity is performing its four enduring strategic roles: preventing conflict, shaping the security environment, prevailing in large-scale combat operations, and consolidating gains. Armies do this by being fully trained to maneuver against other militaries and defend key terrain and populations. The better prepared these ground forces are, the more credible their deterrent effects. The foundational mission of the US Army is primarily to combine fires and maneuver to defeat near-peer threats.[29]

Do the multidomain concepts describe the Army's basic missions—like holding terrain for employment of the land component—as well as the modifications to tactics and formations required to make the Army successful in prosecuting its mission in a fundamentally new way of war or at least against existing but modernizing militaries?

The answer is a qualified yes. MDB and MDO retain a description of the four enduring strategic roles for the Army.[30] They also describe the terrain, formations, and modified tactics required for a lightly modified way of war. The concepts and associated battlefield development plans describe tactics

---

29 U.S. Army Training and Doctrine Command, 'TRADOC Pamphlet 525-3', iii, v, 19.
30 U.S. Army Training and Doctrine Command, 'TRADOC Pamphlet 525-3-1', 24.

associated with the convergence of capabilities across all domains to open windows for maneuver units to exploit, which is also in line with the Army's foundational mission. And finally, the concepts retain a focus on near-peer threats: anticipating a potential war with Russia and the People's Republic of China, with the MDO in particular focusing on the Russian threat and requirement to defend vulnerable NATO allies.

MDO also describes required forces and the need to change the operational focus from the brigade to higher echelons.[31] The qualification comes from General Stephen Townsend, who often said during the development of MDO, "The concept does not describe anything new. U.S. battlefield commanders during the war against the Islamic State of Iraq and Syria regularly converged multi-domain capabilities against enemy forces. However, we will need to do in minutes or seconds what currently takes weeks or months to coordinate." Thus, MDO envisions a war that is largely consistent with current roles and missions but wholly different in scale. The difference in scale is a direct result of the capabilities of the likely opponent. The adversary in Iraq and Syria did not field advanced air defenses, space capabilities, or fires complexes, but Russia and China do.

How to Relate?

Design often starts by delineating a system from its environment and understanding the boundaries and interfaces between the two. Therefore, designing an effective army requires a clear understanding of the boundary and interface between ground forces and anticipated terrain, friendly military counterparts, and adversaries. How well did the multidomain concepts do in defining these relationships?

The concepts effectively describe the relationship between the army, the operational environment, and the rest of the Joint force. As indicated in the title of the second concept, "The U.S. Army in Multi-Domain Operations 2028," the Army was dedicated to working closely with forces in the other four domains. Sections on command-and-control and paragraphs and graphics showing how the forces could work together to create layered options reinforce the intent for comprehensive all-domain coordination.[32] In addition, the battlefield development plans articulated the exact relationships between the multidomain forces in terms of timing, location,

---

31 Ibid, 22.
32 Ibid, 21-26.

network and command-and-control requirements. Furthermore, the Army ensured shared understanding by garnering all-domain participation during experimentation and concept development.

The concepts also effectively described the anticipated terrain in Europe. As discussed earlier, the *Multidomain Operations* concept focuses on a Russian threat in the European theater.[33] While the exact scenario remains classified, concept writers clearly understood the terrain. Standing political direction to defend NATO allies and their territory helped delineate key aspects of the environment. Decades of experience in Europe during the Cold War meant that the Army still had significant institutional knowledge of the terrain, people, and climate. Even if not captured in unclassified concepts, the relationship between the land component and the operational environment was well defined.

However, efforts to describe the relationship between United States, China, Russia and their respective militaries was less effective, particularly, considering the war in Ukraine. On the one hand, both concepts provide battlefield frameworks,[34][35] and the enemy systems encountered in each area within that framework. The *Multidomain Operations* concept explicitly outlines objectives for multidomain forces during competition, conflict, and a return to competition after war ends.[36] MDO directs the Army to deny and defeat a *fait accompli* attack by Russian forces.[37] It even describes how the adversary combines A2AD and maneuver forces to achieve *fait accompli* objectives, and the need to penetrate and disintegrate them to prevent the *fait accompli* attack.

On the other hand, the concepts do not clearly describe what weapons are required for all-domain operations or what the objectives should be vis-à-vis the adversary if the *fait accompli* attack succeeds. More than just the combination of air defenses and offensive fires, A2AD is a linear defense with breadth, depth, and density that make it nearly impossible to flank and penetrate. The stagnation on the Ukrainian battlefield proves as much. Overcoming such comprehensive defenses demands more than just a precision strike network capable of attacking critical nodes in the enemy A2AD complexes, it requires a breakthrough in the speed, concealability, or both of maneuver forces. Here again, the war in Ukraine is informative.

---

33 U.S. Army Training and Doctrine Command, 'TRADOC Pamphlet 525-3-1', 9-14.
34 Ibid, 8.
35 U.S. Army Training and Doctrine Command, 'Multi-Domain Battle', 9.
36 U.S. Army Training and Doctrine Command, 'TRADOC Pamphlet 525-3-1', 24-26.
37 Ibid.

Even armed with the best US tanks, Ukrainian forces cannot maneuver well against Russian A2AD. And the multidomain concepts provide very little insight into this land component gap, offering only that ground forces must be forward stationed or calibrated to prevent the *fait accompli* and minimize the need for new maneuver capabilities.

What to Change?

The multidomain concepts did not describe revolutionary changes in Army missions, friendly forces, or adversaries. Rather than belaboring this point, the following section evaluates the two documents on their ability to outline an evolution in warfare and its associated changes. Did the concepts effectively achieve this goal? Yes.

Appendix B of both concepts captures the list of CRCs required to execute multidomain operations. The list of CRCs in Multi-Domain Battle is extensive and includes a number of information technologies (e.g., self-healing networks, and reliable precision, navigation and timing data).[38] Required capabilities for *Multidomain Operations* are more general and include precision logistics, common operating pictures, and new tools to rapidly converge capabilities.[39]

Both concepts recognize the need for additional fires, sustainment, and command-and-control capabilities. In fact, in development and presentation of the *WayPoint* and *AimPoint* (now *Army 2030*) force packages, Lieutenant General Eric Wesley emphasized the need for command-and-control nodes at echelons above the brigade and enablers such as theater fires and sustainment commands. He also advocated for a rebalancing of sustainment forces between the reserve and active component. At the time, 78 percent of all Army sustainment units resided in the reserves.[40]

Conclusion

For decades, enterprise-level leaders painstakingly built the case for a change in the character of war and for an ensuing military revolution. A large body of thought on the information revolution existed from incredible theoreticians

---

38 U.S. Army Training and Doctrine Command, 'Multi-Domain Battle', 52-61.

39 U.S. Army Training and Doctrine Command, 'TRADOC Pamphlet 525-3-1', B1-B2.

40 'Army Reserve Important for Total Force', *Association of the United States Army*, Oct. 11, 2017, https://www.ausa.org/news/army-reserve-important-total-force.

such as Alan Turing, Claude Shannon, Norbert Wiener, Jon von Neumann, Gordon Moore and Robert Metcalfe. These thinkers early on identified the components of the revolution, namely, networked sensors and compute capabilities used to collect, analyze and act on environmental data. In fact, Norbert Weiner, foresaw that the new epoch would see "the devaluation of the human arm and brain," foreshadowing the replacement of humans on the battlefield and headquarters staffs with an army of autonomous drones and algorithms.

In 2017, the US Army received significant fiscal and symbolic support to paint a bold picture of change and chart a course toward a profoundly different future when Congress authorized establishment of a four-star "futures command" to streamline innovation and acquisition. The underlying conditions seemed conducive to systemic change, providing leaders with the opportunity to anticipate and prepare for the coming storm.

Despite that, when given the chance to imagine something fundamentally new, military thinkers balked. They chose instead to make pragmatic observations regarding enemy weapons and tactics. Based on these observations, concept writers offered recommendations for evolutionary change within the Army and the broader US military. Enterprise leadership wrapped these recommendations in revolutionary language. By adopting this approach, they effectively mitigated internal pushbacks while garnering external support for some of their more contentious recommendations. These leaders created meaningful change in an organization renowned for its conservativism. They fielded new organizations and technologies, published new concepts and doctrines, and procured new munitions.

Looking back, the moment that produced the multidomain concepts is simultaneously a lost opportunity and a success story.

PART 2

PRACTICAL
CONSIDERATIONS

3

Who Left Brigade Combat Teams Out of Multidomain Operations?

Bill Murray

Introduction

The development of AirLand Battle in the late 1970s to early 1980s marked a pivotal moment in the evolution of US Army doctrine. It wasn't simply a top-down directive or a collection of ideas; it was the product of rigorous analysis, intense institutional debates, and—most importantly—a sharp focus on defeating a specific adversary: the Soviet Union. AirLand Battle addressed real problems in real contexts, providing a framework for how the US Army would fight and win in the Cold War era. That doctrine demanded that every capability the US Army developed and employed contribute directly to overcoming the challenges posed by the enemy.[1] In contrast, the current drive to implement Multidomain Operations (MDO) lacks the same institutional rigor, debate, and enemy-focused clarity that made AirLand Battle an enduring concept and doctrine.

Although assessments might differ, the most significant issue with MDO is its detachment from a thorough understanding of the adversary. AirLand Battle was rooted in recognizing specific threats—the size, capabilities, and doctrine of the Warsaw Pact forces, and it was designed to counter them. MDO, on the other hand, often appears overly theoretical, concerned more with integrating various domains—land, air, sea, space, and cyber—than

---

1 Field Manual (FM) 100-5: *Operations* (Washington, DC: U.S. Government Printing Office, 1982), 2-1 to 2-3.

with addressing the actual challenges posed by potential enemies.[2] Instead of being grounded in the realities of the battlefield, it risks becoming a doctrine shaped by what's technologically possible rather than what's tactically and operationally necessary. While multidomain integration is undoubtedly important in modern warfare, it must serve as a means to an end, and not an end in itself. The end must always be defeating the enemy, and that requires a doctrine that prioritizes what the US Army does best: fighting and winning ground combat.

This disconnect is particularly evident in the US Army's growing focus on precision strike capabilities. To be clear, long-range precision fires are a critical asset for the joint force, but they should not come at the expense of the US Army's core mission. Precision strikes alone do not win wars; they shape battles, disrupt adversaries, and help set conditions for success. Warfare's decisive action still takes place on the ground. Historically, the US Army's core strength has been its ability to destroy the enemy in close combat, and its unique value to the joint force lies in its capacity to seize and hold terrain, destroy enemy formations, and impose its will in the unforgiving chaos of ground warfare. The zone between one- and 50-kilometers—the range of direct and indirect fires—is where these fights are won. This zone is the area in which shaping efforts give way to combined arms operations, where soldiers and leaders operate under the harsh realities of heavily contestation, and where the US Army has consistently proven dominant (see Figure 3.1).

To ensure MDO remains grounded in the US Army's core strengths, the focus must shift back to this critical zone of combat. This is where the US Army's basic combat formation, the Brigade Combat Team (BCT) operates, and the BCT's ability to control this space will determine the US Army's relevance in future conflicts. Although the US Army has again designated the division as its unit of action, the Army's divisions provide their subordinate units with tasks and purpose, and set priorities of support, but BCTs execute those orders and conduct the fighting.[3] Brigade Combat Teams need a basic set of tools to accomplish the mission and they cannot rely on the hope that they will be the division's priority for indirect fire or shaping and support

---

2   U.S. Army Training and Doctrine Command, "Multi Domain Operations: Redefining Joint Operations for the 21st Century," TRADOC Pamphlet 525-3-1 (Fort Eustis, VA: U.S. Government Publishing Office, 2018).

3   U.S. Army Training and Doctrine Command, "The US Army in Multi Domain Operations 2028" (Washington, DC: US Army, 2018), 23-25, https://api.army.mil/e2/c/downloads/2021/02/26/b45372c1/20181206-tp525-3-1-the-us-army-in-mdo-2028-final.pdf.

Zone & Range Overview

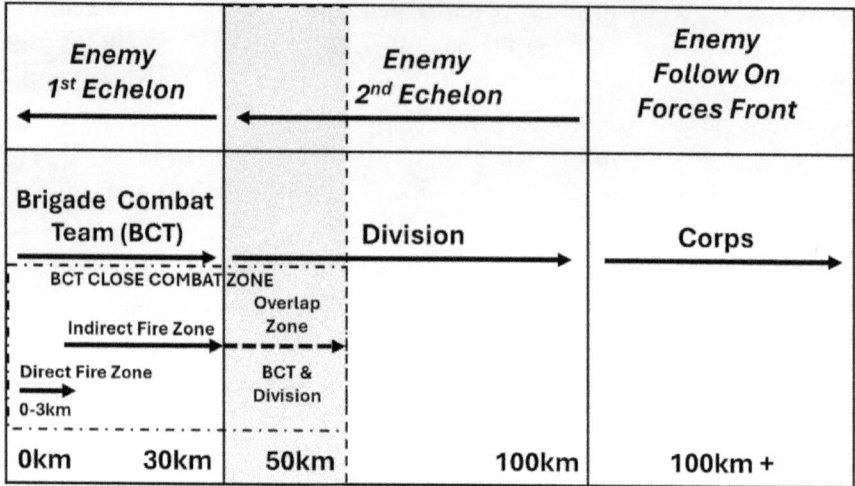

Enemy 1st Echelon		Enemy 2nd Echelon	Enemy Follow On Forces Front
Brigade Combat Team (BCT)		Division	Corps
BCT CLOSE COMBAT ZONE	Overlap Zone		
Indirect Fire Zone			
Direct Fire Zone 0-3km	BCT & Division		
0km 30km	50km	100km	100km +

Overview of BCT, Division, and Corps zones in comparison to enemy echelon zones

Figure 3.1 Zones on the Battlefield
Source: FM 3-0 Operations (US Army) 2022.

capabilities. As a result, the US Army needs robust BCTs, not lean ones. Rather than building a doctrine around the promise of faraway precision strikes, or dizzying cyber effects, generally held at the corps, theater army, or theater fires command level. The US Army must instead prioritize what makes it unique and indispensable to the joint force—brutally dominant close combat.

Achieving dominance in the one-to-50-kilometer zone will not happen by maintaining the status quo. The US Army must invest in building new capabilities within its BCT to meet the demands of 21st century combat. This includes enhancing lethality at the brigade and battalion level by providing those formations with more capable direct and indirect fire weapon systems, enhancing Intelligence Surveillance and Reconnaissance (ISR) capabilities, and building a communication network that will work in any environment. This investment also means innovating in areas like sensor-to-shooter integration to reduce shooter response times and ensure that units have the needed training and relevant doctrine to operate seamlessly across multiple domains. The US Army providing their BCTs these innovations will empower the BCT, and thereby, the Army can ensure that these forces possess the capabilities needed remain dominant in close combat, and yet, still contribute in a meaningful way to the broader multidomain fight.

Moreover, the US Army's MDO doctrine should not center on subordinating the US Army's role to a joint vision of warfare. Rather, US Army doctrine should address sharpening the force's edge where it matters most—on the ground, in the crucible of close combat. Consequently, the US Army must enhance its brigade-level force structure with the capabilities that those forces need to dominate the critical one-to-50-kilometer zone in which close combat occurs in order to win 21st century wars, which is the type of revolutionary doctrine that AirLand Battle brought to the Army in the 1980s.

Why the One-to-50-Kilometer Range Matters in Combat

The one-to-50-kilometer, or close combat zone, is historically the most decisive area in combat.[4] This area includes the space in front and behind of a land force. Having the enemy echeloned to consistently attack or defend at that depth for weeks or months at a time is the most dangerous course of action in any large-scale ground combat conflict. Furthermore, the close combat zone is not linked to a specific enemy or theater, and as a result, the US Army should strive to own this zone in any theater.

The Soviet Army's doctrine and method of attacking in echelons provides a good model that helps conceptually illustrate the close combat zone. Soviet doctrine is characterized by the structured and phased deployment of forces, the goal of which is to enhance their army's operational capability in the close combat zone. The first echelon is designed to move rapidly, often achieving its objectives within the first day or two of the operation.[5] The first echelon of a Soviet attack typically consists of frontline combat units, including reconnaissance forces, mechanized and motorized infantry, and tank formations. The first echelon typically engages into enemy territory at a depth of approximately 10-30 kilometers. These forces are tasked with conducting the initial assault, breaching enemy defenses, securing key objectives, and creating the conditions needed for the commitment of second echelon forces. The first echelon aims to rupture an enemy's front lines through a deep penetration into enemy territory, and subsequently, creating chaos in the enemy's lines, and disrupting the enemy's command and

---

4   Russell F. Weigley, *The Age of Battles: The Quest for Decisive Warfare from Breitenfeld to Waterloo* (Bloomington: Indiana University Press, 1991), 12-15.
5   David C. Isby, *The Soviet Army: Operations and Tactics* (London: Jane's Publishing, 1986), 123-145.

control.[6] Fire support, including artillery and airstrikes, plays a critical role in this phase, as it aims to weaken enemy positions prior to the ground assault.

The second echelon consists of reserve forces that are held back to support and reinforce the first echelon's initial attack. The second echelon operates at a depth of about 30 to 100 kilometers behind the front lines. This distance allows the second echelon to be able to respond flexibly to the developments created by the first echelon. Second, echelon units may include additional armored formations, artillery, and specialized reconnaissance troops. This echelon also provides flexibility to respond to unforeseen developments on the battlefield, whether by reinforcing successful fronts, exploiting emerging opportunities, or countering enemy responses.[7] In addition, the second echelon are tasked with preparing the battlefield for further offensives as the situation evolves.[8] Once the first echelon has secured its objectives, the second echelon's role is to consolidate gains, fortify newly acquired positions, and maintain momentum and finally, the follow on forces' front echelon fills the role of the second echelon.

In the US Army's version of MDO, different echelons engage in combat at varying distances based on their roles. US Army corps, for instance, engage at distances exceeding 100 kilometers, coordinate operations across multiple domains, and integrate joint forces over vast distances and wide areas. US Army divisions, on the other hand, operate within a range of 30- to 100-kilometers, and prioritize executing operational plans, and conducting combined arms maneuvers. Brigade Combat Teams typically operate at distances of 1 to 30 kilometers.[9]

Nevertheless, Army BCTs lack the ability to attack an enemy's second echelon forces, which tend to stage at 30- to 100-kilometers. This structural defect must change. When the original attacking depths were planned for Army BCTs, offensive and defensive strike technology did not exceed 30 kilometers. The technology to attack beyond this range now exists and BCTs need it.

---

6   Ibid.

7   Ibid.

8   Ibid.

9   U.S. Army Training and Doctrine Command, "Multi Domain Operations: Redefining Joint Operations for the 21st Century," TRADOC Pamphlet 525-3-1 (Fort Eustis, VA: U.S. Government Publishing Office, 2018).

Divisions and their BCTs' attack thresholds must overlap. This will facilitate the brigade's ability to focus on second echelon forces in the 30–50-kilometer zone (or the Overlap Zone), while also dominating the close combat zone. Furthermore, a division cannot support more than two BCTs simultaneously in the 30-to-100km range fight in large-scale ground combat, despite the fact that US Army divisions usually command three or more BCTs. Brigade Combat Teams can manage fighting in the additional space and handle the associated increased information requirements if provided with the necessary resources discussed in this chapter, without overstressing their existing capabilities. It is preferable to mitigate this risk by enhancing the capabilities of BCTs, rather than accepting the alternative risk of deploying brigades that are less capable and potentially ill-equipped to succeed in complex operational environments.

In 21st century warfare, the close combat zone is critical for combined arms operations, particularly at the brigade level. This range encompasses two vital zones: the direct fire zone (1-3km) and the indirect fire zone (3-50km), and it allows brigades to engage both the first and second echelon forces. Understanding the dynamics of these ranges is essential, as battles can be won or lost within these critical distances.

Transitioning to the indirect fire zone, which spans from 3- to 50-kilometers, the dynamics shift significantly. This range is essential for delivering fire support and achieving operational goals without exposing ground forces to direct enemy fire. The use of artillery, mortars, and rocket systems within this range allows for concentrated firepower against enemy formations and critical infrastructure. This capability not only disrupts enemy operations but also forces adversaries to adapt their tactics, potentially leading to disarray and confusion.

In the Russo-Ukrainian War, the Ukrainian military's use of UAVs and precision-guided artillery to target Russian command and control nodes, logistics hubs, and troop concentrations within the close combat zone has been a key factor in their ability to disrupt the Russian advance. In the early stages of the war, Ukrainian forces used UAVs to identify and target Russian artillery positions and command centers in the 10- to 20-kilometer range, allowing them to disrupt the Russian ability to coordinate their fires and maneuver their forces.[10] Additionally, Ukrainian forces used precision-guided

---

10 "Ukraine's Drone War: How Ukrainian Forces Are Using Drones to Target Russian Positions," *The New York Times*, June 15, 2022, https://www.nytimes.com/2022/06/15/world/europe/ukraine-drones-russia.html.

munitions, such as the High Mobility Artillery Rocket System (HIMARS), to target Russian logistics hubs and troop concentrations within the 20- to 40-kilometer range, further disrupting the Russian ability to supply and reinforce their troops.[11] This has forced the Russian military to operate in a more dispersed and decentralized manner, making it more difficult to mass their forces and unlock their potential on the battlefield by bringing the full weight of their combat power to bear against the Ukrainians.

To conclude, the close combat zone, or that area (e.g., the length, height, and width) between one- to 50-kilometers on a battlefield, is indispensable for combined arms operations at the brigade level. The success of contemporary military engagements hinges on managing the direct fire zone, where immediate combat takes place, and the indirect fire zone, which allows for sustained support and engagement from a distance. The ongoing conflict in Ukraine illustrates that the ability to integrate and destroy the enemy within these ranges can significantly influence the outcome of battles, underscoring their critical importance in contemporary warfare. While winning battles may not necessarily guarantee the outcome of a war, losing battles can have a profoundly debilitating effect, eroding momentum, undermining strategy, and ultimately increasing the likelihood of defeat, making it a risk that leaders cannot afford to take.

Direct Fire Weapons Systems

The US Army's late 20th century weapon systems—specifically the M1 Abrams tank, M2 Bradley infantry fighting vehicle, and Javelin missile—have provided essential direct fires on the battlefield. However, the evolving nature of warfare, particularly illustrated by the ongoing conflict in Ukraine, has exposed several significant limitations across these systems. To maintain superiority throughout the battlefield's important direct fire zone (i.e., the space from a formation out to approximately three kilometers), brigades require cutting-edge systems that are sufficiently advanced to respond to 21st century threats and challenges.

The M1 Abrams tank, a mainstay of US armored forces, is equipped with a powerful cannon and composite armor, providing a formidable combination of firepower and protection. However, the Abrams tank's

---

11 "HIMARS in Ukraine: A Game-Changer on the Battlefield," *Forbes*, July 20, 2022, https://www.forbes.com/sites/davidaxe/2022/07/20/himars-in-ukraine-a-game-changer-on-the-battlefield/?sh=5a444f6d66f2.

weight and fuel consumption can impede its tactical and operational mobility and limit potential gap (e.g., bridging) crossing opportunities.[12] The Abrams tank's reliance on traditional anti-armor measures makes them less effective against contemporary threats that utilize precision-guided munitions. A lighter tank is desperately needed and adding robotic substitutes will increase a brigade's direct fire capability.

The Javelin missile, while highly capable as a man-portable anti-tank guided missile, also demonstrates the challenges faced by legacy systems. The Javelin's fire-and-forget technology and top-attack capability make it a potent tool against armored threats; however, its limited availability in the field and the need for dismounted operators often constrains its capabilities in fast-paced combat scenarios.[13] In Ukraine, the high demand for Javelins has revealed that while they are invaluable for countering armor, BCTs need complementary systems that can offer greater versatility and immediate response in the direct fire zone.

The ongoing Russo-Ukraine war has highlighted the enduring importance of tanks and anti-tank missile systems in modern combat. The Russian military's initial advances in Ukraine were facilitated by the use of tank columns, working in tandem with aerial assaults, which allowed the Russians to rapidly exploit weaknesses in Ukrainian defenses and gain a significant amount of territory.[14] However, the Ukrainian military's use of anti-tank missile systems, such as the Javelin and Next Generation Light Anti-Tank Weapon (NLAW), proved effective in blunting the Russian armored advance, with reports of significant Russian tanks losses attributed to shoulder-fired anti-tank systems.[15] During the Battle of Kyiv, for instance, Ukrainian forces used Javelin missiles to destroy several Russian tanks, including the advanced T-72B3s, which helped to slow the Russian advance and ultimately forced the Russian withdraw from the city.[16] Similarly, the Ukrainian forces

---

12 "M1A2 Abrams Main Battle Tank," Military.com, accessed January 3, 2025, https://www.military.com/equipment/m1a2-abrams-main-battle-tank.

13 "Javelin," Lockheed Martin, accessed January 3, 2025, https://www.lockheedmartin.com/en-us/products/javelin.html.

14 "Russia's Tank Columns: The Key to Their Initial Success in Ukraine," *The National Interest*, March 10, 2022, https://nationalinterest.org/blog/buzz/russias-tank-columns-key-their-initial-success-ukraine-200636.

15 "Ukraine's Anti-Tank Missiles: The Game-Changer on the Battlefield," *Forbes*, April 20, 2022, https://www.forbes.com/sites/davidaxe/2022/04/20/ukraines-anti-tank-missiles-the-game-changer-on-the-battlefield/?sh=4c944f6d66f2.

16 "The Battle of Kyiv: How Ukraine's Military Held Off the Russian Advance," *The New York Times*, April 5, 2022, https://www.nytimes.com/2022/04/05/world/europe/ukraine-russia-kyiv-battle.html.

use of NLAW missiles attributed to the destruction of numerous Russian armored vehicles, including tanks and infantry fighting vehicles. Moreover, the Ukrainians use of the NLAW in urban environments was particularly effective, where the system's top-attack capability has allowed Ukrainians to target Russian vehicles from unexpected vantage points.[17] As noted by military analyst, Michael Kofman, "the war in Ukraine has shown that tanks and anti-tank missile systems remain critical components of modern ground warfare, and that the side that can effectively employ these systems will have a significant advantage on the battlefield."[18]

To own the direct fire zone, BCTs should integrate advanced systems that enhance mobility, lethality, and survivability. Future platforms might include next-generation armored vehicles equipped with active protection systems (APS) to intercept incoming projectiles and advanced sensors to enhance situational awareness. Furthermore, integrating unmanned ground vehicles (UGVs) and UAVs can provide critical reconnaissance and strike capabilities, facilitating greater battlefield success. Additionally, future direct fire systems should be capable of engaging targets across a spectrum of scenarios, from heavy armor to infantry, while possessing rapid repositioning capacity to avoid counter-battery fire. Tying in sensor platforms like low-earth orbiting (LEO) satellites and UAVs into anti-tank missile systems can increase ranges from two to three kilometers out to, and perhaps beyond, ranges normally within range of indirect fire systems. To be sure, when the maximum range of a weapon system is controlled by line-of-sight, the US Army, and its allies and partners, should invest in systems that increase the line-of-sight.

The development of systems that can integrate with existing command and control networks will also enhance real-time decision-making and operational effectiveness. Ultimately, by investing in advanced systems with these capabilities, BCTs can maintain their competitive advantage and remain prepared to face the evolving challenges of 21st century warfare.

---

17 "NLAW Missiles: The Ukrainian Military's Secret Weapon Against Russian Armor," *The Telegraph*, May 10, 2022, https://www.telegraph.co.uk/news/2022/05/10/nlaw-missiles-ukrainian-militarys-secret-weapon-russian-armour/.
18 Kofman, Michael. "The War in Ukraine: A New Era of Ground Warfare." *Journal of Slavic Military Studies* 35, no. 2 (2022): 155-170.

Indirect Fire Weapon Systems

Legacy indirect fire systems and the M777 Howitzers have long served as critical components of US Army fire support, particularly in BCTs. These systems were designed to provide artillery support over distances ranging from three- to 30-kilometers, perhaps the most vital range for shaping the battlefield for BCT combat in the direct fire zone. However, as evidenced by the ongoing conflict in Ukraine, the limitations of these legacy systems have become increasingly apparent, necessitating the urgent modernization of artillery capabilities to ensure operational effectiveness in contemporary combat scenarios.

Towed artillery has been deemed less capable and outdated due to its limited mobility and vulnerability on the battlefield. As early as the aftermath of World War I, military strategists recognized the limitations of towed artillery, with many considering it to be a relic of a bygone era.[19] The Soviet Army, in particular, was at the forefront of transitioning to self-propelled artillery, recognizing the need for greater mobility and survivability on the battlefield.[20] By the 1960s, the Soviet Army had largely completed its transition to self-propelled artillery, with the majority of its artillery pieces being mounted on armored vehicles.[21] In fact, by the 1970s, the Soviet Army had transitioned nearly all of its artillery to self-propelled systems, rendering towed artillery all but obsolete.[22] This shift towards self-propelled artillery was a deliberate effort to increase the mobility and firepower of Soviet artillery units, and it has since been adopted by most modern militaries.[23] The US Army continues to use towed artillery as its primary means to deliver indirect fire.

The M777 Howitzer is a lightweight, towed artillery piece renowned for its precision and mobility. The range of both systems, even with extended range projectiles, barely strikes into the enemy's second echelon at 30- to 100-kilometers. Capable of firing a variety of munitions, including Global Positioning System (GPS)-guided *Excalibur* shells, the M777 offers a

---

19 J.F.C. Fuller, *The Reformation of War* (London: Hutchinson, 1923), 134-135.

20 Steven J. Zaloga, *Soviet Self-Propelled Guns, 1936-1945* (Oxford: Osprey Publishing, 2018), 12-15.

21 David M. Glantz, *Soviet Military Operational Art: In Pursuit of Deep Battle* (London: Frank Cass, 1991), 56.

22 Nikolai S. Simonov, *Voyennaya Promyshlennost' SSSR v 1920-1950s* (Moscow: Rosspen, 2005), 234-235.

23 Ian V. Hogg, *The Soviet Self-Propelled Guns, 1936-1945* (London: Arms and Armour Press, 1970), 45-50.

maximum range of approximately 30 kilometers with standard projectiles and up to 40 kilometers with precision munitions.[24] Despite its advantages, the M777 possesses limited mobility. The M777 is limited by being a towed howitzer, which reduces its responsiveness to localized attacks and further exposes it to counter-battery fire.

To dominate the three- to-50-kilometer zone (BCT Indirect Fire Zone), BCTs must incorporate advanced capabilities that address the shortcomings of legacy systems. First and foremost, advanced artillery systems must possess improved mobility. The US Army needs motorized artillery to provide a combination of long-range, precision firepower and rapid, mobile artillery support, which will provide its artillery forces with greater flexibility and more capable on the battlefield.[25] Motorized artillery, in particular, offers advantages over traditional tracked artillery, by offering enhanced mobility on improved and unimproved road networks, reducing logistical burdens, and lowering operating costs.

Wheeled artillery offers significant advantages over towed artillery, including improved mobility, increased volume of fire, and enhanced accuracy. For instance, wheeled artillery systems can displace in as little as 30 seconds, compared to towed artillery systems, which can take up to ten minutes to displace.[26] Additionally, wheeled artillery systems can achieve a higher volume of fire, with systems capable of firing up to ten rounds per minute, compared to some towed systems which can fire at a rate of two-to-five rounds per minute.[27] Furthermore, wheeled artillery systems often feature advanced fire control and navigation systems, allowing for more accurate targeting.[28] Overall, the improved mobility, increased volume of fire, and enhanced accuracy of wheeled artillery systems make them a more efficient option.

Viewed as a whole, motorized artillery makes an attractive option for expeditionary operations and rapid-response missions, which is the basic strategy for employing US Army forces.[29] Additionally, the US Army must

---

24 "M777," BAE Systems, accessed January 3, 2025, https://www.baesystems.com/en/product/m777.

25 James F. Dunnigan, *How to Make War: A Comprehensive Guide to Modern Warfare* (New York: William Morrow Paperbacks, 2003), 234-235.

26 David E. Johnson, "Wheeled Artillery: A New Era for Mobile Firepower" (Washington, D.C.: Center for Strategic and International Studies, 2009), 12-15.

27 Hogg, Ian V. The *Illustrated Encyclopedia of Artillery* (London: Arms and Armour Press, 1987).

28 Foss, Christopher F. Artillery of the World. London: Jane's Publishing, 1985.

29 Ian V. Hogg, *The Illustrated Encyclopedia of Artillery* (London: Quarto Publishing, 2011), 156-159.

focus on outfitting BCTs with advanced targeting technologies, including precision-guided munitions and improved fire control systems that utilize real-time data. Doing so will enable artillery units to conduct quicker engagements than they do today, while relying on even more accurate intelligence.

Future systems need longer ranges and more substantial payloads. As the theory goes, precision fires are effective when they are precise, nonetheless, precision is not a guarantee. In that case, more expensive and less precise projectiles are utilized. The US Army needs to pursue the use of thermobaric indirect fire projectiles and focus on employing those systems in mass. HIMARS and Multiple Launch Rocket System (MLRS) already have thermobaric capabilities, but the US Army prioritizes precision over area-of-effect destruction. More thermobaric capabilities are needed to facilitate dominating the close combat zone. In addition, the use of sensor platforms like LEO satellites and UAVs for reconnaissance and target acquisition would enhance the integration of indirect fire support with infantry, armor, and other land forces. Furthermore, implementing automated resupply and logistical support systems can boost the sustainability of artillery units in periods of steady state operations and spikes in artillery-oriented operations. This overhaul of the artillery system would generate continuous fire support without the vulnerabilities associated with traditional supply lines, which have historically been significant battlefield target liabilities.

The current Russo-Ukraine war highlights the importance of long-range direct fire weapons, such as the HIMARS, in 21st century land warfare. The Ukrainian military's use of HIMARS has shown profoundly impactful out to ranges up to 70 kilometers at targeting Russian command and control nodes, logistics hubs, and troop concentrations.[30] In July 2022, for instance, Ukrainian forces HIMARS destroyed a Russian ammunition depot in Nova Kakhovka, which destroyed a large quantity of Russian munitions and equipment.[31] The destruction of the ammunition depot limited the Russian army's ability to resupply and replenish their stocks, hindering their ability to maintain a strong offensive, while the targeting of the command center disrupted the Russian army's ability to coordinate and communicate,

30 "HIMARS in Ukraine: A Game-Changer on the Battlefield," *Forbes*, July 20, 2022, https://www.forbes.com/sites/davidaxe/2022/07/20/himars-in-ukraine-a-game-changer-on-the-battlefield/?sh=5a444f6d66f2.
31 "Ukraine Strikes Russian Ammunition Depot with HIMARS," *The New York Times*, July 12, 2022, https://www.nytimes.com/2022/07/12/world/europe/ukraine-russia-himars-ammunition-depot.html.

making it more difficult for them to respond to Ukrainian counterattacks.[32] The HIMARS strike forced the Russian army to pause their advance and regroup, giving Ukrainian forces time to reinforce their positions and prepare for future battles, ultimately preventing the Russian army from advancing in the region.[33]

In contrast, the Russian military's reliance on towed artillery has often been a liability in combat.[34] For instance, during the battle for the city of Bakhmut in Donetsk Oblast, the Russian army's use of towed artillery, such as the 152mm D-20 gun-howitzer, proved to be a disadvantage.[35] The slow pace of deployment and redeployment of these systems, which required manual towing and setup, limited the Russian army's ability to rapidly respond to changing battlefield conditions and exploit weaknesses in the Ukrainian defenses.[36] Furthermore, the towed artillery's lack of mobility and vulnerability to counter-battery fire made them easy targets for Ukrainian forces, who were able to use UAVs and guided munitions to accurately target and destroy the Russian guns.[37] The Ukrainian military has taken advantage of this vulnerability, using artillery and HIMARS systems to target and destroy Russian towed artillery pieces, which has significantly reduced the Russian military's ability to provide supporting firepower to their ground forces.[38] As noted by military analyst, Phillips O'Brien, "the war in Ukraine has shown that long-range direct fire weapons like HIMARS are a

---

32 Rob Lee, "The Impact of HIMARS on the Russian Army," *Foreign Policy*, July 1, 2022, https://foreignpolicy.com/2022/07/01/himars-russia-ukraine-war/.

33 "Russian Army's Advance in Ukraine Slows After HIMARS Strike," *The New York Times*, June 20, 2022, https://www.nytimes.com/2022/06/20/world/europe/russia-ukraine-war-himars.html.

34 "Russia's Towed Artillery: A Liability in Modern Combat," *The National Interest*, August 15, 2022, https://nationalinterest.org/blog/buzz/russias-towed-artillery-liability-modern-combat-203756.

35 "Russia's Artillery Woes in Ukraine," *The Moscow Times*, August 10, 2022, https://www.themoscowtimes.com/2022/08/10/russias-artillery-woes-in-ukraine-a78131.

36 Michael Kofman, "The Russian Military's Artillery Challenges in Ukraine," *War on the Rocks*, September 1, 2022, https://warontherocks.com/2022/09/the-russian-militarys-artillery-challenges-in-ukraine/.

37 "Ukraine's Drone Warfare: A Game-Changer on the Battlefield," *The Economist*, October 1, 2022, https://www.economist.com/the-economist-explains/2022/10/01/ukraines-drone-warfare-a-game-changer-on-the-battlefield.

38 "Ukraine's Counter-Battery Fire: How They're Taking Out Russian Artillery," *The Telegraph*, September 10, 2022, https://www.telegraph.co.uk/news/2022/09/10/ukraines-counter-battery-fire-how-theyre-taking-out-russian-artillery/.

game-changer on the modern battlefield, while towed artillery is increasingly obsolete and vulnerable to counter-battery fire."[39]

While the M777 howitzer and M109 Paladin have provided capable artillery support in the past, the realities of 21st century land warfare necessitate an urgent upgrade of these systems. The lessons learned from the ongoing conflict in Ukraine highlight the need for BCTs to adopt enhanced artillery solutions that can operate in the indirect fire zone. By investing in advanced artillery systems with wheeled mobility, state-of-the-art targeting capabilities, and innovative logistical support, then US forces can improve their battlefield prowess in 21st century conflicts.

ISR Systems

ISR systems are fundamental to the impact that the US Army's BCTs can deliver on the battlefield. These systems encompass a range of platforms and technologies, including UAVs, unmanned ground vehicles, and satellite capabilities, which collectively enhance situational awareness and decision-making capabilities.

Division-Level Systems

Divisions, on the other hand, use the MQ-1C *Gray Eagle* for ISR. The *Gray Eagle* maintains an advanced sensor suite and the ability to stay aloft for up to 25 hours. The *Gray Eagle* provides division commanders and their staffs with real-time ISR capabilities, thereby affording them with the tools to make better informed decisions.[40] The *Gray Eagle's* capabilities include a high-resolution electro-optical/infrared (EO/IR) sensor, a synthetic aperture radar (SAR) system, and a communications relay system, allowing it to provide persistent surveillance and communications support to the division's BCTs. However, the *Gray Eagle* also has some shortcomings, including its limited weapons payload capacity and vulnerability to air defenses, which limits its impact in contested environments.

---

39 O'Brien, Phillips. "The War in Ukraine: A New Era of Artillery Warfare." *Journal of Military and Strategic Studies* 20, no. 2 (2022): 1-15.
40 US Army, "MQ-1C Gray Eagle," US Army Fact Files, https://odin.tradoc.army.mil/WEG/Asset/MQ-1C_Gray_Eagle_American_Medium-Altitude.

Brigade Combat Team-Level Systems

BCTs used the RQ-7 *Shadow* extensively throughout the Global War on Terror, but on March 19, 2024, the US Army deleted the program without fielding a replacement program.[41] This situation has created a massive problem for BCTs by generating a capability gap between what a BCT is required to do and their means to accomplish those requirements. This problem is compounded exponentially when considering that the fact that the US Army eliminated the majority of BCT cavalry scouts in 2024, leaving the force without all-weather ISR capabilities that it has enjoyed since its formation.[42]

Company-Level Systems

The Army's company-level cannot be overlooked either. Each BCT's company possesses the RQ-20 Puma UAV, which come with a range of 20 to 30 kilometers in line-of-sight conditions, and a beyond line-of-sight transmission upward of 60 kilometers with satellite augmentation.[43]

Looking Forward

The elimination of BCT-level UAVs, coupled with the elimination of cavalry formations has left the Army BCTs woefully under-resourced when it comes to UAVs and the ability to conduct proximal reconnaissance and security operations. To support MDO and conduct operations within the close combat zone, BCTs require advanced, and legacy, ISR systems that can closely integrate with joint and coalition forces. BCTs must be able to use sensor platforms like LEO satellites for targeting and communications. LEO systems have the potential to revolutionize the way BCTs conduct targeting and ISR operations. By providing persistent surveillance over a wide area, LEO satellites can enable BCTs to track enemy movements, identify high-value targets, and conduct battle damage assessments in real time, further increasing the impact that BCTs provide to Army, joint, and coalition forces

41 U.S. Army, "Shadow UAS Retires After Decades of Service," Army.mil, March 19, 2024, https://www.army.mil/article/275946/shadow_uas_retires_after_decades_of_service.

42 U.S. Army, "Army Force Structure Transformation," Army.mil, February 27, 2024, https://api.army.mil/e2/c/downloads/2024/02/27/091989c9/army-white-paper-army-force-structure-transformation.pdf.

43 AeroVironment, "Puma AE: Unmanned Aircraft System," AeroVironment, Inc., https://www.avinc.com/uas/puma-ae.

on 21st century battlefields. In addition, these capabilities are particularly useful providing support to direct and indirect fire missions. This is because LEO satellites provide high-quality targeting data for precision-guided munitions, artillery, and mortar systems, and therefore accelerates improved accuracy at the BCT level.

Furthermore, LEO satellites can also deliver the information that BCT commanders, and their staffs, need to develop real-time situational awareness, which can assist them in making better informed decisions about ongoing operations, and also quickly respond to emerging threats. What's more, LEO satellites' advanced sensors and communication relay capabilities also facilitate intelligence sharing and targeting data sharing across forces and command and control systems. Additionally, LEO satellites can operate at high altitudes, above the weather and out of range of most enemy air defenses, making them a survivable and reliable asset. Overall, the integration of LEO satellites into BCT operations has the potential to provide a significant, positive impact on BCT targeting and ISR operations, allowing them to provide an outsized impact on rapidly changing, and increasingly lethal, 21st century battlefields.

Looking toward the future, BCTs need UAVs in mass, especially when considering their loss of cavalry formations. In the close combat zone, UAVs may play a critical role in supporting land operations by offsetting the loss of cavalry by generating real-time surveillance and reconnaissance information. As these systems support the BCTs ability to detect and track enemy movements, identify high-value targets, and conduct battle damage assessments, BCTs will provide more positive battlefield outcomes for the divisions to which they are attached. Considering direct fire, future UAVs could be used to provide targeting data for precision-guided munitions, such as tank rounds or anti-tank missiles, allowing BCTs to engage enemy armor, infantry, engineer, and artillery formations with improved accuracy and therefore greater lethality.

Additionally, BCT level UAVs could be used to provide overwatch and surveillance for dismounted infantry patrols, enabling them to detect and engage enemy forces from more advantageous ranges and from more advantageous locations. For indirect fire support, future UAVs could be used to provide targeting data for artillery and mortar systems, allowing BCTs to conduct precision strikes against enemy positions and command centers.

Furthermore, ground drones could be used to conduct reconnaissance and surveillance in urban and complex terrain, providing critical information to support indirect fire missions and reduce the risk of collateral damage. By integrating these capabilities, BCTs could use UGVs and UAVs to support a range of direct and indirect fire missions, including precision strikes, suppressive fire, and area denial, ultimately enabling them to target and destroy enemy forces in the one-to-50-kilometer zone with greater speed, precision, and effectiveness. Even with increased technological capabilities, BCTs need cavalry scout formations. Weather and electronic warfare interfere with even the most technologically advanced ISR systems, except for the soldier on the ground.

The current Russo-Ukraine war has highlighted the critical importance of ISR platforms, such as UAVs in modern combat. The Ukrainian military's use of UAVs, such as the Bayraktar TB2, has proven highly capable in providing real-time intelligence on Russian troop movements and positions, allowing them to target Russian forces with precision-guided munitions.[44] In March 2022, Ukrainian forces used a Bayraktar TB2 to identify and target a Russian column near the city of Kharkiv, resulting in the destruction of several Russian tanks and armored vehicles.[45] Similarly, the use of smaller UAVs, such as the RQ-20 Puma, has allowed Ukrainian forces to conduct reconnaissance and surveillance in urban areas, providing critical intelligence on Russian troop movements and strongpoints.[46] The Russian military has also made extensive use of UAVs, including the Orlan-10, to conduct reconnaissance and surveillance, but the Ukrainian military's use of UAVs has been more effective due to their greater numbers and more advanced capabilities.[47] As noted by military analyst Samuel Bendett, "the war in Ukraine has shown that UAVs and drones are now a critical component of modern warfare,

---

44 "Ukraine's Drone War: How Ukrainian Forces Are Using UAVs to Target Russian Positions," *The New York Times*, June 15, 2022, https://www.nytimes.com/2022/06/15/world/europe/ukraine-drones-russia.html.

45 "Bayraktar TB2: The Drone That's Changing the Face of Warfare," *Forbes*, April 20, 2022, https://www.forbes.com/sites/davidaxe/2022/04/20/bayraktar-tb2-the-drone-thats-changing-the-face-of-warfare/?sh=4c944f6d66f2.

46 "RQ-20 Puma: The Drone That's Helping Ukraine's Military in Urban Warfare," *The Telegraph*, May 10, 2022, https://www.telegraph.co.uk/news/2022/05/10/rq-20-puma-drone-thats-helping-ukraines-military-urban-warfare/.

47 "Russia's Orlan-10: The Drone That's Providing Critical Intelligence for Russian Forces," *The National Interest*, July 20, 2022, https://nationalinterest.org/blog/buzz/russias-orlan-10-drone-thats-providing-critical-intelligence-russian-forces-203456.

---

providing real-time intelligence and surveillance capabilities that are essential for combat operations."[48]

Enhanced capabilities such as multi-sensor fusion, which combines data from various ISR platforms to provide a comprehensive operational picture, are essential. This approach not only improves situational awareness but also facilitates better decision-making in rapidly changing environments. Future ISR systems should incorporate resilience against these tactics, employing technologies such as low-probability-of-detection (LPD) radars and secure communication links to ensure the continuous flow of information even in challenging operational contexts.[49]

Communications Network and Systems

To maximize capabilities, the US Army needs advanced communications networks and systems. The MDO framework emphasizes the importance of speed and agility in decision-making so that US forces, and their partners and allies, can outpace enemy responses, and so that they can quickly adapt to the dynamic conditions of 21st century battlefields.[50] Enhanced ISR capabilities and targeting systems require an agile and simple communications network and systems. Currently, the US Army has neither, and in an era where the United States has led the world in the development of cutting-edge networks and communications systems, this is unacceptable. The Army's legacy communication systems are 10 to 30 years behind current civilian capabilities, and the few networks and systems that look remotely modern are only in the hands of a fraction of the US Army.[51] There is an urgent need for communications modernization to support MDO and fight within the close combat zone.

BCTs need modern and capable communication systems that can provide integrated connectivity across domains to maximize their capabilities to reinforce MDO. This necessitates the fielding of secure, high-bandwidth

---

48 Bendett, Samuel. "The War in Ukraine: A New Era of Drone Warfare." *Journal of Slavic Military Studies* 35, no. 2 (2022): 171-185.

49 "The Future of Communications in Satellite-Denied Environments," QinetiQ, accessed January 3, 2025, https://www.qinetiq.com/en-au/news/the-future-of-communications-in-satellite-denied-environments.

50 Joint Chiefs of Staff, *Joint Publication 3-0: Joint Operations* (2017), https://www.jcs.mil/Portals/36/Documents/Doctrine/pubs/jp3_0.pdf.

51 "Study: U.S. Military Communications Technology and Cyber Defense Challenges Remain," PR Newswire, December 10, 2020, https://www.prnewswire.com/news-releases/study-us-military-communications-technology-and-cyber-defense-challenges-remain-301190660.html.

communication networks capable of transmitting data, voice, and video in real time. Advanced systems such as the Tactical Network Transport (TNT) and the Integrated Tactical Network (ITN) represent significant upgrades over legacy systems, offering improved bandwidth, interoperability with joint, coalition, and partnered forces, as well as network resilience against jamming and other cyber threats.[52] These advanced systems are designed to support integrated operations and facilitate the flow of information among ground forces, air support, and other domains. Furthermore, BCTs should focus on enhancing situational awareness through improved data-sharing capabilities. These systems require the use of LEO satellites, which is not common across all the Army's BCTs. The use of these satellites must become standardized across the US Army. Using any future communication network without LEO satellites would be like using a cell phone without a cell phone or wi-fi network.

The amount of data that modern armies use is massive, and one of the only ways to solve data management issues is with cloud-based solutions. This will also help support the use of artificial intelligence in the ISR and targeting processes. Cloud-based solutions and advanced data analytics will help commanders make better informed decisions due to having accurate and timely information on hand.[53] The ability to share intelligence quickly across units is critical for successful operations in the close combat area, particularly as threats become more sophisticated and adaptive.

The ability to operate in contested environments is another essential capability that BCTs must develop. Modern communications systems must be resilient against electronic warfare and cyber operations, as demonstrated by Russian tactics in Ukraine, which have included jamming and spoofing communications to disrupt Ukrainian command and control. During the Russian invasion of Ukraine in 2022, Russian forces successfully jammed Ukrainian communication systems in the Kharkiv region, disrupting their ability to coordinate artillery strikes and troop movements. This allowed Russian forces to gain an advantage and capture key cities, including Izyum.

To counter these threats, BCTs should implement technologies such as frequency-hopping spread spectrum (FHSS) communications and employ

---

52 "ITN Radio: Seamless Communication," GovConWire, accessed January 3, 2025, https://www.govconwire.com/articles/itn-radio-seamless-communication/.

53 U.S. Army Command and General Staff College, "Command Post Computing Environment (CPCE) Overview," last modified January 2023, https://www.dote.osd.mil/Portals/97/pub/reports/FY2020/army/2020cpce.pdf.

secure satellite communication (SATCOM) capabilities that allow for robust connectivity even in denied environments.

The Russo-Ukraine war highlights the critical importance of reliable and secure communications in modern combat, with cellular networks and satellite communications playing a vital role in facilitating command and control, intelligence sharing, and coordination between units. The Ukrainian military's use of cellular networks, such as those provided by Kyivstar and Vodafone Ukraine, has allowed them to maintain communication with their forces in the field, even in areas where traditional communication infrastructure has been damaged or destroyed.[54] The Russian military's attempts to disrupt Ukrainian cell networks through electronic warfare and cyber-attacks have led to the widespread adoption of satellite communications, particularly *Starlink*. *Starlink* is a LEO satellite, which has provided a secure and reliable means of communication for Ukrainian forces.[55] Between May 24 and June 1, 2022, Russian forces launched a series of intense artillery and infantry assaults on Ukrainian positions in Bakhmut, but Ukrainian forces were able to use Starlink to maintain communication and coordinate their defense, ultimately repelling the Russian attack and inflicting significant casualties on the Russian military. Specifically, Ukrainian forces used Starlink to transmit critical information, such as artillery coordinates and troop movements, allowing them to launch precision strikes against Russian positions and destroy several Russian tanks and infantry fighting vehicles. As a result, the Russian attack on Bakhmut was repelled, and Ukrainian forces were able to reinforce their defenses.[56]

The use of Starlink has also allowed Ukrainian forces to access critical intelligence and surveillance data, such as that provided by UAVs, in real time, enabling them to make more informed and timely decisions on the battlefield.[57] As noted by military analyst, Peter Singer, "The war in Ukraine has shown that access to reliable and secure communications, particularly

---

54 "Ukraine's Cell Networks: A Lifeline for Ukrainian Forces," *The New York Times*, April 10, 2022, https://www.nytimes.com/2022/04/10/world/europe/ukraine-cell-networks.html.

55 "Russia's Electronic Warfare: Disrupting Ukrainian Communications," *The National Interest*, May 20, 2022, https://nationalinterest.org/blog/buzz/russias-electronic-warfare-disrupting-ukrainian-communications-202456.

56 Jack Watling and Nick Reynolds, "Ukrainian Artillery: Innovation in the Face of Adversity," Royal United Services Institute (RUSI), June 2022, https://rusi.org/explore-our-research/publications/occasional-papers/ukrainian-artillery-innovation-face-adversity.

57 "UAVs and Starlink: A Powerful Combination for Ukrainian Forces," *The Telegraph*, July 10, 2022, https://www.telegraph.co.uk/news/2022/07/10/uavs-starlink-powerful-combination-ukrainian-forces/.

through satellite networks like Starlink, is now a critical component of modern warfare, allowing forces to maintain situational awareness and coordinate their actions in real time."[58]

While legacy communications systems served their purpose, the realities of 21st century land warfare necessitate a comprehensive modernization of BCT communications capabilities. The lessons from the Ukraine conflict illustrate the importance of advanced, interoperable, and resilient communication systems in supporting MDO and enabling operations within the close combat area. By investing in modern communication technologies that enhance situational awareness, support rapid decision-making, and ensure connectivity in contested environments, BCTs can significantly improve their operational capabilities and readiness for future conflicts.

Conclusion

The development of AirLand Battle serves as a paradigm of US Army doctrine; one that was grounded in a deep understanding of the adversary and focused on addressing force structure changes inside of divisions and BCTs to make the US Army more dominant in ground combat. In contrast, the current pursuit of MDO risks losing sight of the US Army's core strengths and the fundamental importance of ground combat. To ensure MDO remains relevant and delivers results, the US Army must refocus its efforts on dominating the critical close combat zone, where land warfare is fundamentally won and lost. To achieve this objective, the US Army must prioritize the development of critical capabilities at the BCT-level, focusing on four key areas: direct and indirect fires, sensor-to-shooter capabilities, Intelligence, Surveillance, and Reconnaissance (ISR), and communications.

This will require renewed emphasis on the US Army's core competencies, including its ability to seize and hold terrain, destroy enemy formations, and impose its will in the unforgiving chaos of ground warfare. By doing so, the US Army can maintain its position as a vital component of the joint force, one that brings unique capabilities and strengths to the table. The future of warfare will undoubtedly be characterized by complexity, uncertainty, and a high degree of interconnectedness across multiple domains. However, amidst this complexity, the fundamental importance of ground combat and the US Army's role in it must not be forgotten. By refocusing its efforts on the

---

58 Singer, Peter. "The War in Ukraine: A New Era of Space-Based Warfare." *Journal of Strategic Studies* 45, no. 3 (2022): 347-363.

close combat zone and prioritizing the development of BCTs, the US Army can ensure that it remains a dominant force on the battlefield, capable of adapting to and overcoming the challenges of modern warfare.

MDO should not sacrifice the US Army's unique value on the altar of jointness but should rather sharpen its edge on the ground in the face of a determined enemy. By learning from the successes of AirLand Battle and reorienting the approach to MDO, the US Army can ensure that its doctrine remains grounded and focused on the ultimate goal of winning ground combat and defeating the enemy.

4

A Deficit of Technique: Addressing Tactical Shortcomings in Multidomain Operations Doctrine

Bryan Quinn

The United States is facing a rapidly evolving security landscape. In Europe, an unfavorable military balance is emerging, driven primarily by Russia's accelerating capacity to rearm while Europe struggles to match pace. Concurrently, threats to US interests have become increasingly interconnected, as demonstrated by the strengthening ties between Moscow, Beijing, Pyongyang, and Tehran, all seemingly aligned in opposition to the West. Compounding these challenges is a rise in technological complexity, driven by artificial intelligence (AI) and a meaningful pivot towards data centricity, which threatens to fundamentally alter the battlefield. This confluence of factors is transforming Europe and the Levant into a proving ground where the character of warfare is evolving in real-time.

To address this evolving landscape, the US Army has transitioned multidomain operations (MDO) from future concept to formal doctrine and has begun developing accompanying tactics to operationalize its principles. While theory and technology often dominate academic discourse, the third crucial element of military adaptation— technique or tactics—is frequently overlooked. Closing the gap between concepts and tactics remains one of the most pressing challenges facing efforts to operationalize MDO.

Despite its integration into US doctrine and accelerating adoption by other Western militaries, MDO currently faces a deficit of technique.[1] This deficiency prevents the connection between the concept's principles and tenets and actionable guidance for field forces, leading to a lack of coherence between the operational concept and execution. Without clear, prescriptive tactical guidance to translate grand concepts into battlefield action, militaries risk fielding updated concepts disconnected from the practices of combat forces or battlefield realities. Correcting this disconnect requires increasing detail in tactical doctrine aligned with core multidomain principles, supported by critical enabling capabilities such as modernized command and control. By doing so, militaries can align concepts, tactics, and materiel more closely, increasing military readiness to succeed in conflict.

The central focus of this chapter is examining the relationship between a military's overarching theory for defeating its enemies, or a theory of victory, and the techniques it requires to do so. That is, between the German attempt at *Bewegungskrieg* ("war of movement") in 1918 and the infiltration tactics of the *Sturmtruppen* (stormtrooper tactics) or the US Army's Unified Land Operations (ULO) concept (i.e., theory of victory) and combined arms maneuver (i.e., technique).[2]

The importance of aligning theory with practice and concept with action cannot be overstated. Neglecting tactical development in developing new concepts may prevent success in future conflicts. In his assessment of General William E. DePuy's 1976 operations manual, Paul Herbert notes, "When well-conceived, doctrine can instill confidence throughout an army… [and] can have the most profound effect on its performance in war."[3] It follows, therefore, that doctrine lacking in clarity or coherence could just as profoundly undermine a military's readiness for the next war.

---

1 'Multidomain Operations,' NATO Allied Command Transformation, accessed Jan. 10, 2025, https://www.act.nato.int/activities/multidomain-operations/; Alexandra Schulte, 'Empty Promises: A Year Inside the World of Multidomain Operations,' *War on the Rocks*, Jan. 2024, https://warontherocks.com/2024/01/empty-promises-a-year-inside-the-world-of-multidomain-operations/.

2 Department of the Army, *Army Doctrinal Publication (ADP) 3-0: Unified Land Operations* (Washington, DC: Headquarters, Department of the Army, 2011); Karl-Heinz Frieser, *The Blitzkrieg Legend: The 1940 Campaign in the West*, trans. John T. Greenwood (Annapolis, MD: Naval Institute Press, 2005), 614.

3 Paul H. Herbert, *Deciding What Has to Be Done: General William E. DePuy and the 1976 Edition of FM 100-5, Operations* (Fort Leavenworth, KS: US Army Command and General Staff College, 1988), 3.

The Foundations of Military Adaptation: Concepts, Technique, and Materiel

Since the Vietnam War, the US Army has revised its capstone doctrine, Field Manual 3-0 (then FM 100-5) *Operations*, seven times. Each change introduced a new concept of how the Army should approach warfare, including Active Defense, AirLand Battle, and Full-Spectrum Operations (FSO). Each change represented an adaptation in the military's operational concept or theory of victory, designed to address a novel operational environment.

MDO is the most recent adaptation of the Army's concept, pivoting away from ULO's focus on joint operations in the land domain.[4] MDO's development was meant to better prepare the Army to succeed against an emboldened adversary, equipped with its own adaptation in technology and strategy, primarily advances in anti-access and areal denial (A2AD) systems.[5]

Yet, military change consists of more than an updated view of the problem and its corresponding operational concept. To better understand the complexity of organizational change, military adaptation can be understood through three primary elements: a military's theory of victory, its technique or tactics, and its materiel.[6]

A theory of victory is a military's hypothesis for the best way to succeed in conflict, given everything that can be known about friendly forces, the enemy, and the environment.[7] The articulation and distillation of this theory is a military's operational concept, which provides the overarching idea or vision for what war will look like and how it will be fought, tying tactics and materiel to strategic objectives.[8] Military doctrine—expressed through field manuals, publications, regulations, and other institutional documents—is largely constructed around these concepts, which is "the core of doctrine," reflecting "the way the Army fights its battles and campaigns."[9]

The second element of military adaptation is the tactics and techniques that detail how a military operates on the battlefield consistent with the principles of the operational concept. As FM 3-90 *Tactics* articulates, "tactics

---

4 Department of the Army, *ADP 3-0*, 2011, 5.

5 Department of the Army, *Field Manual (FM) 3-0: Operations* (Washington, DC: Headquarters, Department of the Army, October 2022), 6-15 – 6-17.

6 Robert I. Sickler, 'The Technology Triad: Reimagining the Relationship Between Technology and Military Innovation' (PhD diss., Arizona State University, 2021), 57.

7 Ibid.

8 Ibid, 65.

9 Department of the Army, *FM 100-5: Operations* (Washington, DC: Headquarters, Department of the Army, 1982), 2-1.

are the employment, ordered arrangement, and directed actions of forces in relation to each other,"[10] such as the series of physical actions that an artillery battery may take to emplace or displace from a firing position or support ground forces. Tactics translate a military's theory of victory into battlefield action and guides how field forces fight and employ weapon systems in war, connecting physical material with concept.

The final element of adaptation is a military's materiel, or equipment, which includes the weapons, tools, and technologies that military forces use in war, such as the cannon itself.

Together, these three elements of adaptation interact in a complex relationship and provide a framework for understanding the different elements involved in military evolution.[11] This framework draws from previous studies on technology and innovation, notably Andrew Ross' military innovation triad, Braden Allenby and Daniel Sarewitz's technological taxonomy, and, most significantly, Robert Sickler's technology triad.[12] Each author articulates a socio-technical model that, to varying degrees, links ideas, action, and materiel.

As Sickler explains, each element of adaptation changes at a different rate.[13] A theory of victory can change rapidly based on an updated assessment of the threat environment, the introduction of a new weapon, or change in political guidance, while tactics and material can take longer to adapt.

Even this framing, however, fails to fully capture the relationship's complexity as all three elements evolve simultaneously, though at varying rates. As concepts clash with reality and materiel moves from the laboratory to the battlefield, each interacts with the enemy and the environment, resulting in continuous interaction with one another. The tank and the machine gun had as profound an impact on concepts and tactics just as the Active Defense and AirLand Battle concepts had on influencing the US Army's 'Big Five' weapon systems.[14] Consequently, each element is evolving,

---

10 Department of the Army, *FM 3-90: Tactics* (Washington, DC: Headquarters, Department of the Army, May 2023), 1-1.

11 Sickler, 'The Technology Triad,' 57.

12 Andrew L. Ross, 'On Military Innovation: Toward an Analytical Framework,' SITC 2010, no. Policy Brief 1 (Sep. 1, 2010); Braden R. Allenby and Daniel R. Sarewitz, *The Techno-Human Condition* (Cambridge, Massachusetts: MIT Press, 2011); Sickler, 'The Technology Triad.'

13 Sickler, 'The Technology Triad,' 9.

14 Ben Jensen, *Forging the Sword: Doctrinal Change in the US Army* (Washington, DC: National Defense University Press, 2017), 17; Rodney D. Fogg, 'From the Big Five to Cross-Functional Teams: Integrating Sustainment into Modernization,' Army.mil, accessed Jan. 10, 2025, https://www.army.mil/article/227832/from_the_big_five_to_cross_functional_teams_integrating_sustainment_into_modernization.

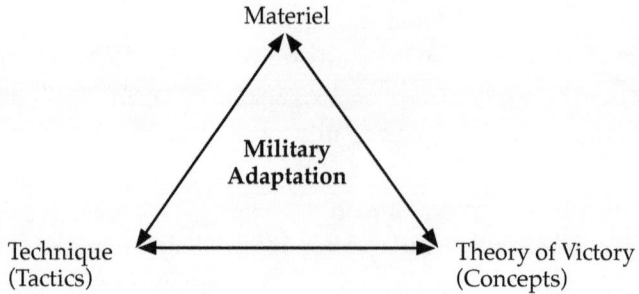

Figure 4.1 Military Adaptation Framework
Source: Author.

not only in accordance with their own internal logics and limitations, but also in dynamic, two-way interactions with the other elements, forming a complex system as illustrated in Figure 4.1.[15]

The US Army's Pentomic transition following the Korean War is an effective example of the interplay between concepts, tactics, and material. In 1953, President Dwight Eisenhower's "New Look" policy prioritized massive nuclear retaliation, raising doubts about the relevance of ground forces in a future dominated by nuclear deterrence.[16] To align with this strategy, remain relevant, and address Soviet numerical superiority, the Army developed a new theory of victory.[17] Army planners hypothesized that technological advancements in battlefield firepower, such as tactical nuclear weapons, could compensate for the stark military imbalance between Soviet and NATO ground forces. Eisenhower's guidance and the Department of Defense's subsequent work on nuclear deterrence spurred what would become known as the *First Offset Strategy*.[18]

This new concept marked a stark departure from previous methods of warfare, requiring a fundamental and commensurate rethinking of tactics and technique. Throughout the 1950s, Army officers turned the new operational concept into "concrete, practical methods for warfighting" through doctrinal

---

15 Sickler, 'The Technology Triad,' 10.
16 Andrew J. Bacevich, *The Pentomic Era: The Army Between Korea and Vietnam* (Washington, D.C.: National Defense University Press, 1986), 49-51.
17 Ibid.
18 Claudette Roulo, "Offset Strategy Puts Advantage in Hands of US, Allies," *Department of Defense*, 28 January 2015. Available at: https://www.defense.gov/News/News-Stories/Article/Article/603997/offset-strategy-puts-advantage-in-hands-of-us-allies/.

changes, reflected in the 1954 edition of Field Manual 100-5.[19] Recognizing the vulnerability of concentrated formations to nuclear strikes, for instance, tactics emphasized survivability on the nuclear battlefield and required field forces to disperse rapidly to reduce the risk of presenting a high-payoff target to the enemy.[20] To adapt in line with this Pentomic tenet, the Army transformed the structure of its field forces to improve self-sustainability and organic fires by combining the regimental and battalion echelons.[21] Offensively, penetration became the cornerstone of action, driving a number of specific changes in technique and tactics to ensure a formations' ability to exploit battlefield successes.[22] These changes included a pivot away from flanking maneuvers, instead favoring tightly controlled columns that would attack rapidly to execute frontal penetration following a tactical nuclear strike.[23]

To support the new operational concept, the Army also identified new material requirements, including low-yield nuclear weapons.[24] Achieving the dispersion and concentration described in tactics required increasing the mobility of ground forces, expanding the mechanization of ground vehicles, and incorporating new capabilities like the helicopter. While the Pentomic Era was not without flaws, it serves as a prime example of comprehensive military adaptation, where an updated operational concept was combined with specific, actionable tactics and the necessary materiel developments to effectively execute its theory of victory.

As demonstrated by the Army's Pentomic Era, meaningful military adaptation requires change across all three elements – concepts, tactics, and materiel. Adapting one element without a commensurate change in the others can create an imbalance between concept and battlefield realities. For instance, fielding a new armored vehicle without developing the accompanying technique for its employment or integrating it into the broader operational concept may, at best, constitute an incremental improvement in materiel. The performance of France's armored forces in 1940 provide a stark illustration of this imbalance, where superior tanks such as a the Char B1 proved ineffective due to outdated tactics.[25] Similarly, adopting a

---

19 Ibid, 60-64.
20 Ibid, 103-106.
21 Ibid.
22 Ibid, 108-109.
23 Ibid.
24 John H. Cushman, *Command and Control of Theater Forces: Adequacy* (Washington, D.C.: National Defense University Press, 1986), 20-23.
25 Frieser, *The Blitzkrieg Legend*, 595-598.

new operational concept without equipping the military with the necessary tools or concrete instructions for applying its conceptual principles risks leaving the military unable to translate ideas into action and ill-prepared to succeed on the battlefield.

In contrast to the military adaptation of the 1950s, where there was a clear connection between the Pentomic concept and its supporting technique and tools, consider the Army's failed attempt to rapidly modernize through its Future Combat Systems (FCS).[26] This program, inaugurated in 2003, focused primarily on fielding advanced equipment at the brigade-level including a networked set of 18 interdependent sensors, robots, drones, and manned ground vehicles.[27] The FCS program was broadly considered a failure and cancelled in 2009 after nearly 200 billion dollars and little delivery on promised improvements in capabilities.[28]

Why did FCS fail to achieve the "array of deployable, agile, versatile, lethal, survivable, and sustainable formations" that then-US Army Chief of Staff General Eric Shinseki originally envisioned?[29] Among other reasons, a significant factor was that the program largely focused on "leap ahead" technologies and material modernization, a single aspect of adaptation, without consideration of how the equipment would be employed on the battlefield or support a greater operational concept.[30] As one study would later highlight, "…greater specificity was needed to describe to engineers what exactly TRADOC wanted the brigade to do, how it would fight, how integrated systems would interact, and how the network would operate."[31]

Updating only one or two elements, without considering how all three elements interact, risks creating a maladapted military, where concepts and theories are disconnected from battlefield tactics, and/or materiel realities. To avoid repeating similar mistakes in adapting the military for MDO, concepts and materiel cannot be developed without supporting developments in technique.

---

26 Christopher G. Pernin, 'Lessons from the Army's Future Combat Systems Program' (Santa Monica, CA: RAND Corporation, 2012), 2.

27 Ibid, 1.

28 US Government Accountability Office, 'Defense Acquisitions: Future Combat System Risks Underscore the Importance of Oversight,' GAO-07-672T (Washington, DC: Government Accountability Office, 2007), xvii, https://www.gao.gov/products/gao-07-672t.

29 Pernin, 'Lessons from the Army's Future Combat Systems Program,' 55.

30 Ibid, xvii.

31 Ibid, xxii.

The Prescriptive Gap: Bridging Theory and Practice

Doctrine began as a codification of battlefield tactics. It served as a drill manual for infantry formations and provided a unifying reference to standardize actions on the battlefield. While large formations no longer march in lockstep across open fields, doctrine still provides the single authoritative source describing how to employ military force. Given the increased complexity of integrating effects across domains and services, the need for detailed tactical doctrine is arguably even greater for MDO.

Concepts and tactics—two essential components of military adaptation—are intertwined in modern doctrine. As a former Combined Arms Center (CAC) Commander, Lieutenant General John Cushman, observed, "Doctrine in one sense can be called the soul of a military... [However,] there is another kind of doctrine, more narrowly defined. It has to do with how to fight."[32] Thus, military doctrine must articulate both the military's theory of victory and the specifics of how to implement it to fulfill its purpose.

However, doctrine has trended away from tactical specificity in favor of a broader operational focus since the 1976 edition of the US Army's operations manual. The introduction of the operational level of war in 1982 and its expansion in 1986, alongside the release of AirLand Battle in 1982 further broadened the scope of doctrine and crowded out tactical detail.[33] To compensate for this shift in focus, subsequent updates to FM 100-5 (and FM 3-0) have relied on subordinate manuals, such as FM 3-90 *Tactics* and FM 4-0 *Sustainment*, to provide more precise tactical and functional guidance on how to implement the operational concept.[34]

Now formally integrated into FM 3-0 *Operations*, MDO fulfills the foundational doctrinal requirement noted by Lieutenant General Cushman and provides broad guidance for applying its principles. Specifically, MDO is the "combined arms employment of joint and Army capabilities to create and exploit relative advantages that achieve objectives, defeat enemy forces, and consolidate gains."[35] The concept emphasizes the need to overcome the enemy's layered stand-off capabilities such as "integrated fires complexes

---

32 Cushman, *Command and Control of Theater Forces*, 85.
33 Michael P. Coville, 'Tactical Doctrine and FM 100-5' (Fort Leavenworth, KS: US Army Command and General Staff College, 1991), 24–27.
34 Huba Wass de Czege, Memorandum for Record, SUBJECT: Transfer of FM 100-5 Proponency from Department of Tactics to School of Advanced Military Studies. Sep. 6, 1984.
35 Department of the Army, *FM 3-0*, 2022, 1-2.

and air defense systems" so that "maneuver forces can exploit the resulting freedom of action."[36]

To achieve these goals, FM 3-0 advises commanders to pursue "positions of relative advantage" through "decision dominance and overmatch," to "[create] multiple dilemmas" for the enemy and ensuring joint force "lethality."[37] Yet, how to achieve any of the principles detailed within this manual is left unaddressed by the operational concept, leaving the practical steps for implementing these ideas to tactical doctrine. Dwight Phillips underscores this assessment, noting that "however dissatisfying for an operations doctrine, MDO provides few cogent examples of what these terms practically mean for unit operations and tactics."[38]

Consistent with Lieutenant General Cushman's second "kind of doctrine," tactical doctrine should provide detailed guidance on implementing MDO and clearly articulate how to achieve objectives in alignment with the concept's principles and tenets. Following the publication of FM 3-0, the Army has begun revising its tactical manuals to close the gap between theory and action. For example, FM 4-0 Sustainment—a subordinate manual—should articulate how the US Army organizes, trains, and employs sustainment formations at each echelon to provide commanders with freedom of action and operational reach in line with MDO's tenet of endurance.[39]

Given the importance FM 3-0 places on targeting to integrate joint capabilities, FM 3-60 *Army Targeting* should detail how multidomain task forces (MDTFs), fires brigades, and division and corps fires elements synchronize cross-domain and cross-service effects to maximize MDO's tenets of convergence and depth.[40]

However, current tactical doctrine suffers from three interrelated shortcomings that collectively undermine the military's ability to effectively conduct MDO. First, "updated" sustainment and tactics manuals reveal either significant similarities with outdated doctrine or, even worse, no substantive changes at all. For example, FM 3-90 introduces a new tactical framework, ostensibly designed to address MDO's imperatives of continuously

---

36 Ibid, ix.

37 Ibid, 1-3 – 1-5.

38 Dwight Phillips, *Multidomain Operations: Passing the Torch* (The Hague: The Hague Centre for Strategic Studies, Nov. 2023), accessed Jan. 10, 2025, https://hcss.nl/wp-content/uploads/2023/11/Multidomain-Operations-Passing-The-Torch-HCSS-2023.pdf.

39 Department of the Army, *FM 4-0: Sustainment* (Washington, DC: Headquarters, Department of the Army, Jul. 2024).

40 Department of the Army, *FM 3-0*, 2022, 1-3, 3-2 – 3-8.

anticipating transitions, sustaining the main effort, and consolidating gains.[41] Yet, the framework, "find, fix, finish, follow through," is essentially a rebrand of the "find, fix, finish, exploit, analyze, and disseminate" (F3EAD) targeting cycle that has existed for decades. F3EAD, in turn, traces its roots back to John Boyd's "observe, orient, decide, and act" (OODA) loop, which he introduced almost 40 years ago.[42] Similarly, the updated sustainment manual exactly replicates the "principles of sustainment framework" with the one exception of replacing the heading "Unified Land Operations" with "Multidomain Operations."[43] This superficial adjustment underscores a lack of substantive adaptation on the part of tactical doctrine. Meanwhile, the Army's manual for tactics mentions MDO fewer than ten times, primarily in introductory paragraphs that reiterate the language of the operational concept.

It appears that "updated" tactical doctrine merely dress previous publications in multidomain verbiage, a fallacy that Ben Jensen refers to as "old wine in a new bottle."[44] Without an observable and commensurate change in tactics, any adaptation remains superficial and incomplete. If left unaddressed, these shortcomings in tactical doctrine will impede the effective implementation of MDO, while undermining the authority of those respective manuals. Further, an overreliance on previous doctrine risks sending soldiers into battle armed only with the knowledge of how they've always fought with tactics of outdated concepts.

Second, today's tactical doctrine lacks sufficient depth and detail. The most recent FM 3-0 continues the long trend of strengthening its focus on conceptual theory at the expense of tactical detail; concerningly, subordinate doctrines fail to fill this critical gap. For example, FM 4-0 acknowledges the vulnerability of sustainment nodes and routes due to enemy reconnaissance and long-range fires. However, it provides little detail for countering such threats beyond general recommendations for dispersing or improving camouflage.[45] The updated tactical doctrine provides even fewer details on MDO targeting than does the theoretical, conceptual exposition of the concept.

Targeteers and fires officers are consequently left ill-equipped to (a) handle new processes that should account for cross-service coordination,

---

41 Department of the Army, *FM 3-90*, 2023, 1-6 – 1-7.
42 John R. Boyd, *A Discourse on Winning and Losing*, ed. and comp. Grant T. Hammond (Maxwell AFB: Air University Press, 2018), 1-2.
43 *FM 3-0*, 2022, 3-2 – 3-8.
44 Ben Jensen, *Forging the Sword: Doctrinal Change in the US Army*, 10.
45 Department of the Army, *FM 4-0*, 2024, 98, 175.

Range of Doctrine

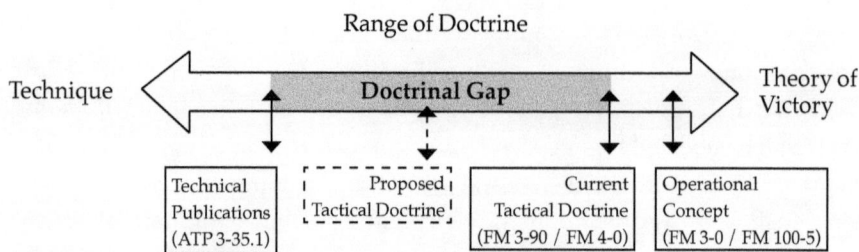

Figure 4.2 Range of Doctrine
Source: Department of the Army, *ADP 1-01: Doctrine Primer* (Washington, DC: Headquarters, Department of the Army, 2019), 2-3 – 2-5.

(b) manage procedures for layering or converging multidomain effects, or (c) operate the new technical systems required for these tasks. This lack of tactical, technical detail creates a gap between the doctrine's conceptual emphasis on MDO and soldiers' preparedness for the practical demands of sustaining MDO. Current doctrine struggles to implement the theory of victory as techniques that can be applied, synchronized, and executed on the battlefield. Each doctrinal level—the future operating concept (in this case, 525-3-1; now the Army Warfighting Concept), the military's operational concept as articulated in FM 3-0, and subordinate, tactical doctrine including FM 4-0, 3-60, or 3-90—simply re-articulates the broader operational-level framework without additional prescriptive details that explain how a soldier should go about the business of winning.

At the same time, technical publications (e.g., ATP 3-35.1) often receive iterative updates oblivious to conceptual changes, resulting in a growing gap between forward-looking concepts and slow-changing techniques. Figure 4.2 illustrates all three aspects of the challenge: (1) the undue similarities between operational concept and current tactical doctrine, (2) the doctrinal gap that this creates when tactical doctrine lacks sufficient specificity, and (3) with no tactical doctrine to bridge the gap, the ever-widening distance between the theory of victory and technique. Consequently, field forces currently lack tactics that are compatible with MDO, creating misalignment and incoherence between the operational concept and its execution. Grand descriptions of how the enemy will be defeated cannot substitute for tactical detail.

Third and finally, to compensate for a lack of concrete battlefield guidance, updated tactical doctrine overly relies on impractical materiel solutions. These shortcomings are particularly evident in tactical sustainment doctrine. For example, to solve the challenges of sustaining initial entry forces and other multidomain formations, FM 4-0 suggests emergent capabilities like the Joint Tactical Autonomous Aerial Resupply System (JTAARS) – an experimental unmanned platform.[46] However, it does so without considering the real-world limitations of these systems: JTAARS and other similar aerial platforms only offer a fraction of the payload or resupply capacity of even a light logistics truck.[47]

The failure of commercial companies like Amazon to deliver on promises of deploying autonomous vehicles at scale, despite over a decade of investment, highlights the significant challenges of fielding such technology.[48] The war in Ukraine further demonstrates the limited functionality and viability of capabilities like JTAARS in denied environments on the modern battlefield. Yet, tactical doctrine readily suggests this system as a solution to sustainment challenges without considering the limitations that will prevent its effectiveness on the battlefield. This overemphasis on technology is a red herring, diverting attention and effort away from developing practical and actionable tactical doctrine.

Such theoretical solutions echo the flaws of past programs like Future Combat Systems, discussed earlier, where "The goals that Shinseki laid out… were not only operationally but also technically ambitious. In hindsight, few of these early performance requirements were realistic, and none apparently had been seriously vetted for technical feasibility, let alone affordability."[49] While emerging technologies may help mitigate some technical challenges, they cannot by themselves bridge the gap between concepts and actionable tactics or replace the need for practical, sound techniques. The reliance on theoretical solutions risks leaving sustainment operations underprepared for the realities of contested battlefields. Technology must complement, rather than substitute for, the development of effective techniques and tactics to ensure sustainment can meet operational demands.

---

46 Department of the Army, *FM 4-0*, 2024, 17.
47 Brett Davis, 'UAS of All Sizes Aim for Army Programs at Fall AUSA Conference,' Inside Unmanned Systems, accessed Jan. 10, 2025, https://insideunmannedsystems.com/uas-of-all-sizes-aim-for-army-programs-at-fall-ausa-conference/.
48 Noor Al-Sibai, "Amazon's Drone Delivery Program Is an Utter Failure," Futurism, last modified Jan. 8, 2025, accessed Jan. 10, 2025, https://futurism.com/the-byte/amazon-drone-delivery-fail.
49 Pernin, 'Lessons from the Army's Future Combat Systems Program,' 55.

Reimagining Sustainment: Strengthening the Link Between Concepts, Tactics, and Materiel

To reduce doctrine's dependence on technological solutions and correct the imbalance between concepts, tactics, and materiel, militaries must develop more prescriptive tactical doctrine. This effort should focus on drawing clear and logical linkages between the tenets of the operational concept and practical actions required on the ground. To enable these tactics, militaries must develop truly joint and interoperable command and control systems rather than focusing on unproven or impractical technological solutions.

For an operational concept that relies on the reinvigorated role of sustainment, logistics is an area that demands more rigorous thinking. An important example of this requirement is Joint Forcible Entry Operations (JFEO), which is the "seizing and holding of a military lodgment in the face of armed opposition or forcing access into a denied area."[50] As a central component of MDO, JFEO is envisioned as a way to seize positional advantages, create multiple dilemmas, and defeat adversarial A2/AD or stand-off approaches.[51] JFEO's success hinges on a sustainment framework capable of overcoming the logistical complexities and vulnerabilities inherent in such operations.

Yet JFEO's sustainment tactics remain largely unchanged from those outlined in Unified Land Operations and continues to heavily rely on aerial resupply – despite the impracticality of aerial resupply in the environment assumed by the operational concept. This reliance on aerial resupply directly contradicts other recommendations derived from the concept—such as autonomous vehicles and joint logistics over the shore (JLOTS)— demonstrating the disconnect between concept and tactics.[52]

To ensure the sustainment of ground forces, militaries must correct this doctrinal deficiency by developing a tactical sustainment framework that aligns with MDO's tenets of endurance and depth. This framework should operationalize the sustainment principles of (a) extending operational reach, (b) prolonging endurance, and (c) emphasizing dispersion by (1) increasing ground lines of communication and (2) expanding the number of staging bases.

Central to this sustainment framework are two key assumptions about the expected operating environment already captured in the operational

---

50 Department of the Army, *FM 3-0*, 2022, 6-18.
51 Ibid, 3-15.
52 Department of the Army, *FM 4-0*, 2024, 90; Department of the Army, *FM 3-0*, 2022, 1-19.

concept and validated by observations from the war in Ukraine. First, relying on large-scale aerial resupply is increasingly impractical due to enemy anti-access weapons and electronic warfare.[53] Therefore, ground forces must be sustained through ground lines extending from support areas to the deep maneuver area. Second, bases of supply, because of their vulnerability and high payoff nature, are likely to be sought out, targeted, and rapidly destroyed in future conflict.[54] As a result, ground forces can no longer rely on a handful of large main supply routes. Both assumptions support the need for a more effective sustainment framework that prioritizes resilience and decentralization over the efficiency emphasized in current doctrine.

Assuming that a specific battlefield allows it, sustainment resiliency and responsiveness must be increased by increasing the number of ground lines of communication to (a) allow for expanded routing options for field, combat, and company trains and (b) shorten supply lines. Updates in tactical doctrine can support this by expanding the role and presence of deliberate intermediate staging bases and dynamic logistics release points. Staging bases increase entry points to the operating area and ensure sustainment capacity is kept further from anticipated threats but close enough to provide immediate support.[55] Meanwhile, release points offer a faster and more flexible method of distributing supplies by forward positioning or caching more supplies forward for easier access by front line troops.

This framework mirrors a modern mesh network characterized by interconnections between nodes and the absence of dependency on any single node, eliminating single points of failure.[56] In the event of any given node's disruption, the network's ability to reconfigure allows for greater dynamic distribution. Similarly, with an increased number of supply bases and shorter supply lines, commanders and logisticians can quickly disable or reenable lines of communication as they are threatened or disrupted, actively diverting critical supplies through other lines as necessary.[57] This approach creates a flexible distribution network capable of withstanding

---

53 Ronald R. Ragin and Christopher G. Ingram, 'Theater Sustainment Transformation: Lessons from the Russia-Ukraine War,' Army.mil, Apr. 23, 2024, accessed Jan. 10, 2025, https://www.army.mil/article/274914/theater_sustainment_transformation_lessons_from_the_russia_ukraine_war.
54 Ibid.
55 Department of the Army, FM 4-0, 2024, 87.
56 Stephen Rayment, 'Capacity of Wireless Mesh Networks,' in *Emerging Technologies in Wireless LANs: Theory, Design, and Deployment*, ed. Benny Bing (Cambridge: Cambridge University Press, 2008), 220-227.
57 Bryan J. Quinn, 'Sustaining Multidomain Operations: The Logistical Challenge Facing the Army's Operating Concept,' *Military Review*, Mar.-Apr. 2023.

the challenges of contested logistics and achieving the dispersion called for in tactical doctrine.

By implementing a sustainment framework that is aligned with the tenets and principles of the operational concept, instead of relying on impractical capabilities like JTAARs, this approach reduces the risks associated with sustainment gaps caused by large, vulnerable supply bases and extended lines. Without sufficient sustainment resiliency in contested environments, combat forces will become disconnected from the support area and struggle to maintain operational tempo or achieve relative advantages envisioned by MDO. Historical examples, such as the failure of the British 1st Airborne Division during Operation Market Garden in World War II or, more recently, the Russian airborne troops' (VDV) attempt to seize the Hostomel airport in the early days of the Russia-Ukraine War highlight the consequences of initial entry forces losing connectivity to their bases of supply.[58]

Implementing such a robust approach to sustaining MDO introduces risks to efficiency. To mitigate, this change in tactics must be supported by corresponding materiel adaptation. To effectively implement this sustainment framework, militaries must also improve their ability to orchestrate operations through command and control (C2) systems. These systems consist of the communication networks, hardware, and software applications necessary to enable commanders to plan, direct, and coordinate people and things. A modernized C2 system should foster shared understanding – the ability to see ourselves and the enemy accurately—and enable decision superiority—the ability to use that information to decide and act faster than the adversary. Effective C2 capabilities are essential to realizing MDO, as they underpin the implementation of advanced logistics, fires, intelligence, and other necessary systems.

The requirement for modernized C2 is already acknowledged in sustainment doctrine, which highlights the criticality of "precision logistics" and "predictive sustainment."[59] These concepts aim to provide a "layered, agile, and responsive sustainment capability" enabled by real-time common operating pictures and predictive decision support systems.[60] While FM 4-0 envisions this capability as a means of reducing demand, modernized C2

---

58 Liam Collins, Michael Kofman, and John Spencer, 'The Battle of Hostomel Airport: A Key Moment in Russia's Defeat in Kyiv,' *War on the Rocks*, Aug. 10, 2023, accessed Jan. 10, 2025, https://warontherocks.com/2023/08/the-battle-of-hostomel-airport-a-key-moment-in-russias-defeat-in-kyiv/.
59 Department of the Army, *FM 4-0*, 2024, 14.
60 Ibid, 17.

should instead enable more effective and resilient sustainment consistent with the proposed framework.

Implementing such a sustainment solution requires a capable and modernized C2 capability that does not exist today, much less one capable of effectively operating with allies and partners. Combatant commands, the organizations charged with managing the nation's crises and conflicts, manually count ships, airplanes, and tanks.[61] Even the Army's technical manual on supply distribution acknowledges this issue, bluntly stating, "Unfortunately, present systems do not completely satisfy the requirements of force tracking, and much of the process must be accomplished manually."[62] The fact that this issue is highlighted in a manual published as recently as 2023 underscores how embarrassingly short C2 falls in relation to MDO's vision.

The challenge with improving C2 is the unrealistic belief that it can leap from its current manual state, incapable of even providing shared understanding, to providing decision advantage. C2 deficiencies—including data, processes, and interoperability—must be corrected in a more iterative and deliberate manner to enable a future where software can recommend optimal logistics routes and sustainment nodes or manage the complexity of MDO.

Militaries must first address their data. In the US military, this means addressing data that is currently fragmented across services, Joint Staff, the Office of the Secretary of Defense (OSD), national repositories, and open-source platforms. Much of this data is unstructured, inaccessible, and not machine-readable. Resolving this issue requires a deliberate effort to identify and integrate authoritative data sources into advanced systems that can form a common operational picture. It is not enough to simply collect the data; the next step is to effectively leverage the data to gain decision-making advantages. This involves digitizing existing manual decision-making tools found in doctrine, such as decision support templates and synchronization matrices, to begin automating procedural aspects of C2.

Lastly, any C2 system capable of supporting MDO must prioritize interoperability with other services, partners, or alliance systems. Integrated Deterrence, codified in the 2022 National Defense Strategy (NDS), calls for

---

61 Peter Andrysiak and Bryan Quinn, 'Empowering the Combatant Commands Is Critical for the Future Fight,' *War on the Rocks*, Dec. 10, 2024, accessed Jan. 10, 2025, https://warontherocks.com/2024/12/empowering-the-combatant-commands-is-critical-for-the-future-fight/.
62 Department of the Army, *Army Techniques Publication (ATP) 3-35: Army Deployment and Redeployment* (Washington, DC: Headquarters, Department of the Army, Mar. 2023), 5–8.

greater integration with allies and partners to counter threats like Russia and China.[63] A concept as simple as burden sharing, called for in the 2025 National Defense Strategy development guidance, is not possible without the ability to communicate.[64] Without the ability to share a common operational picture, combined operations, let alone conduct MDO, will not be a possibility.

Ultimately, improving capabilities central to the operational concept, such as C2, establishes a closer and more logical connection between materiel, concepts, and tactics while reducing reliance on more impractical technological solutions like JTAARs.

Closing Comments: Overcoming the Tactical Deficit

It's clear that FM 3-0 was written without fully appreciating that it would only have a limited impact on tactics without the commensurate work of translating the concept into actionable tactics, a result of assuming a linear or hierarchical relationship between tactics and concepts rather than the complex reality. Three years after its publication, and despite a handful of updates to doctrine, MDO's tactical deficit persists. In the place of a deliberate effort to connect the two elements—concepts and tactics—a dangerous misconception is gaining traction; the belief that tactics, by rule, will follow concepts. The prevailing belief assumes that with MDO formalized, it is the responsibility of the warfighter to fill in the details.[65] While the US Army's Capability Development Integration Directorates (CDIDs) and Centers of Excellence (COEs) are working to update tactics, many assume that "MDO's success will be realized through the dedication and rigorous training of crews, companies, battalions, and staff echelons..."[66]

While concepts, tactics, and materiel admittedly progress at different rates, the belief that tactics can "be figured out later" abdicates institutional responsibility to sufficiently consider tactics and the practical implementation of materiel in developing concepts. This attitude allows the authors of doctrine to avoid key challenges, such as how to sustain ground forces or converge

---

63 US Department of Defense, *National Defense Strategy of the United States of America 2022* (Washington, D.C.: US Department of Defense, 2022), 8-11.
64 US Department of Defense. *Memorandum Directing the Development of the 2025 National Defense Strategy.* Memorandum from the Office of the Secretary of Defense, Washington, DC, May 1, 2025. https://media.defense.gov/2025/May/02/2003703230/-1/-1/1/MEMORANDUM-DIRECTING-THE-DEVELOPMENT-OF-THE-2025-NATIONAL-DEFENSE-STRATEGY.PDF.
65 Phillips, *Multidomain Operations: Passing the Torch*, 2023.
66 Ibid.

fires, assuming soldiers in the field will work out these critical details. While operational units should be able to refine and perfect tactics based on real-world experience, these organizations have neither the conceptual expertise nor time required to develop MDO's tactics. This approach, placing the full burden of operationalization on tactical units, may further sever the link between the two elements and risk the creation of concepts wholly divorced from tactical realities.

Current tactical shortcomings aside, the Army is already moving out with its next conceptual framework—*Transformation in Contact*—before the ink has dried on its current one.[67] This new framework attempts to address the complexity between the elements of adaptation by emphasizing bottom-up innovation from tactical units.[68] However, the development of new tactics remains unclear as authors of the emerging concept still appear to assume a linear relationship between adaptation elements (i.e., tactics will follow concepts) and rely on the timely delivery of new technological capabilities.

For every example where success hinged on fielding advanced capabilities faster than an adversary, like the English longbowmen at Crecy, there are just as many instances where new capabilities were introduced without considering their tactical employment.[69] Just as the Germans did, the French possessed the tank in 1940 but failed to capitalize on the capability's full potential.[70] Thus, simply fielding new weapons is not enough. This sentiment is echoed by Army Futures Command, General James Rainey, who acknowledges that "new technology is not transformational by itself. To fully exploit the technology's potential, we must change how we operate, organize, and equip with it."[71] To fully exploit the technology's potential, the Army must also not forget the practical instructions of how to connect materiel solutions to broad concepts.

General William Momyer commented that, "We find ourselves constantly in a dilemma as to whether too much detail has been presented or whether we have become so terse that the meaning of doctrine is clouded

---

67 James Rainey, 'Continuous Transformation: Transformation in Contact,' *Military Review*, Aug. 2024.
68 Ibid.
69 Bernard Brodie and Fawn McKay Brodie, *From Crossbow to H-Bomb*, rev. and enl. ed. (Bloomington: Indiana University Press, 1973), 39–40.
70 Marc Bloch, *Strange Defeat: A Statement of Evidence Written in 1940* (New York: Norton, 1999), 36-38.
71 Rainey, 'Continuous Transformation: Transformation in Contact,' 2.

and darkness descends upon the reader."[72] Between the two options, MDO is trending towards the latter.

To move beyond doctrinal darkness and broad concepts to prescriptive, actionable solutions that enable field forces to succeed in contested, multidomain environments, militaries must improve tactical doctrine. While the introduction of MDO into doctrine marked an important step in shifting away from ULO, it remains largely a semantic change. Without the appropriate tactics and techniques—detailed in subordinate doctrinal publications—and the necessary equipment to operationalize its principles, the changes in FM 3-0 lack the prescriptive "how" to effectively conduct MDO. Without addressing these doctrinal gaps, MDO will struggle to achieve strategic objectives in future conflicts.

---

[72] S.A. Mackenzie 'Strategic Air Power Doctrine for Small Forces,' (Canberra: Air Power Studies Centre, 1994), 2.

5

Without a Why There Is No How
How Bureaucratic Complexity Detached MDO from Strategic Reality

Robert Rose

In 2023, I led a working group within Operations Group at the US Army's National Training Center (NTC) to update its training scenario in response to the newly published doctrine of Multidomain Operations (MDO). We concluded that the NTC should make just two changes to align its training scenarios with MDO: we replaced the use of decisive operations and shaping operations with main effort and supporting effort in the orders that we provided to the training unit; and second, we highlighted times on the division's synchronization matrix specifying when brigades could expect convergence, a key new term in MDO.

As defined in ADP 3-0, *Operations* "Convergence is an outcome created by the concerted employment of capabilities against combinations of decisive points in any domain to create effects against a system, formation, or decision maker, or in a specific geographic area."[1]

For the brigades training at the NTC, it was clear that they did not understand how convergence applied to them. Maybe, it was a period when they could take advantage of having the priority of fires, or potentially,

---

1 Army Doctrinal Publication 3-0, *Operations* (Washington, DC: US Government Publishing Office, March 2025), 27.

the cyber disruption of enemy forces.[2] Considering that the training brigades were already aware of when they had priority of fires and that they could request cyber effects, they did not change how they operated. The NTC routinely updated its training scenario based on observations of contemporary conflicts, but those refinements were not due to MDO or instructions from the Training and Doctrine Command (TRADOC).

If MDO produced few adjustments at the NTC, it is unlikely that many other Army units drastically changed how they prepared for war due to MDO's inclusion into Army doctrine. In its current form, MDO fails to provide clear guidance to the force for how to defeat its adversaries. FM 3-0, *Operations* explains that "Multidomain operations are the combined arms employment of joint and Army capabilities to create and exploit relative advantages that achieve objectives, defeat enemy forces, and consolidate gains on behalf of joint force commanders."[3] Stripped of inflationary language, this could be written as "MDO uses stuff well, to do stuff." Such a broad premise is difficult to criticize.

One criticism came from Brigadier General (Retired) Huba Wass de Czege. Wass de Czege was the lead author of AirLand Battle, the Army's 1982 doctrine. In a critique of the initial MDO concept, he explained, "An operating concept, like the logic of a campaign at war, needs to be the product of design based on a specific mission and context."[4] Although it may have started to address a specific situation, by the time MDO passed from initial concept to Army doctrine, it had lost its context.[5]

Doctrines need context to provide purpose. They need to be grounded in a specific theory of victory against clear adversaries and translate that theory of victory into how units will fight and train for that fight. To develop such a doctrine, armies need to integrate doctrinal development with war planning, operations, and training. With that close integration, armies can ensure their doctrines meet their operational requirements and rapidly adapt their doctrines to changing strategic environments. However, the US Army

---

2 When I asked a visiting doctrine writer to further define convergence and how it was not simply a synonym of synchronization, he responded that he hoped NTC would help the Army better understand convergence.

3 FM 3-0, 47.

4 Huba Wass de Czege, *Commentary on "The US Army in Multi-Domain Operations 2028"* (Carlisle, PA: Strategic Studies Institute, 1 April 2020), 8, https://press.armywarcollege.edu/cgi/viewcontent.cgi?article=1908&context=monographs.

5 Robert G. Rose, "Returning Context to Our Doctrine," *Military Review* (October 2023), https://www.armyupress.army.mil/journals/military-review/online-exclusive/2023-ole/returning-context-to-our-doctrine/.

has evolved a complex structure that silos thinking and creates bureaucratic inertia to slow doctrinal adaptation and disconnect it from strategic realities.

Over time, bureaucracies tend towards increasing complexity.[6] While often well-meaning, each additional layer reduces marginal productivity.[7] In the Army, additional four-star commands, such as TRADOC or Army Futures Command, create vast new staffs and the ensuing staff processes of decision boards, coordination meetings, working groups, progress reviews, etc. While each individual staff member may be hardworking, the requirement for disaggregated staffs to coordinate slows adaptation, detaches doctrine from operational requirements, and induces theories towards vagueness to achieve bureaucratic consensus.

Harold Seidman, a scholar of federal bureaucracy, described the quest for coordination as the "twentieth-century equivalent of the medieval search for the philosopher's stone."[8] Our complicated bureaucratic structure brought that quest into the twenty-first century by producing a doctrine based around integrating, synchronizing, federating, synergizing, and converging multiple domains.[9]

The Twisted Roots of MDO

The development of MDO revealed how the requirement to achieve consensus across a complex bureaucracy results in ineffectual doctrines. The seed of the concept was planted by US Deputy Secretary of Defense Robert Work's desire for a largely technologically driven "third off-set strategy." In 2015, Work explained, "The real essence of the third offset strategy is to find multiple different attacks against opponents across all domains..."[10]

In nurturing that idea, Lieutenant General H.R. McMaster, director of the US Army Capabilities and Integration Center (ARCIC), focused on

---

6   Michel Crozier, *The Bureaucratic Phenomenon* (Chicago: University of Chicago Press, 1964).

7   For an explanation of how organizations tend towards complexity and how that reduces marginal productivity, see Joseph A. Tainter, *The Collapse of Complex Societies* (Cambridge, UK: Cambridge University Press, 1988).

8   Harold Seidman, *Politics, Position, and Power: The Dynamics of Federal Organization* (New York: Oxford University Press, 1970), 164.

9   In searching for synonyms for coordinate, TRADOC Pamphlet (TP) 525-3-1, *The US Army in Multi-Domain Operations 2028* (Fort Eustis, VA: TRADOC, February 2020) used "synchronize" six times, "integrate" seventy-three times, "converge" ninety-four times, "federate" three times, and "synergy" twenty-three times.

10 Robert Work, "Deputy Secretary of Defense Speech," Army War College Strategy Conference, Carlisle, Pennsylvania, Apr. 8, 2015, https://www.defense.gov/News/Speeches/Speech-View/Article/606661/army-war-college-strategy-conference/.

how the Army could defeat a Russian invasion of the Baltic in 2030.[11] With this specific context, it was a promising start to develop clear thoughts on a future war. However, over the course of the concept development and eventual doctrine adoption, this context was diluted to secure consensus. By 2017, when General David Perkins, the TRADOC commander, wrote about the concept, he explained that it was not tied to a specific context writing "Whereas in AirLand Battle, the terrain, politics, and enemy were known, today, multiple adversaries of varying and growing capabilities are actively achieving their objectives under the threshold of armed conflict."[12] It became a broad idea to address multiple adversaries revealed by the first Multidomain Task Force being oriented on the Pacific rather than the Baltic.[13]

The broadening of the concept began when the Army partnered with the Marine Corps on MDO's initial development. The Marines desired a maritime rather than Russian focused concept.[14] Likely because of that early partnership, MDO focused on the problem of overcoming anti-access and area denial (A2AD) systems. At the time, A2AD was a preoccupation of the Navy in a conflict with China. Admiral John Richardson, the Chief of Naval Operations, even had to write an article criticizing the terms overuse, saying "A2AD is a term bandied about freely, with no precise definition, that sends a variety of vague or conflicting signals, depending on the context in which it is either transmitted or received."[15]

While A2AD originally was seen as a problem to overcome in fighting China, MDO made A2AD the primary problem in fighting any adversary.[16] While this approach made MDO more attractive to joint partners, it was not clear that all our adversaries pursued A2AD as their primary approach to war. For example, Michael Kofman argued, "[A2AD] is not a concept in Russian

---

11 For a history of the development of the MDO concept, see J.P. Clark's chapter "The Army in Multi-Domain Operations: The Intellectual Journey of an Organization" in this volume.

12 David G. Perkins, "Multi-Domain Battle: Driving Change to Win in the Future," *Military Review* (July-August 2017), https://www.armyupress.army.mil/Journals/Military-Review/English-Edition-Archives/July-August-2017/Perkins-Multi-Domain-Battle/.

13 Sean Kimmons, "Army updates future operating concept," US Army, Dec. 6, 2018, https://www.army.mil/article/214632/army_updates_future_operating_concept.

14 Kelly McCoy, "The Road to Multi-Domain Battle: An Origin Story," The Modern War Institute, Oct. 27, 2017, https://mwi.westpoint.edu/road-multi-domain-battle-origin-story/.

15 John Richardson, "Chief of Naval Operations Adm. John Richardson: Deconstructing A2AD," *The National Interest*, Oct. 3, 2016, https://nationalinterest.org/feature/chief-naval-operations-adm-john-richardson-deconstructing-17918.

16 Stephen Townsend, "Forward," TP 525-3-1, *The US Army in Multi-Domain Operations 2028* (Fort Eustis, VA: TRADOC, February 2020), iii, https://adminpubs.tradoc.army.mil/pamphlets/TP525-3-1.pdf.

military thought, and there is no Russian strategy bearing that name."[17] In the Russo-Ukraine War, Russia has not fought with A2AD approaches. It has not meaningfully disrupted arms shipments to Ukraine. Additionally, its much-vaunted cyber capabilities have had minimal impact on combat operations in Ukraine, and on civil society in Europe or the US

As the first large-scale war in Europe since World War II, the Russo-Ukraine War should provide many observations to update our doctrine. MDO did not predict the density of drone usage, the effectiveness of various countermeasures against long-range strike, or the difficulties in returning maneuver to the battlefield after an opponent has established a defense-in-depth. TRADOC published a revision to MDO in March 2025, but it provided only minor changes. While it does offer additional context for how Russia and China fight, that context does not modify the overall doctrine.[18]

In December 2024, TRADOC issued an assessment of the operational environment from 2024 to 2034. The assessment reinforced the assumptions behind MDO. It continued to discuss the problem of A2AD, but bizarrely, provided Russia's employment of mines as "best exhibiting area denial efforts in the Russia-Ukraine war." Whereas previously A2AD had meant adversaries' capabilities to disrupt strategic deployments, now it included tactical obstacles.[19] Underlying the Army's inertia to adapt its doctrine to strategic realities, the assessment assumed that "Geopolitical events that fundamentally reshape US adversaries' approaches to LSCO are unlikely."[20]

Doctrine needs to adapt to strategic realities, both due to changes in our adversaries and allies. Our allies in the Baltic, witnessing Russian crimes against occupied areas and Ukraine's difficulties in a counterattack, are developing the Baltic Defense Line to allow them to defend forward and "protect every inch of NATO territory."[21] However, MDO provides little guidance on how to reinforce such a forward defense. Our doctrine

---

17 Michael Kofman, "It's Time to Talk about A2/AD: Rethinking the Russian Military Challenge," *War on the Rocks*, Sep. 5, 2019, https://warontherocks.com/2019/09/its-time-to-talk-about-a2-ad-rethinking-the-russian-military-challenge/.

18 FM 3-0, 122-125.

19 Gary N. Brito, "Forward," TRADOC Pamphlet 525-92, *The Operational Environment 2024-2034: Large-Scale Combat Operations* (Fort Eustis, VA: TRADOC, Dec. 2024), 16, https://adminpubs.tradoc.army.mil/pamphlets/TP525-92.pdf.

20 TRADOC Pamphlet 525-92, 6.

21 Marta Kepe, "From Forward Presence to Forward Defense: NATO's Defense of the Baltics," RAND, Feb. 14, 2024, https://www.rand.org/pubs/commentary/2024/02/from-forward-presence-to-forward-defense-natos-defense.html.

does not align with how our allies will fight, because we detached doctrinal development from operational planning.

The Divergence of Doctrinal Development from Strategic Realities

When the US Army established TRADOC in 1973 and tasked it to create the Army's post-Vietnam War doctrine, it divided doctrinal development from training and operational planning. While TRADOC ran institutional training, Forces Command (FORSCOM) oversaw training for operational units.[22] Meanwhile, regional planning and command of units deployed to conflict fell under the Army Service Component Commands (ASCCs) subordinate to the Geographic Component Commands (GCCs). Due to these bureaucratic divides, TRADOC became detached from operational planning and large-scale exercises.

While TRADOC might cite AirLand Battle as an early success, in many ways, it is the exception that proves the problem. General Don Starry, who commanded TRADOC during the development of AirLand Battle, imported his concept of *Central Battle* to produce the doctrine. Before TRADOC, Starry commanded US V Corps, which stared down Soviet forces in Central Europe. Starry led the development of the Central Battle as the concept for how V Corps would fight those Soviet forces. Due to this unique experience, he could steward a doctrine optimized for that operational environment.[23] He grounded the doctrine in strategic requirements to deter enemy decision-makers that if they attacked, the US and its allies could threaten the liberation of Soviet satellite states. He explained that to preserve the territory of NATO, a defense should be "well forward and proceed aggressively…" He forecasted a clear timeline for how the Warsaw Pact would employ its forces and used that to develop a theory of victory based on interdicting its follow-on echelons.[24]

---

22 TRADOC Military History & Heritage Office Staff, *Victory Starts Here: A Short 50-Year History of the US Training and Doctrine Command* (Fort Leavenworth, KS: Combat Institute Press, 2023), 1.

23 John L. Romjue, *From Active Defense to AirLand Battle: The Development of Army Doctrine, 1973-1982* (Fort Monroe, VA: TRADOC, June 1984), 23-25, https://www.tradoc.army.mil/wp-content/uploads/2020/10/From-Active-Defense-to-AirLand-Battle.pdf.

24 Donn A. Starry, "Extending the Battlefield," *Military Review* (March 1981), https://www.armyupress.army.mil/Portals/7/online-publications/documents/1981-mr-donn-starry-extending-the-battlefield.pdf.

While Starry developed a doctrine tied to strategic realities, he straddled TRADOC with developing concepts that attempted to predict technologies for eight to ten years in the future.[25] Thus, TRADOC started basing its thinking on forecasted technologies instead of the requirements of the contemporary strategic environment.

In 2018, the US Army established another four-star command, Army Futures Command (AFC) and tasked it to lead the Army's technological modernization. The Army also transferred the role of ARCIC to the AFC's Futures and Concept Center (FCC).[26] With AFC focused on technology and leading concept development, TRADOC's downstream doctrines will likely continue to center on integrating expected technologies rather than being grounded in specific strategic contexts.

Case Studies of how to, and not to, Adapt Forces to Win

To illustrate how bureaucratic complexity leads to ineffective doctrines, let us first compare French and German doctrinal development before World War II. The Battle of France provides a valuable case study because the opposing sides arrayed nearly identical combat power, but Germany's approach to fighting yielded a spectacular victory. We can then analyze Israel, which provides an example of initial bureaucratic simplicity producing an effective doctrine in the 1960s, but then, its increasing bureaucratic complexity resulted in an imprecise doctrine that yielded military embarrassment against Hezbollah in 2004. Nonetheless, Israel provides a case for optimism. After that crisis, it reversed its bureaucratic bloat.

France's World War II Doctrinal Failure

In defeating Germany in World War I, France displayed a notable capability for innovation. France led Germany in motorization, built over 4,300 tanks compared to Germany's few dozen, and achieved air superiority with its capable aircraft. Towards the end of the war, it demonstrated tactical finesse in defeating the Spring Offensive and rapidly counterattacking with combined

---

25 *Victory Starts Here*, 8.
26 Jen Judson, "Army Futures Command drafting next operating concept," *Defense News*, Jul. 31, 2023, https://www.defensenews.com/land/2023/07/31/army-futures-command-drafting-next-operating-concept/.

arms during the Hundred Days Offensive.[27] France seemed that it would continue to innovate. In the early 1920s, a survey of French military writing observed that two-thirds of articles foresaw the likelihood of a more mobile war enabled by air and mechanized coordination.[28] As early as 1922, General Jean-Baptist Estienne called for the creation of an armor division.[29]

Yet, in 1939, when France again went to war with Germany, its doctrine was little changed from the end of World War I.[30] During the Battle of France, the Allies employed the same number of divisions as Germany and had the advantage of being in the defense.[31] Yet, France suffered a humiliating defeat. A major factor in its defeat was that its doctrine had remained static. France did not adapt its doctrine to address its strategic situation, changes in technology, or the approach that Germany took to fighting. France recognized its strategic problem, and its Army contained innovative thinkers, but the complex structure of its military bureaucracy impeded reform.

Recognizing the Strategic Problem

During the interwar period, France worried that Germany could launch a surprise attack to seize the majority of French iron mines, coal mines, and industry, which were perilously near the German border. France needed to defend against such a German *fait accompli*, while also preparing for a longer war.[32] Therefore, France invested in the Maginot Line to defend against a surprise attack and buy time to mobilize its resources.

However, Germany could bypass the Maginot Line and attack through Belgium as in World War I. France's border with Belgium extends 620 kilometers, but if France could defend forward in central Belgium, it would reduce by half the length of its defensive lines. Unfortunately, since Belgium maintained neutrality, France could not fortify Belgium before a war. If

---

27 James S. Corum, "A Clash of Military Cultures: German & French Approaches to Technology Between the World Wars" (Washington, DC: Joint Doctrine Division Support Group, 1997), 1-3.
28 Faris R. Kirkland, "Governmental Policy and Combat Effectiveness: France 1920-1940," *Armed Forces & Society*, Vol. 18, No. 2 (Winter 1992), 175-191, https://www.jstor.org/stable/45305303.
29 P. A. Bourget, *Le General Estienne, Penseur, Ingenieur et Soldat* (Paris: Berger-Levrault, 1956).
30 Ministère de la Guerre, *Instruction Provisoire sûr l'Emploi Tactique des Grande Unités* (October 1921).
31 Robert A. Doughty, *The Seeds of Disaster: The Development of French Army Doctrine, 1919-1939* (Mechanicsburg, PA: Stackpole Books, 1985), 5.
32 Robert A. Doughty, *The Breaking Point: Sedan and the Fall of France, 1940* (Mechanicsburg, PA: Stackpole Books, 1990), 4.

Germany invaded Belgium, France would have to rapidly come to its aid before Belgium was overrun.

France recognized this problem. In 1936, General René Tournès described the debate regarding the development of mechanized forces to quickly reinforce Belgium: "Seven mechanized divisions which had prepared in peacetime could render most valuable service in the first days of mobilization. They could go to the aid of the Belgian army, well entrenched on the fortified plateau of Herve east of Liège but exposed on its southern flank in the Ardennes..."[33]

France's Bureaucratic Inertia

While Tournès displayed that France recognized its problem, the French army delayed developing a method to rapidly reinforce Belgium and fight in an initially open, fluid battlefield. Tournès continued: "The problems raised by the development of mechanized divisions have not been solved any more than those raised by motorized divisions -- indeed they are even more experimental and controversial. In consequence, the French High Command has so far created only one experimental unit..."[34] In the three years between Tournès's remarks and the outbreak of the war, France still had not organized a mechanized division. During the same period, Germany expanded from three to ten *panzer* divisions.[35] Two months after Germany's dramatic success with Panzer divisions in Poland, France finally decided to establish a mechanized division.[36]

While some argue that France's sluggishness to adapt was due to a failure of imagination,[37] the French army had been innovative in World War I and possessed many forward thinkers who contributed a lively debate to their journals. Simply saying that the French lacked imagination fails to explain their failure. French officers had forward-looking concepts, but the

---

33 René Tournès, "The French Army, 1936," Foreign Affairs, April 1936, https://www.foreignaffairs.com/articles/france/1936-04-01/french-army-1936.

34 Tournès, "The French Army, 1936."

35 Alban Merglen, "La Responsabilité Du Maréchal Pétain Dans Le Désastre Militaire de Mai-Juin 1940," *Guerres Mondiales et Conflits Contemporains*, No. 184 (1996), 147–49, http://www.jstor.org/stable/25732385.

36 Eugenia C. Kiesling, "If It Ain't Broke, Don't Fix It': French Military Doctrine Between the World Wars," *War in History*, Vol. 3, No. 2 (April 1996), 222, https://www.jstor.org/stable/26004549.

37 Elizabeth Kier, *Imagining War: French and British Military Doctrine between the Wars* (Princeton, N.J.: Princeton University Press, 1997).

army did not implement them. To implement such ideas, armies need a simple, unified structure for thinking about war.

During World War I, when the French army displayed adaptability, first Marshal Joseph Joffre and then Marshal Ferdinand Foch unified the roles of Commander-in-Chief and Chief of the Army General Staff.[38] They were empowered with the wartime staff of the *Grand Quartier Général (GQG)* to address all aspects of war to meet national policy. However, in 1919, the GQG was dissolved, and the French army devolved from wartime unity to bureaucratic division.[39]

In peacetime, no one commanded the French army. After the experience of Napoleon III, the Third Republic was chronically concerned about a "man on horseback" overthrowing the republic.[40] During the interwar period, the civilian minister of war should have provided unity, but ministers regularly deferred to military experts.[41] The Superior Council of National Defense brought together ministers and military leaders and was supposed to set overall strategy and military policy, but it rarely met. A separate Permanent Committee of National Defense would conduct a war.

In 1932, France created the High Military Committee between the army, air force, and navy. That committee focused on inter-service cooperation, but it did not create a unified concept for how France would win its wars to bridge the competing views of the services. The Chief of the General Staff of National Defense led that committee, but he only had powers of coordination, not command over the services.[42] After the war, when the Vichy government conducted the Riom trial to investigate Frances's defeat, it apportioned blame on General Maurice Gamelin, as the Chief of the General Staff. He replied "The title is one thing. The power is another."[43]

In addition to a chief of the general staff, the French army had an Inspector General who was in charge of "all questions concerning organization, training, and mobilization."[44] Furthermore, the separate branch

---

38 Elizabeth Greenhalgh, "Marshal Ferdinand Foch versus Georges Clemenceau in 1919," *War in History War*, vol. 24, no. 4, 2017, https://www.jstor.org/stable/26393387.

39 Benoît Lagarde, "Grand Quartier Général, 1914-1918," Service Historique de la Défense (2011), https://www.servicehistorique.sga.defense.gouv.fr/sites/default/files/notices_files/SHDGR_REP_16NN.pdf.

40 Greenhalgh, "Marshal Ferdinand Foch versus Georges Clemenceau in 1919."

41 Doughty, *The Seeds of Disaster*, 133.

42 Philippe Vial, "1932-1961: Unifying Defense," *Inflexions*, Vol. 21, No. 3, https://doi.org/10.3917/infle.021.0011.

43 Doughty, *The Seeds of Disaster*, 120-125.

44 Doughty, *The Seeds of Disaster*, 123.

departments (infantry, artillery, cavalry, etc.) largely operated independently of the general staff, making their own decisions on personnel, material, training, organizations, and doctrine. The general staff's third bureau oversaw operations and training, but it did not have authority over the branches.[45] The French War College also took a lead role in doctrinal developments.[46] No single individual or institution could resolve doctrinal disagreements or rapidly adapt doctrine to strategy.[47]

While in 1935, Gamelin established that the high command was the sole arbiter for doctrine, this action only further slowed approval and restricted debate.[48] Any adjustments to doctrine had to be coordinated amongst departments resulting in doctrinal inertia. Efforts to adapt doctrine were caught up in committee meetings and waiting periods for comments from the departments. This structure impeded innovation. Officers published articles that revealed an increasing professional focus on bureaucratic infighting and minutia.[49] Proposed changes to doctrine tended to be conservative and in line with existing concepts to maximize consensus.[50] For example, the doctrine on the *Regulation of Tank Units* began as a draft in 1938, but it was not published until the spring of 1940, days before Germany's attack.[51]

In 1934, Charles De Gaulle wrote a book arguing for the creation of a mechanized force of six mechanized divisions as a rapid response force. In the book, he reveals frustration with bureaucratic inertia, writing "the conditions in which the state functions today allow no one the authority or the time to carry through such an undertaking." He describes the situation as a "paralysis" and a "sclerosis of power."[52]

---

45 Martin S. Alexander, *The Republic in Danger: General Maurice Gamelin and the Politics of French Defence, 1933-1940* (Cambridge: Cambridge University Press, 1992), 118-119.
46 David Campbell and Jesse McIntyre III, "A Policy of Defeat," *Small Wars Journal*, Jul. 4, 2018, https://smallwarsjournal.com/jrnl/art/policy-defeat.
47 Douglas Porch, "Military 'Culture' and the Fall of France in 1940: A Review Essay," *International Security*, Vol. 24, No. 4, 2000, http://www.jstor.org/stable/2539318.
48 Williamson Murray, "Armored warfare: The British, French, and German experiences," *Military Innovation in the Interwar Period* (Cambridge, UK: Cambridge University Press, 1998), 31.
49 Kirkland, "Governmental Policy and Combat Effectiveness: France 1920-1940."
50 Doughty, *The Seeds of Disaster*, 193.
51 Doughty, *The Seeds of Disaster*, 135.
52 Charles de Gaulle, *The Army of the Future* (Philadelphia: J. B. Lippincott Company, 1941), 178.

Methodical Battle's Shortcomings

With its slowness to adapt its doctrine, the French army was ill-prepared to meet the strategic situation of the late 1930s. Its approach to fighting assumed an initial defensive campaign, in which France would have time to mobilize resources before conducting well-synchronized, methodical attacks for shallow objectives. Their primary regulation stated that "One should... fight the battle only in a planned manner and only after bringing up all available fire support."[53] A German observer of French training explained that "French tactics are essentially characterized by systematization, which seeks to anticipate and account for any eventuality in the smallest detail."[54] French generals argued on the eve of World War II that a successful attack required "three times as much infantry, six times the artillery, and twelve times the ammunition."[55]

However, in assuming time to mobilize and synchronize combat power, the French army was unprepared to rapidly respond to Germany's remilitarization of the Rhineland in 1936[56] and was not ready to support France's commitment to Czechoslovakia or Poland. When Germany invaded Poland and France declared war on Germany in response, the French army did little to assist the beleaguered Poles. Potentially, France could have overrun Germany's thin forces on the border and seized Germany's industrial heartland in the Rhur Valley, but its doctrine was not suited for such a quick offense. After the war, General Alfred Jodl would state that "if [Germany] did not collapse already in the year 1939 that was due only to the fact that during the Polish campaign, the approximately 110 French and British divisions in the West were held completely inactive against the 23 German divisions."[57] Germany could concentrate on Poland before turning west.

When Germany attacked in the west, France moved its primary forces 175 km through Belgium to establish a defense in Holland along the Scheldt River. Meanwhile, Germany only had to move 90 km.[58] Instead of the solidified, continuous front that French doctrine expected, French forces

---

53 Ministère de la Guerre, *Instruction provisoire sur l'emploi tactique des grandes unites*, 130.
54 Elizabeth Kier, *Imagining War: French and British Military Doctrine between the Wars*, 194.
55 Louis Chauvineau, *Une invasion est-elle encore possible?* (Paris: Berger-Levrault, 1939), 101.
56 Stephen A. Schuker, "France and the Remilitarization of the Rhineland, 1936," *French Historical Studies*, Vol. 14, No. 3 (Spring 1986), https://www.jstor.org/stable/286380.
57 *Trial of the Major War Criminals Before the International Military Tribunal*, Vol. 15 (Nüremberg: International Military Tribunal, 1948), 350, https://www.loc.gov/rr/frd/Military_Law/pdf/NT_Vol-XV.pdf.
58 Doughty, *The Breaking Point*, 9.

collided with Germans in unplanned, fluid engagements. The commander of the Northeastern Front, general Alphonse George, had observed how German mechanized forces had bypassed Polish forces and directed his units to organize a defense-in-depth. If they were bypassed, he instructed them to hold in place, establish a perimeter defense, and prepare to support counterattacks. In spite of this order, his subordinate commanders fought as proscribed in doctrine and attempted to withdraw to preserve a continuous front.[59] Armies that have trained to fight one way, cannot instantly adopt a new approach.

Even though the French army found itself in a situation that had been predicted, it was unprepared to fight a chaotic, rapid war of movement. It collapsed. Its doctrine had been correct for the battlefields of 1918, it might even have been correct in 1943, Field Marshal Bernard Montgomery would succeed with an approach very similar to methodical battle,[60] but the doctrine was not aligned with France's requirements in 1940.

How Germany Aligned its Doctrine to its Strategic Situation

Unlike the French, during the interwar years, the Germans had a simplified military structure that supported the development of a doctrine aligned with Germany's strategic situation. A key source of Germany's doctrinal adaptability was the general staff. The network of rigorously selected and educated general staff officers scattered throughout the echelons of the German army provided an informal means of sharing lessons and implementing new ideas. Those officers cut across bureaucracy and flattened the already simple structure of the German army.[61]

German Interwar Doctrinal Development

After World War I, the Allies placed restrictions on the German army which forced it to create a streamlined organization of only 100,000 soldiers. To fill this shrunken army, the leader of the army from 1919 to 1926, General

---

59 Maurice Gamelin, *Servir*, Vol. 3 (Paris: Plon, 1946), 487-91.
60 Williamson Murray, "British Military Effectiveness in World War II," in Millett and Murray, *Military Effectiveness*, Vol. 3, *The Second World War*, 119
61 Corum, "A Clash of Military Cultures: German & French Approaches to Technology Between the World Wars," 22.

Hans von Seeckt prioritized retaining general staff officers.[62] Von Seeckt promoted debate on the lessons of World War I and formed 57 committees with 400 officers to study all aspects of the war.[63]

Overseeing this learning process, Von Seeckt provided a unified vision. He saw the need for Germany to be able to wage a quick, aggressive war against Poland and France to prevent a war of exhaustion.[64] Instead of delaying the start of a war to mobilize reserves, he argued for a professional army that could be stationed on the frontier in a high state of readiness. Such an army could achieve strategic surprise against an opponent and defeat it before it could fully mobilize and establish an effective defense. To prevent the enemy from establishing a defense, he emphasized infiltration tactics to achieve an initial breakthrough followed by a continuous and aggressive attack to pursue and annihilate the enemy.[65]

Because the air force was then subordinate to the army, Von Seeckt could align air force doctrine with the army. He prioritized gaining air superiority by initially concentrating against enemy airfields. Afterward, the air force would disrupt the enemy's troop mobilization and then provide reconnaissance and ground support. He viewed strategic bombing as a costly diversion of aircraft.[66] He established a clear, practicable concept to integrate air and land forces due to the bureaucratic simplicity of not having a separate air service.

The Training-Doctrine Feedback Loop

While Von Seeckt provided a cohesive concept, he did not produce a detailed series of doctrinal manuals. Instead, he focused on rigorous, realistic training to allow units to create shared understanding of how they would fight. Exercises concentrated on training for meeting engagements and a war of movement that aligned with Von Seeckt's concepts.[67]

---

62 James S. Corum, *The Roots of Blitzkrieg: Hans Von Seeckt and German Military Reform* (Lawrence, KS: Kansas University Press, 1992), 33-34.

63 Corum, *The Roots of Blitzkrieg*, 37-38.

64 Seeckt's approach shared much with Prussia's traditional approach of seeking "Kurz and Vives" (short and lively) wars as a solution to its strategic context. Located in the center of Europe without natural obstacles on its borders, it feared France, Russia, and Austria threatening it with overwhelming combat power on multiple fronts, see Robert M. Citino, *The German Way of War: From the Thirty Years War to the Third Reich* (Lawrence, KS: The University Press of Kansas, 2005).

65 Robert A. Doughty, *The Seeds of Disaster*, 114.

66 Hans von Seeckt, *Thoughts of a Soldier* (London, UK: Ernest Benn, 1930), 61-62.

67 Corum, *The Roots of Blitzkrieg*, 89.

Before it attacked France, Germany demonstrated how rapidly it could learn, adapt, and produce an effective fighting force. In October 1939, after it overran Poland, the general staff led a campaign of self-criticism to improve the army for the fight against France. After collecting lessons, it shared them across the army and directed all divisions to conduct realistic, large-scale training.[68] By March 1940, the entire army retrained to meet the demands of an invasion of France.

Germany's doctrine was optimized to defeat Poland, France, and the smaller states of Europe, but not for wars with continental-sized powers on the scale of the Soviet Union and the United States. Germany's ability to adapt how it fought deteriorated as Adolf Hitler created an increasingly complex bureaucracy, which broke the air force from the army, weakened the general staff, and injected Nazi leaders into decisions.[69] In the end, it is unlikely any approach to fighting could have made up for the material imbalance between the Axis and Allied powers.

Israel's Doctrinal Success and Failure

In 1948, the Israel Defense Forces (IDF) were born in war. In forming a new organization driven to deal with a clear encompassing threat of national survival, they developed a streamlined military bureaucracy. Israel created a simple military structure with a single service united, a single commander, and a single staff.[70] Its general staff combined strategic planning, wartime operations, doctrinal development, and training in a single organization. It provided a unity of military thought and preparations.[71]

Within this simplified organization structure, Israel encouraged innovation, free thinking, and open discourse through a flattened hierarchy and culture of informality. Today, across the current Israeli Defense Forces, which number 600,000 when mobilized, there is one three-star and 24 two-star generals. It's air force, which is larger than the British Royal Air Force, is commanded by a two-star general. By reducing the rank structure and

---

68 Williamson Murray, "The German Response to Victory in Poland: A Case Study in Professionalism," *Armed Forces & Society*, Vol. 7, No. 2 (1981), 285–98, http://www.jstor.org/stable/45346229.

69 Carey Brewer, "The General Staff of the German Army," *Proceedings*, Vol. 62, No. 2 (February 1956).

70 Edward N. Luttwak and Eitan Shamir, *The Art of Military Innovation: Lessons from the Israel Defense Forces* (Cambridge, MA: Harvard University Press, 2023), 5.

71 Luttwak and Shamir, *The Art of Military Innovation*, 34.

intermediate echelons between decision-making, the IDF reduced the need to coordinate between staffs.[72]

While the German army relied on its general staff officers to foster the free flow of ideas, the IDF promoted an informal culture of free discourse. Since the days of the socialist Palmach militia, Israeli forces displayed a lack of deference compared to other militaries. Most visibly, the IDF did not have dress uniforms, avoided drill and ceremony, and called senior leaders by their nicknames. With such a culture, subordinates felt free to debate, innovate, and employ flexible tactics.[73]

Israel's Doctrinal Alignment before the Six-Day War

With an organization optimized for learning, the IDF aimed towards a clear strategic problem of defeating the Arab states that threatened its existence. With a comparatively small population compared to its opponents, the IDF established a reserve system inspired by Swiss and Finnish models. While it could mobilize its population for a short war to equal the numbers of the Arab states, it could not fight a prolonged war of exhaustion with so much of its manpower pulled out of its economy.

In 1953, David Ben-Gurion, the long serving minister of defense, formulated a coherent concept of defense, which Israel would follow for the next three decades. He stated that "If [the Arab states] attack us in the future, we do not want the war to be waged in our country, but rather, in the enemy's country, so that we will not be on the defensive, but rather on the offensive. This war is waged not by border settlements, but rather by mobile forces equipped with rapid vehicles and strong firepower."[74] His vision of a rapid war using maneuver approaches provided a clear vision to drive the IDF's doctrinal development.

With its unified structure, the IDF could ensure that all services supported this concept. Dan Tolkosky, the head of the Air Corps in 1953, focused its doctrine and aircraft development on employing short-range fighter-bombers to destroy enemy aircraft on the ground in an all-out surprise

---

72 Luttwak and Shamir, *The Art of Military Innovation*, 33-35.
73 Eitan Shamir, *Transforming Command: The Pursuit of Mission Command in the US, British, and Israeli Armies* (Stanford: Stanford University Press, 2011), 84-85.
74 Gabi Siboni, Yuval Bazak, and Gal Perl Finkel, "The Development of Security-Military Thinking in the IDF," *Strategic Assessment*, Vol. 21, No. 1 (April 2018), 2, https://www.inss.org.il/wp-content/uploads/2018/05/The-Development-of-Security-Military-Thinking-in-the-IDF.pdf.

attack against their airfields to ensure air superiority before transitioning those same aircraft to close support of the army.[75]

With war planning, doctrine, and training closely aligned, the IDF conducted large-scale training exercises to develop techniques to fight in the specific scenarios in which units would be called to fight. For an example of how this system allowed for rapid adaptations, in 1964, the general staff identified that the Egyptian and Syrian armies had switched to the latest Soviet doctrine. The general staff published revised training guidance so that all units trained against their adversaries' latest methods. By the Six-Day War in 1967, IDF units had optimized their techniques to deliver a devastating defeat to those Egyptian and Syrian forces.[76]

Israeli's Doctrinal Failure against Hezbollah

However, nearly 40 years later, the IDF suffered embarrassment against the much weaker Hezbollah. The chiefs of the Israeli intelligence services told their prime minister that "the war was a national catastrophe."[77] During the war, the IDF appeared to be confused by its doctrine.[78]

Starting in the 1970s, the IDF had developed a more complex bureaucracy, with an expanded staff that emphasized technological capabilities at the expense of doctrine and training.[79] It broke the general staff's unity of planning, operations, doctrine, and training. A separate planning directorate came into being, which siloed strategic thought.[80] Meanwhile, different branches began to develop independent doctrines, and in 1993, the IDF established the Institute for Operational Doctrinal Research (MALTAM), which acted as a think tank on doctrinal concepts separate from the general staff.[81]

---

75 Luttwak and Shamir, *The Art of Military Innovation*, 25.
76 Siboni, Bazak, and Finkel, "The Development of Security-Military Thinking in the IDF," 11.
77 "Secret Meeting," Ynetnews.com, Mar. 30, 2007, ynetnews.com/Ext/Comp/ArticleLayout/CdaArticlePrintPreview/1,2506,L-3383151,00.html.
78 Matt M. Mathews, "We Were Caught Unprepared: The 2006 Hezbollah-Israeli War," The Long War Series Occasional Paper, No. 26 (Fort Leavenworth, KS: Combat Studies Institute Press, 2008), https://www.armyupress.army.mil/Portals/7/combat-studies-institute/csi-books/we-were-caught-unprepared.pdf.
79 Emanuel Wald, *The Wald Report: The Decline of Israeli National Security Since 1967* (Boulder, CO: Westview, 1992).
80 Siboni, Bazak, and Finkel, "The Development of Security-Military Thinking in the IDF," 11.
81 Tamir Hayman, "Learning in the General Staff," *Dado Center Journal*, No. 8 (September 2016), https://www.idf.il/en/mini-sites/dado-center/vol-8-the-general-staff-part-a/learning-in-the-general-staff/.

MALTAM pushed for Systemic Operational Design (SOD), which in contrast to Israel's previously threat-based approach to doctrine, employed a complicated methodology for analyzing problems that drew from "postmodern French philosophy, literary theory, architecture, and psychology." The doctrine's primary author stated that it was "not intended for ordinary mortals."[82] Instead of focusing on developing tactical techniques, IDF officers became enmeshed in theoretic debates on ambiguous terms. It created confusion when applied in combat.[83]

Furthermore, the IDF no longer invested in large-scale, realistic training to ensure training, doctrine, and operational plans aligned. Such training could have helped validate SOD and potentially turned it into something actionable. Official guidance prioritized small unit training, which meant that reserve units did not train in large formations for four to six years and divisions almost never trained in the field.[84]

However, after the war, the IDF reformed. Israel commissioned the Winograd Report, which was highly critical of the IDF and enumerated training, doctrinal, and organizational improvements.[85] The next chief of the general staff, Lieutenant General Gabi Ashkenazi eliminated SOD and focused on a threat-aligned doctrine.[86]

Fixing Doctrine Development

Israel's experience demonstrates that armies can come back from a doctrinal malaise. For the US military, the ideal way to unify doctrinal thinking would be to create a single service with a single general staff. As in the German and Israeli examples, a general staff would bring together supreme command, war planning, training, and doctrine. Realistically, service pride would make such a proposal infeasible.

A more feasible change would be to reform the Planning, Programming, Budgeting, and Execution (PPBE) process. Initially developed in 1961 to ensure a centralized means for the Department of Defense (DoD) to align budget

82 Yotam Feldman, "Dr. Naveh, Or, How I Learned To Stop Worrying and Walk Through Walls," *Haaretz*, Oct. 27, 2007, https://www.haaretz.com/2007-10-25/ty-article/dr-naveh-or-how-i-learned-to-stop-worrying-and-walk-through-walls/0000017f-db53-df9c-a17f-ff5ba92c0000.
83 Hayman, "Learning in the General Staff."
84 Mathews, "We Were Caught Unprepared: The 2006 Hezbollah-Israeli War," 27-28.
85 "The Winograd Report: The Main Findings of the Winograd Partial Report on the Second Lebanon War," *Wall Street Journal*, 2007, https://www.wsj.com/public/resources/documents/winogradreport-04302007.pdf.
86 Hayman, "Learning in the General Staff."

with strategy, in the 1970s, it decentralized management to the services who would submit budget proposals for evaluation by the Office of the Secretary of Defense (OSD) and submission to Congress.[87] To fight for their priorities, services regularly submit budgets without clear guidance from OSD. Due to this interservice budget brawl, services are incentivized to establish organizations such as Army Futures Command (AFC) to create capability-based concepts in advance of OSD's strategic guidance. The process has led to bureaucratic bloat as OSD, the Joint Staff, and services created duplicative staffs, which are bogged down in coordination process.[88] By returning PPBE to OSD, we could ensure that strategic requirements drive the budget and reduce the pressure for services to develop capabilities-based concepts.

Within the Army, the Army Staff should write the Army's capstone doctrine as did the German and Israeli general staffs. To unify thinking on doctrine, training, and operational requirements, we should remove concepts from AFC and doctrine from TRADOC and centralize them within an empowered Army Staff. The Germans and Israelis did not have to write a doctrine optimized for threats across the globe, but we do. Therefore, the Army Service Components Commands (ASCCs) should provide regional annexes to our doctrine. Those annexes should make up the preponderance of the doctrine and specify how units would defeat specific adversaries. The main body of the document would capture the commonalities and how the Army would think through mobilization and deployment to those theaters.

Furthermore, to reduce bureaucratic complexity and unify planning, operational command, training, and doctrine, the Army should eliminate Forces Command (FORSCOM) and assign all units to regional ASCCs. To ensure doctrine, training, and capability development align with operational requirements, all commands that do not lead forces in combat should be reduced below the four-star level of command. In criticizing how little agency regional commands currently possess, Major General Peter Andrysiak, the chief of staff of European Command, wrote that Combatant Commands control just 0.7 percent of the Department of Defense's budget.[89]

---

87 Robert Hale et al., "Defense Resourcing for the Future," Commission on Planning, Programming, Budgeting, and Execution Reform (Mar. 6, 2024), 24, Commission-on-PPBE-Reform_Full-Report_6-March-2024_FINAL.pdf.

88 Clark A. Murdock et al., "Beyond Goldwater-Nichols (BG-N): Defense Reform for a New Strategic Era - Phase 1 Report." Center for Strategic and International Studies (Mar. 1, 2004), 7, https://www.csis.org/analysis/beyond-goldwater-nichols-phase-i-report.

89 Peter Andrysiak and Bryan Quinn, "Empowering the Combatant Commands is Critical for the Future Fight," *War on the Rocks*, Dec. 10, 2024, https://warontherocks.com/2024/12/empowering-the-combatant-commands-is-critical-for-the-future-fight/.

As an example of how to align units to regional commands, the Army should assign V Corps with subordinate divisions and brigades to US Army Europe and Africa. Those units would train in scenarios that would refine concepts for how to reinforce the Baltic Defense Line and ensure interoperability with the North Atlantic Treaty Organization. I Corps and its subordinate units already have developed such a relationship with US Army Pacific. By providing units with specific context to prepare for, they can refine techniques and drive bottom-up doctrinal feedback.

The Army can enable units to better develop techniques by reforming its industrial-age personnel system. The Army manages its personnel through a complicated bureaucratic approach, which requires most soldiers to move every couple of years.[90] This churn prevents units from conducting long-term learning and creating the institutional knowledge to refine techniques.[91] The Army's approach mirror's France's approach to its personnel, which treated them like cogs in a machine. Conversely, Germany and Israel prioritized unit cohesion and assigned personnel to the same units for most of their career. Over years of exercises, units could continuously improve their techniques and establish a strong shared understanding of how they would fight.

Finally, to make our doctrines more adaptable, we should stop giving them grandiose names. The French had Methodical Battle; the Germans and the Israelis (at least before SOD) had no such grand names for their doctrines. Imposing names provide doctrines with added inertia. Pushing a change through the bureaucratic process requires consensus over a new name and becomes a marketing campaign. Maybe we should just publish annual doctrines titled by the year. Instead of relying on a catchy name, a doctrine's quality should come from providing guidance that prepares units to solve specific strategic problems.

---

90 Robert G. Rose "Ending the Churn: To Solve the Recruiting Crisis, the Army Should be Asking Very Different Questions," Modern War Institute, Feb. 9, 2024, Ending the Churn: To Solve the Recruiting Crisis, the Army Should Be Asking Very Different Questions - Modern War Institute.
91 Based on my reviews of division and brigade SOPs collected by the National Training Center and Mission Command Training Program, the vast majority lacked detail, heavily copied from doctrine, and did not present a coherent vision of how the unit would fight.

PART 3

TENSIONS WITH CONTEMPORARY CONFLICT AND FUTURE DEVELOPMENTS

The Muddle of Multidomain Operations
Why and How the Concept Should Be Reined In

Heather Venable[1]

In preparation for teaching an upcoming course on contemporary warfighting, a group of field grade officers and civilian professors recently tried to determine what exactly multidomain operations (MDO) is and how much it differs from previous concepts. After some heated debate, the only real agreement we arrived at was that the US military now sought to integrate fires rather than simply deconflict them. That a group of committed professionals struggled to identify the essence of MDO says much about the extent to which the most prominent operational concept to emerge in the US military in the last decade raises more questions than it answers.[2]

Having strayed too far from its original purpose, the US military should refocus the concept of MDO on a very specific problem: how to suppress or destroy fifth-generation air defenses to enable freedom of action, while protecting its personnel and materiel from an opponent's long-range fires. The flaws in this concept also suggest a need to clean house when it comes

---

1 Disclaimer: The views expressed are those of the author and do not necessarily reflect the official policy or position of the Department of the Army, Department of Defense, or the US Government.

2 COL Marco J. Lyons, USA, & COL David E. Johnson, USA, Ret., "People Who Know, Know MDO: Understanding Army Multidomain Operations as a Way to Make It Better," Land Warfare Paper 151, November 2022, AUSA, https://www.ausa.org/publications/people-who-know-know-mdo-understanding-army-multidomain-operations-way-make-it-better.

to the intellectual foundation of Western military thinking, which seeks to obfuscate more than it does to clarify.[3]

Initially, MDO properly focused on synergizing the traditional domains of land, sea, and air fully with the newer cyber and space domains given the recognition that acquiring air superiority would be far more difficult in anti-access and anti-denial (A2AD) environments.[4] As the Army and Marine Corps' first white paper on what it then called multi domain battle, or MDB, explained in 2017, the US now had to contend with challenges to superiority in every domain. Losing superiority in the air domain, however, was particularly critical.[5] What began in January 2017 as a 14-page white paper that attempted to update combined arms for the 21st century, had a year later morphed into 100 pages of tedious, illogical, and largely unreadable prose when the Army published its "The US Army in Multidomain Operations 2028."[6] That focus has been lost as the concept has morphed into seeking to be a comprehensive approach to future war including contesting peer competitors' efforts to create "layered stand-off in the political, military, and economic realms."

This document is also unnecessarily confusing. In explaining how "[m]ulti-domain formations can conduct independent maneuver, employ cross-domain fires, and maximize human potential," for example, the document employs circular logic. In effect, the Army states that multidomain formations can employ cross-domain fires. What is the difference between the two? In the glossary, the document defines cross-domain as "having an effect from one domain into another." It defines MDO as "conducted across multiple domains and contested spaces." There is no meaningful difference

---

3 Luke O'Brien, "The Doctrine of Military Change: How the US Army Evolves," July 25, 2016; https://warontherocks.com/2016/07/the-doctrine-of-military-change-how-the-us-army-evolves/; Steve Leonard, "Broken and Unreadable: Our Unbearable Aversion to Doctrine," May 18, 2017; https://mwi.westpoint.edu/broken-unreadable-unbearable-aversion-doctrine/.
4 US Air Force, "Air Superiority 2030 Flight Plan Enterprise Capability Collaboration Team," May 2016; https://www.af.mil/Portals/1/documents/airpower/Air%20Superiority%202030%20Flight%20Plan.pdf; Todd Smith, "This 3-star Army general explains what multidomain operations mean for you," Army Times, 11 August 2019; https://www.armytimes.com/news/your-army/2019/08/11/this-3-star-army-general-explains-what-multidomain-operations-mean-for-you/.
5 "United States Army-Marine Corps White Paper Multidomain Battle: Combined Arms for the 21st Century," January 18, 2017; https://imlive.s3.amazonaws.com/Federal%20Government/ID19831026121158752582394001180406110474/Multidomain_Battle_White_Paper_Final_18_Jan_2017.pdf.
6 TRADOC, "The US Army in Multidomain Operations, 2028," 6 December 2018, iii; https://api.army.mil/e2/c/downloads/2021/02/26/b45372c1/20181206-tp525-3-1-the-us-army-in-mdo-2028-final.pdf.

between these definitions. The glossary not only defines cross-domain but also cross-domain fires, cross-domain synergy, and cross-domain maneuver to create word soup.[7]

Confusion reigns in the national security community about whether the concept offers anything that is meaningfully new, as seen at a recent one-day conference focused on the topic.[8] When one individual suggested that MDO could be seen as far back as Operation Iraqi Freedom, another pushed back, suggesting the concept is far newer. If one can see significant multidomain operations at work in Operation Iraqi Freedom, it is unclear why there is a need for a new concept. Indeed, coalition partners engaged in Operation Iraqi Freedom enjoyed significant superiority in multiple domains, thus the campaign does not accord with the need for multidomain operations in the first place.

Given the concept's amorphous and expanding nature, the notion of MDO unsurprisingly has been loosely and widely applied to the Russo-Ukraine War. One analyst has argued that the Russo-Ukraine War has had a "significant effect on its trends (e.g. kinetic military operations with non-kinetic cyber-attacks; missile and artillery strikes based on information collected by space satellites."[9] This kind of analysis lumps together a vague reference to multiple domains supporting each other, but this type of employment has been used as early as Operation Allied Force in 1999.[10]

Others refer vaguely to multidomain operations to use the Russo-Ukraine War to validate MDO. Importantly, analysts have warned about using current conflicts to validate developing concepts given the dangers of confirmation bias, especially when the US military has largely observed the war at a safe distance due to prohibiting its servicemembers going to Ukraine.[11]

---

7 "Army in Multidomain Operations 2028," GL-3 and GL-7.

8 The Hague Center for Strategic Studies Symposium and NATO HQ SACT, "Rethinking Fire and Manoeuvre across physical and non-physical aspects of domains." September 28,2023; https://hcss.nl/news/symposium-rethinking-fire-and-manoeuvre-across-physical-and-non-physical-aspects-of-domains/.

9 Futoshi Takabatake, "NATO's Approach to Multidomain Operations: From the Perspective of the Economics of Alliances," *Defence and Peace Economics* 35 (July 2023: 3): 281–94: 281-82.

10 For cyber operations in Operation Allied Force see Fred M. Kaplan, *Dark Territory: The Secret History of Cyber War* (New York: Simon and Schuster, 2016).

11 John A. Nagl, "Learning from Russia's war on Ukraine," Military Times, September 1, 2023; https://www.militarytimes.com/opinion/2023/09/01/learning-from-russias-war-on-ukraine/; David Johnson, "The Army Risks Reasoning Backwards in Analyzing Ukraine," *War on the Rocks*, June 14, 2022; https://warontherocks.com/2022/06/the-army-risks-reasoning-backwards-in-analyzing-ukraine/.

The disagreement over MDO not only reflects understandable and evolving debate but also the tendency of the US military to over-engineer its doctrine in a way that obscures and complicates, rather than clarifies. The Army, perhaps more so than any other service, has excelled in obfuscation. As one Army officer has argued, the Army excels at taking "prosaic ideas and dress[ing] them in the multisyllabic finery of sophistication."[12]

This tendency is evident in the very definition of MDO. The Army's Field Manual 3-0 *Operations,* published in the fall of 2022, defines MDO as "the combined arms employment of joint and Army capabilities to create and exploit relative advantages that achieve objectives, defeat enemy forces and consolidate gains on behalf of joint force commanders."[13] This definition is unsatisfactory in that it is impossible to determine why one needs a new concept since the US has been pursuing "relative advantage" ever since it adopted maneuver warfare after the Vietnam War and it has employed combined arms even longer. The confusing phrasing of "achieve objectives, defeat enemy forces and consolidate gains" is also a muddled and disorganized mess of what exactly a military seeks to achieve.

These definitions also raise the question of whether or not multidomain is just another way to say combined arms, as suggested by the early white paper. Of course, the concept of combined arms stretches back thousands of years. In discussing multidomain operations, one scholar has even argued that the Athenians' disastrous expedition to Sicily in 413 B.C. owes much to the success of multidomain operations, with the key lesson being that the best way to achieve victory is by "establishing superiority in a combination of domains offers the freedom of action necessary to attain mission success."[14] What is less clear is why the US military requires hundreds of pages of pamphlets, white papers, and PowerPoint slides to learn such a lesson.

Meanwhile, the US Army and NATO resoundingly sound the rhetorical drum on the topic. NATO extravagantly describes the concept as a "paradigm shift" that "empowers the Alliance to strategically influence events, synchronize efforts with external stakeholders, and present formidable

---

12 Major Daniel J. Kull, USA, "The Myopic Muddle of the Army's Operations Doctrine," *Military Review,* May 24, 2017; https://www.armyupress.army.mil/Journals/Military-Review/Online-Exclusive/2017-Online-Exclusive-Articles/Myopic-Muddle-of-Army-Ops-Doctrine/.
13 Quoted in "Defense Primer: Army Multidomain Operations (MDO)," Congressional Research Service, updated October 1, 2024; https://crsreports.congress.gov/product/pdf/IF/IF11409.
14 Jeffrey M. Reilly, "Multidomain Operations A Subtle but Significant Transition in Military Thought," *Air & Space Power Journal* 30, no. 1 (2016): 61-73: 62-63.

challenges to adversaries."[15] It is difficult to reconcile this rhetoric with the reality of Europe's limited military capabilities. A recent British report warns not only that its Army is too small to meet its commitments but also that NATO must "re-establish a credible deterrence posture towards Putin's Russia."[16] In this context, MDO appears to be an attempt to window dress a broken force. Some Army officers even dismiss those who critique the concept as members of "entrenched bureaucracies" afraid of new ideas.[17]

To resolve this conceptual conundrum in updating combined arms, Western doctrine should be scoped more narrowly within the operational level of war given the primary challenge MDO seeks to solve: how to penetrate an opponent's fifth-generation air defenses. In this way it neatly mirrors the joint development of the Army and the Air Force's AirLand Battle concept. AirLand Battle was developed in response to the seemingly striking operational success of the Egyptian military against Israel during the 1973 Arab-Israeli War as it advanced under an umbrella of ground-based air defenses.[18] Airland Battle correctly sought to solve a specific, tailored problem: that of defeating the "second echelon."[19] MDO must be a truly coalition product developed together rather than within services. Most importantly, though, the muddle over MDO may reveal a fatal flaw in Western military thinking; perhaps most damning being the unchecked willingness to fall in line with any concept the US unveils. It is time to return

---

15 "Multidomain Operations in NATO—Explained," October 5, 2023; https://www.act.nato.int/article/mdo-in-nato-explained/.

16 International Relations and Defence Committee, "Ukraine: a wake-up call for the UK and NATO," 26 September 2024; https://committees.parliament.uk/committee/360/international-relations-and-defence-committee/news/202994/ukraine-a-wakeup-call-for-the-uk-and-nato/.

17 Maj. Kyle David Borne, US Army, "Targeting in Multidomain Operations," *Military Review*, May-June 2019, 60-67 :60; https://www.armyupress.army.mil/Journals/Military-Review/English-Edition-Archives/May-June-2019/Borne-Targeting-Multidomain/.

18 LTC Nathan Jennings, PhD, USA, & LTC Kyle Trottier, USA, "The 1973 Arab-Israeli War: Insights for Multidomain Operations," AUSA, December 16, 2022; https://www.ausa.org/publications/1973-arab-israeli-war-insights-multidomain-operations. Ironically, this Israeli operational success was an Israeli strategic failure. See George W. Gawrych, 'The 1973 Arab-Israeli War: The Albatross of Decisive Victory,'" Combat Studies Institute, US Army Command and General Staff College, Fort Leavenworth (1995), 80-81.

19 TRADOC Military History and Heritage Office, "AirLand battle emerges: Field Manual 100—5 Operations, 1982 and 1986 editions: TRADOC 50th anniversary series," May 22, 2023; https://www.army.mil/article/266846/airland_battle_emerges_field_manual_100_5_operations_1982_and_1986_editions_tradoc_50th_anniversary_series.

to the enduring principles of war, stripping out the excessive bureaucratese of US military thinking, concept and doctrine development, and planning.[20]

The Evolution of Multidomain Operations from the Lens of Bureaucratic Politics

The bureaucratic politics model helps to illuminate why the Army is so invested in MDO. This model asserts that nations do not consist of a unitary actor who always makes the most rational decisions. Rather, organizations "compete in attempting to affect" choices.[21] The pivot to Asia has accorded U.S Indo-Pacific Command with increased importance. Given the theater's vast stretches of water, an admiral has always commanded this theater.[22] The bureaucratic politics model explains how naturally the Army seeks to preserve and even possibly magnify its influence in this region given the importance of the China threat.

In some ways, though, the Army is late to the game. Even during the Global War on Terror and the resultingly intense focus on counterterrorism, US officials recognized the need to prepare to operate within an intensified anti-access/anti-denial environment and to better integrate the increasingly important space and cyber domains. The Quadrennial Review of 2010 highlighted how the Navy and the Air Force had begun wrestling with China's anti-access/anti-denial challenge, which initially gained traction as AirSea Battle.[23] It also warned that the previous eight years had "stressed the ground forces disproportionately, but the future operational landscape could also portend significant long-duration air and maritime campaigns for which the US Armed Forces must be prepared."[24] Any ground service reading that sentence would naturally worry about subsequent budget cuts. The Army,

---

20 Ben Zweibelson, *Beyond the Pale: Designing Military Decision-Making Anew* (Maxwell Air Force Base: Air University Press, 2024), xxvii.
21 Graham T. Allison and Morton H. Halperin, "Bureaucratic Politics: A Paradigm and Some Policy Implications," *World Politics*, 24(S1), 40-79: 42.
22 US Indo-Pacific Command, "Previous Commanders," https://www.pacom.mil/About-USPACOM/USPACOM-Previous-Commanders/.
23 US Department of Defense, Quadrennial Defense Review Report, February 2010, 32; https://dod.defense.gov/Portals/1/features/defenseReviews/QDR/QDR_as_of_29JAN10_1600.pdf; Andrew F. Krepinevich, "Why AirSea Battle," Center For Strategic and Budgetary Assessments, 1; https://csbaonline.org/uploads/documents/2010.02.19-Why-AirSea-Battle.pdf.
24 "Quadrennial Review," vi.

and the Marine Corps to a lesser degree, also believed that it had been left out of the concept, which it argued did not focus enough on land.[25]

The Army and the Marine Corps subsequently began seeking to solve the same problem, publishing a white paper about multidomain battle in 2017.[26] While it may be that the Air Force and the Navy had excluded the Army and the Marine Corps in 2010 or that they believed China's anti-access and area-denial challenge to be primarily a sea- and air-domain challenge given the Pacific Ocean's immense range, it is still notable that the Army's foray into thinking about anti-access and area-denial notably excluded the two services who had done the most thinking already. Ironically, MDO was born out of interservice rivalry and the worry about budgets as much as it emerged from institutions seeking genuinely multidomain solutions.

A joint military, much less a multidomain one, must develop its concepts together rather than developing service-specific understandings of what the concept entails. Nonetheless, this challenge continues to exist today to some extent in the various programs contributing to joint all-domain command and control.[27] Since developing multi domain battle, the Army has incorporated the Navy and the Air Force, although the Navy has demurred at times, believing it has been doing multidomain operations all along. In part, the Navy's belief reflects the understandable confusion over what multidomain operations entails.[28]

Despite some reticence, all services have accepted the concept to varying extents. In the meantime, the concept has morphed from MDO into joint all-domain operations, also referred to as JADO, although the Army seems to prefer multidomain operations as does NATO.[29] By contrast Air Force doctrine uses joint all-domain operations.[30] Any concept will go through

---

25 Sam LaGrone, "Pentagon Drops Air Sea Battle Name, Concept Lives On," January 20, 2015; https://news.usni.org/2015/01/20/pentagon-drops-air-sea-battle-name-concept-lives.

26 "People Who Know, Know MDO"; "United States Army-Marine Corps White Paper Multidomain Battle."

27 "How Project Convergence and JADC2 aim to accelerate the decision loop," Defense News, undated; https://www.c4isrnet.com/native/Raytheon-intelligence-space/2021/09/27/how-project-convergence-and-jadc2-aim-to-accelerate-the-decision-loop/.

28 Sydney J. Freedberg, Jr., "All Services Sign On To Data Sharing – But Not To Multidomain," Breaking Defense, February 8, 2019; https://breakingdefense.com/2019/02/all-services-sign-on-to-data-sharing-but-not-to-multidomain

29 "Defense Primer;" "Multidomain Operations in NATO—Explained.

30 LeMay Center; https://www.doctrine.af.mil/; accessed December 30, 2024; Air Force Doctrine Publication 3-99 and Space Doctrine Publication 3-99, "The Department of the Air Force Role in Joint All-Domain Operations," November 19, 2021; https://media.defense.gov/2022/Jan/19/2002924106/-1/-1/0/SDP%203-99,%20THE%20DAF%20ROLE%20IN%20JOINT%20ALL-DOMAIN%20OPERATIONS.PDF.

multiple iterations, and name changes might be construed as showing the concept's maturation. Given General Martin Dempsey's musing back in 2011 as to what should come after jointness, though, the rise of the term "Joint All-Domain Operations" suggests a confusing return back to what came before multidomain operations.[31]

Most recently, the US is proceeding to rename JADC2–another variation of the idea focused more on the technological connectivity between the services–as CJADC2, with the C denoting coalition even if the actual word used is "combined." One might wonder why the term needs to acknowledge jointness and combined arms, but the "C" actually stands for "combined," highlighting the inscrutable language that dominates the Department of Defense, largely serving only to confuse and obfuscate. Indeed, a *Breaking Defense* journalist snarkily commented that the DoD had just added "another letter to the already murky acronym."[32]

More recent internal Army documents highlight how the organization has doubled down on land-centric thinking, which undermines a culture in which true multidomain thinking can thrive. A Chief of Staff paper from 2021 entitled "Army Multidomain Transformation" envisions how "[g]round forces will decisively shape the first battle," stubbornly eschewing a joint perspective.[33] The same paper magnifies this land-centric focus by reinterpreting key historical operations. Regarding seizing control of the Mediterranean in World War II, for example, the paper explains how "Allied landpower enabled the transformation of the Mediterranean into an Allied "lake" by the middle of 1943." This interpretation ignores the significant contributions of air and seapower in providing the superiority necessary for successful land operations in the first place. It also ignores the underlying political reason the Allies agreed to invade North Africa in the first place.[34] Lacking air superiority, the British convinced the United

---

31 Reilly, "Multidomain Operations," 61.
32 Jasprett Gill, "Return of CJADC2: DoD officially moves ahead with 'combined' JADC2 in a rebrand focusing on partners," Breaking Defense, May 16, 2023; https://breakingdefense.com/2023/05/return-of-cjadc2-dod-officially-moves-ahead-with-combined-jadc2-in-a-rebrand-focusing-on-partners/.
33 Headquarters, Department of the Army, "Army Multidomain Transformation Ready to Win in Competition and Conflict," Chief of Staff Paper #1, Unclassified Version, March 16, 2021, 1, https://api.army.mil/e2/c/downloads/2021/03/23/eeac3d01/20210319-csa-paper-1-signed-print-version.pdf.
34 Chief of Staff Paper, 5. For the critical contributions of air and sea power, see Phillips Payson O'Brien, *How the War Was Won: Air-Sea Power and Allied Victory in World War II* (Cambridge University Press, 2019).

States that their coalition could not directly attack Germany in France but, instead, needed to take a far more indirect approach.

The document similarly offers a highly revisionist interpretation of Operation DESERT STORM to offer a simplistic and flawed interpretation of its contribution to achieving air superiority. Ignoring the months of preparation that enabled achieving air superiority over Iraq, the paper claims that Operation EAGLE ANVIL single-handedly enabled the Air Force to gain air superiority:

> On January 17th, 1991, US Army attack helicopters from
> the 101st Aviation Regiment struck deep into enemy
> territory to destroy Iraqi early-warning radar sites. Within
> minutes, Army forces had established a 10-kilometer-wide
> penetration in the enemy's air defense … US Air Force
> fixed wing aircraft immediately exploited this penetration,
> and flew on to strike strategic targets in Baghdad,
> signaling the beginning of Operation DESERT STORM.[35]

Most importantly, this account in a document designed to further multidomain operations excludes the joint nature of this mission given that USAF MH-53 helicopters also participated in Operation EAGLE ANVIL. In fact, the original plane for the operation only envisioned using Air Force Special Operations helicopters.[36] Determined to position itself as *the* key player in the future fight, this document revises far more sophisticated joint understandings of operational success to marginalize other services' contributions.

The document envisions a similar role for itself in the future fight, describing how in 2035 by "exercising dominant land power – will enable the Joint Force to penetrate complex, high-end adversary defensive systems."[37] Landpower advocates traditionally defend the institution's importance by arguing quite reasonably that it is "the man on the scene with the gun" who is in control.[38] But this Army document effectively ignores the nation's joint

---

35 Chief of Staff Paper, 8.
36 Col. Paul E. Berg and Kenneth E. Tilley, "Task Force Normandy: The Deep Operation that Started Operation Desert Storm," https://www.armyupress.army.mil/Books/Browse-Books/iBooks-and-EPUBs/Task_Force_Normandy/.
37 Chief of Staff Paper, 6.
38 J.C. Wylie, *Military Strategy: A General Theory of Power Control* (Rutgers University Press, 1967; Reprint, Naval Institute Press, 2014), 72.

long-range capabilities to put itself at the center of a mission long dominated by the Air Force.[39] The previous commanding general of United States Army Pacific recently echoed these comments in 2024, arguing, "the land has always been decisive – and it remains so."[40] His speech set off memes around Twitter.

In this vein, it is useful to return to the bureaucratic politics model to understand one reason the US Army may have prioritized long-range fires as its number one modernization goal. In an Army arguing loudly for a dominant role in a region characterized by vast ocean, long-range fires keeps the Army relevant. As such, many view this priority as a parochial grab for power.[41] A truly joint service would recognize that no single service can solve the A2AD problem; indeed, that was the initial justification for multidomain battle in the first place.

Varieties of Domain Operations

Is there, then, something worth keeping in MDO? As a starting point, it is useful to distinguish between single, cross-domain, and multidomain operations. In many ways, single and cross-domain operations unsurprisingly go back centuries. In single domain operations, one domain has an effect in that same domain. For example, an aircraft seeks to achieve air superiority by shooting down another aircraft. By contrast, cross-domain operations use one domain to have an effect in another, such as an aircraft striking a target on land.[42]

By contrast, multidomain operations not only seek to integrate a greater number of domains but greatly intensify the domains acted upon to meet the original white paper's intent of creating combined arms for the 21st century. The ultimate intent of this integration is to "create synergistic effects in

39 Congressional Research Service, "US Army Long-Range Precision Fires: Background and Issues for Congress," March 16, 2021, 1; https://crsreports.congress.gov/product/pdf/R/R46721.

40 US Army Pacific, "Gen. Charles Flynn Opening Remarks at LANPAC24," May 14, 2024; https://www.usarpac.army.mil/Our-Story/Our-News/Article-Display/Article/3775247/gen-charles-flynn-opening-remarks-at-lanpac24/.

41 Dave Deptula, LTG USAF (Ret) as quoted in Theresa Hitchens, "Long-Range All-Domain Purpose Prompts Roles & Mission Debate," BreakingDefense.com, July 9, 2020; https://breakingdefense.com/2020/07/long-range-all-domain-prompts-roles-missions-debate/; Matthew Cox, "Army Chief Defends Long-Range Missile Effort After Air Force General's Public Attack," military.com, April 13 2021, https://www.military.com/daily-news/2021/04/13/army-chief-defends-long-range-missile-effort-after-air-force-generals-public-attack.html; Sydney J. Freedberg, "Army Says Long Range Missiles Will Help Air Force, Not Compete," July 16, 2020. https://breakingdefense.com/2020/07/army-says-long-range-missiles-will-help-air-force-not-compete/.

42 "Army in Multidomain Operations 2028," x.

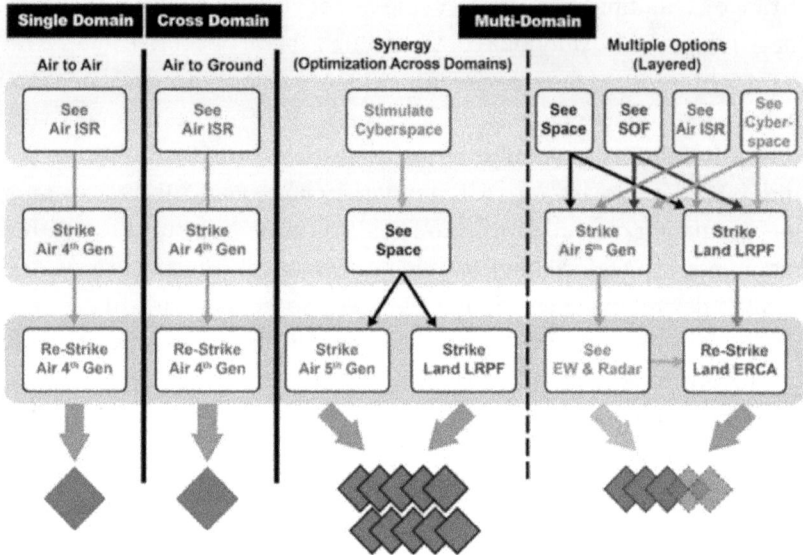

Figure 6.1 Converging Capabilities to Generate Cross-Domain Synergy and Layered Options
Source: TRADOC Pamphlet 525-3-1.

windows of convergence."[43] In terms of desired effect, the Army particularly stresses the need to plan these operations to attack an opponent's air defense as well as target long-range fires. Thus, an example of MDO might consist of "synchronized cyberwarfare, space warfare, and EW effects" to support an F-35 striking [an] air defense asset" but also targets required to support other domains.[44]

The below table conceptualizes single, cross, and multidomain operations (Figure 6.1) even if the accompanying text in the TRADOC publication provides no explanation of the table itself, leaving readers trying to decipher it on their own.[45] The final red squares on the bottom of the image hints at how concept writers envision "layered" effects having a force-multiplying effect, at least in theory. And, for those nations unable to script a highly orchestrated approach, they may be able to also gain more effects by integrating into these windows of opportunity.

---

43 Borne, "Targeting," 67.
44 Borne, "Targeting," 61.
45 "Army in Multidomain Operations 2028," 21.

Implementation Challenges in the Context of War Theory

The promise of multidomain operations is that it orchestrates the employment of the traditional domains of air, ground, and sea with the newer domains of space and cyber to create more significant effects when employed together. In theory, this approach makes the sum much greater than its individual parts, helping the US to solve its mass problem in the most significant way since the development of precision weapons.[46] Even if considered more appropriately within the operational level of war, the concept faces enormous challenges to be realized, especially because of the significant planning challenges required to create the localized windows of superiority that are so essential to the concept.[47] To accompany its stable of small but exquisite military capabilities, the US has now developed an exquisite notion of planning, hoping to orchestrate these key operations in hours rather than the current timeline of days or weeks.[48] Current combat operations, meanwhile, highlight the almost real-time targeting cycles of far less exquisite technology like artillery.[49]

Further challenging this goal is that the US planning timelines for the various domains are of greatly differing amounts, as seen below (Figure 6.2). Much hope is placed in the notion that artificial intelligence will resolve these problems, enabling seamless and smart planning even as sceptics point out potential problems.[50] The newer domains of cyber and space unfortunately require longer planning cycles than the traditional domains. Mission command also rests uneasily within this construct given the need to carefully

---

46 David Alman, "Bending the Principle of Mass: Why That Approach No Longer Works for Airpower," December 15, 2020; https://warontherocks.com/2020/09/bending-the-principle-of-mass-why-that-approach-no-longer-works-for-airpower/.

47 Sean Kimmons, "With multidomain concept, Army aims for 'windows of superiority,'" US Army Website, November 14, 2016; https://www.army.mil/article/178137/with_multi_domain_concept_army_aims_for_windows_of_superiority.

48 US Army, "The US Army in Multidomain Operations, 2028; Slide 10; https://www.ncoworldwide.army.mil/Portals/76/courses/mlc/ref/Multidomain-Operations.pdf.

49 Sam Cranny-Evans, "Russia's Artillery War in Ukraine: Challenges and Innovations," August 9, 2023; https://www.rusi.org/explore-our-research/publications/commentary/russias-artillery-war-ukraine-challenges-and-innovations.

50 David K. Spencer, Stephen Duncan, Adam Taliaferro, "Operationalizing artificial intelligence for multidomain operations: a first look," Proceedings, Volume 11006, May 10, 2019; https://www.spiedigitallibrary.org/conference-proceedings-of-spie/11006/2524227/Operationalizing-artificial-intelligence-for-multidomain-operations--a-first/10.1117/12.2524227.short?SSO=1&tab=ArticleLinkCited; Cameron Hunter and Bleddyn E. Bowen, "We'll Never Have a Model of an AI Major-General: Artificial Intelligence, Command Decisions, and Kitsch Visions of War," *Journal of Strategic Studies* 47 (1): 116–46.

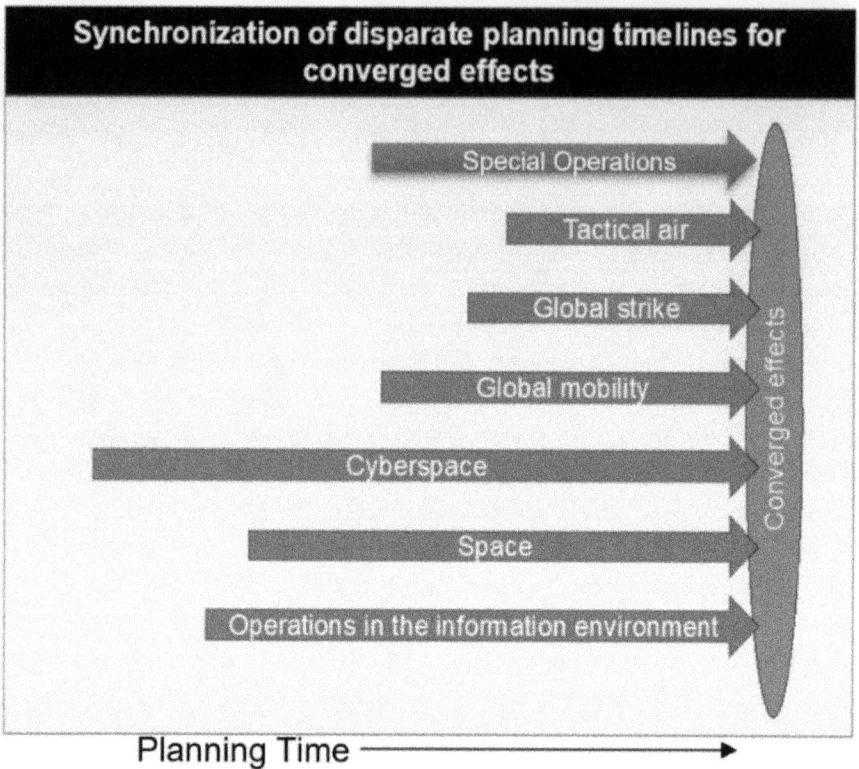

Figure 6.2 Synchronization of disparate planning timelines for converged effects
Source: TRADOC Pamphlet 525-3-1.

script the application of capabilities across various domains in environments characterized by limited communications.

Paring Down MDO and the Conflict Continuum

By applying the concept of MDO across the conflict continuum, the US Army makes the concept about solving any and every potential future problem, with the challenges of competition particularly not as well served as those of conflict. Army doctrine, for example, identifies the Russian operational center of gravity during competition as being the "close integration of information warfare, unconventional warfare, and conventional warfare."[51] Given the

---

51 "Army in 2028," 11.

very kinetic example of multidomain operations the Army provides for the concept, it is difficult to understand how this approach helps solve the extent to which adversaries seek to undermine the nation's political and economic systems and alliances, which the doctrine refers to repeatedly.

By contrast, doctrine defines the operational center of gravity for conflict neatly into the problem-set MDO seeks so solve. Doctrine explains more understandably that the Russian center of gravity in conflict is its mid- and long-range fires.[52] The danger is that further editions of this doctrine will only expand this focus across the conflict continuum as the Army adds new layers resulting in added complexity that seeks distractions into Army doctrine,[53] thereby losing focus on the most pressing problem: destroying or suppressing fifth-generation air defenses while protecting one's own personnel and materiel from long-range fires. As such, it is important to keep MDO pared down to the precise problems it seeks to solve rather than seeking to be a one-stop shop for the changing character of war.

The Emperor's New Clothes: The Bankruptcy of Western Military Concept Development and Doctrine

While some military officials acknowledge the West may have provided problematic advice to Ukraine regarding the failed offensive of the summer and fall of 2023, few have questioned Western doctrine itself.[54] Rather, they have assumed that, were NATO to be in the position Ukraine is, that it could effectively use its doctrine with more advanced capabilities especially airpower, to maneuver. Their main error centers on "miscalculate[ing] the extent to which Ukraine's forces could be transformed into a Western-style fighting force in a short period—especially without giving Kyiv air power integral to modern militaries."[55]

---

52 "Army in 2028," 13.

53 Capt. James Tollefson, "Fixing Army Doctrine: A Network Approach," *Military Review*, January-February 2018; https://www.armyupress.army.mil/Journals/Military-Review/English-Edition-Archives/January-February-2018/Fixing-Army-Doctrine-A-Network-Approach/.

54 Tom Porter, "Ukrainian troops are abandoning US tactics in their counteroffensive because they haven't worked," Business Insider, August 3, 2023; https://www.businessinsider.com/western-trained-ukrainian-troops-are-abandoning-us-tactics-report-2023-8.

55 "Miscalculations, Divisions Marked Offensive Planning by US, Ukraine," *The Washington Post*, December 4, 2023; https://www.washingtonpost.com/world/2023/12/04/ukraine-counteroffensive-us-planning-russia-war/.

Lesser emphasis rests on the quality of Western advice. It is also interesting to note that Western advisers advocated a sole focus while Ukraine sought to launch three simultaneous attacks. In effect, Ukraine sought to force the enemy to respond to multiple dilemmas, one of the central ideas of MDO.[56] Meanwhile, Western advisors steeped in maneuver warfare mythology faulted the Ukrainian military for trying a frontal assault on a force that had months to dig in.[57]

It is time to consider whether US military doctrine is highly flawed, and US military officers are far too optimistic regarding their own intellectual foundation. Regarding the offensive, for example, the US intelligence community far better understood the tremendous defensive efforts Russia had invested in fortifications prior to the offensive.[58] Some have even asserted that Ukraine does better when it ignores or does not even consult with the West.[59] If the muddle of the Army's MDO is any indication, it may be time to start from scratch in simplifying and then strengthening the intellectual foundations of US warfighting. It is not too late, as one analyst warns, to recognize the "dangers of an overwrought, hyper-intellectualized view of battlefield maneuver."[60] One might even possibly correct the analyst by changing "hyper-intellectualized" to "pseudo-intellectualized."

Conclusion

The concept of MDO is probably here to stay given the US Army has bought into this concept wholesale and incorporated it into doctrine, which has since been accepted by NATO. But to gain more meaningful buy-in outside that institution, it and the larger joint and coalition force should work to narrow the concept because it currently means too many things to too many people. It must be distilled to the three-part operational problem that joint force it is trying to solve: in an operational environment threatened by (1) anti-access/anti-denial weapons systems and (2) long-range fires, (3) how can the US and its allies find ways to maneuver?

---

56 "Miscalculations."
57 Ibid.
58 Ibid.
59 Sinéad Baker, "Retired NATO commander says Ukraine's Kursk invasion proves it can succeed without Western advice," Business Insider, September 3, 2024; https://www.businessinsider.com/ukraine-kursk-shows-ability-without-much-west-advice-former-nato-2024-9.
60 Harry B. Halem, "Positional Warfare: Prospects for Ukraine in 2024-27," *Marine Corps Gazette*, September 2024, 40-46: 40.

This approach means stressing what is transformative about the concept, rather than getting distracted by pointing out supposed uses of it in the past. It also means shedding the concept's blob-like character, which has become a substitute for truly thinking about future warfare as a whole, let alone mentioning its larger strategic-level economic and diplomatic implications. There also needs to be more serious analysis focused on the planning and synchronization challenges of implementing such plans that require careful orchestration, especially in the context of the unforgiving fog and friction of war. MDO is no exception to the basic rule that no plan survives first contact with the enemy, yet the concept hedges its bets precisely on the elusive dream of a perfectly synchronized plan fit for Carnegie Hall.

7

Addressing the Flaws in Multidomain Operations
The Pursuit of Dominance in 21st Century Warfare

Amos C. Fox

Strategic theorist J.C. Wylie posits that narrow strategic theories and corresponding doctrines inhibit success in an adversarial environment.[1] Wylie asserts that theories and doctrine should possess the conceptual breadth to make them truly useful.[2] In the spirit of Wylie's vision, this chapter illuminates three flaws in Multidomain Operations (MDO) doctrine. The purpose of doing so is to help correct the doctrine and make it something more than a strike-centric strategy replete with hollow defense jargon, and light on meaningful directions and pathways to tangible outcomes.

MDO doctrine, embodied in the US Army's Field Manual (FM) 3-0 *Operations*, possesses many flaws, two of which are discussed in detail here. First, given the weight placed on dominance within the MDO doctrine, dominance is insufficiently described in FM 3-0 and consequently misapplied throughout the doctrine. This flaw, however, should be rectified by the incorporation of a combat power application mechanism called *Zones of Proximal Dominance* (ZoPD).

Second, MDO's insistence on *persistence* and *convergence*, which are underwritten by assumptions about a never-ending stream of astronomically

---

1 J.C. Wylie, *Military Strategy: A General Theory of Power Control* (Annapolis, MD: Naval Institute Press, 2014), x.
2 J.C. Wylie, *Military Strategy*, x.

expensive precision strike and long-range fire munitions, has consistently been proven false during the wars of the 21st century. The US ran out of precision munitions on multiple occasions during the height of fighting non-state actors in Iraq, Syria, Afghanistan, and The Philippines between 2016-2017.[3] Likewise, Jack Watling and Nick Reynolds, among many other analysts, highlight how both Russia and Ukraine exhausted their precision munitions and ballistic artillery at multiple points within the Russo-Ukrainian War.[4]

Between the US's own munitions shortages during its campaigns to defeat the Islamic State in Iraq and Syria and both Russia and Ukraine's ammunition shortages throughout the Russo-Ukrainian War, it is hard to believe that a military doctrine can be underpinned by such an outlandish assumption. Thus, MDO's second flaw is that the doctrine does not appropriately account for hard constraints, consumption rates, and frontage problems when considering how to organize for combat and how to apply combat power.

If left unaddressed, MDO is at best wishful thinking. At worst, if these problems are left unattended, they make MDO little more than a strike strategy, loosely augmented by other elements of combat power and supporting elements. Moreover, if left unaddressed, when the flag goes up, MDO serves as a recipe for continued strategic, operational, and tactical failure in Western military operations.

Dominance

MDO's first flaw is an insufficient description of dominance. Though metatheoretical, a clear understanding of dominance is critical because of the central position that the concept and the doctrine (i.e., dominance) holds within MDO. To be sure, Field Manual (FM) 3-0 mentions dominance (and its variant forms) no less than 25 times but fails to define the time. One is left to infer a definition from the manual's various uses of the term. For instance, FM 3-0 mentions dominance like a commonly understood idea and that all the US and its allies and partners—or any actor; state or non-state

---

3 Amos Fox, "Precision Paradox and Myths of Precision Strike in Modern Armed Conflict," *RUSI Journal* Vol. 169, no. 1-2: 62-74. DOI: 10.1080/03071847.2024.2343717.
4 Jack Watling and Nick Reynolds, "Russian Military Objectives and Capacity in Ukraine Through 2024, *RUSI*, February 13, 2024, https://www.rusi.org/explore-our-research/publications/commentary/russian-military-objectives-and-capacity-ukraine-through-2024.

actor—needs to do is show up, deliver overwhelming firepower, and presto – they have achieved dominance.[5]

Furthermore, the MDO concept, TRADOC Pamphlet 525-3-1, states that dominance is one of the four emerging trends shaping emerging operational environment.[6] It goes on to state specifically that US dominance in emerging operational environments is *not* assured, but then insufficiently elaborates on the impact dominance plays within MDO.[7]

To account for dominance, MDO should begin with several first-order questions, including but not limited to the following. What is the character of dominance? How is dominance measured? What are the modulating features of dominance? And, can a simple theory help to express dominance in a meaningful and tangible way for policymakers, strategists, analysts, scholars, and military practitioners? The remainder of this section answers each of those questions in detail.

The Character of Dominance

Dominance is conditional, meaning that it requires resource stabilization plus resource overmatch vis-à-vis the adversary for a period sufficient to either force an adversary to change their plan or force the adversary to acquiesce. In that respect, dominance is fleeting, fragile and prone to shock and surprise. Hence, a proportional relationship exists between resource expenditure and the ability to secure or preserve dominance; that is, the greater one's expenditure in resources, the less likely they can achieve and maintain dominance relative to their opponent (see Figure 7.1).

Measuring Dominance

Dominance, being resource-dependent, is measurable and can be forecasted by zones, degrees, and duration. This methodology is useful for both assessing and anticipating the potential for friendly and adversary dominance. To be sure, those metrics can assist in forecasting when, where and for how long an actor may—or may not—possess dominance or be capable of persistence in each domain or across multiple domains.

---

5   Field Manual 3-0 *Operations* (Washington, DC: Government Printing Office, 2022), 154.
6   Training and Doctrine Command (TRADOC) Pamphlet 525-3-1, *The U.S. Army in Multi-Domain Operations* (Washington DC: Government Printing Office, 2017), vi.
7   TRADOC Pamphlet 525-3-1, vi.

Scale of Relative Dominance

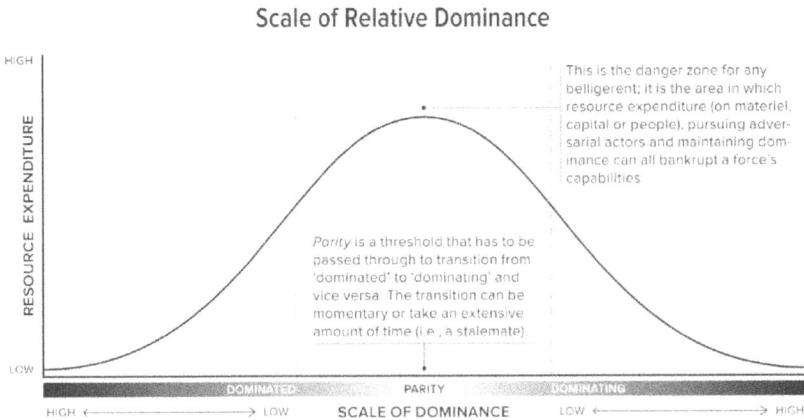

This is the danger zone for any belligerent; it is the area in which resource expenditure (on materiel, capital or people), pursuing adversarial actors and maintaining dominance can all bankrupt a force's capabilities

Parity is a threshold that has to be passed through to transition from 'dominated' to 'dominating' and vice versa. The transition can be momentary or take an extensive amount of time (i.e., a stalemate).

RESOURCE EXPENDITURE

HIGH

LOW

DOMINATED   PARITY   DOMINATING

HIGH ← → LOW   SCALE OF DOMINANCE   LOW ← → HIGH

Figure 7.1 Scale of Relative Dominance
Source: Author.

Controlling Features

Because of resource interdependence, anything an actor does to degrade or disrupt an opponent's resources tempers its ability to perpetuate dominance at a specific point in both time and space. This endeavor includes not only those actions directed at disrupting adversaries' resources, but also anything that expends that actor's own resources in ways for which it was not prepared.

For instance, an actor might launch a surprise, spoiling, or counterattack to catch an opponent off-balance, and thereby disrupt the adversary's bid to grab dominance. Alternatively, an actor might strike indirectly by avoiding an adversary's main effort and focus a full-throated attack on a supporting force or an unanticipated flank to cause an opponent to divert resources and attention from its main effort, hence disrupting the pursuit of dominance. Ukraine's attack into Russia and subsequent seizure of Kursk in August 2024 is an instructive example of this idea. Seeking to take the pressure off areas in and around the Donbas, Ukraine boldly attacked into Russia in early August and gained control of the city of Kursk.[8] This caused the Kremlin to divert significant combat power away from the Donbas front to address the military problem in Kursk, making combat operations in the Donbas

---

8 "Kursk Offensive: A Timeline of Ukraine's Attack and Russia's Fightback," *Reuters*, 12 March 2025, https://www.reuters.com/world/europe/ukraines-attack-russias-fightback-kursk-2025-03-12/.

more manageable for Kyiv.[9] Moreover, Kyiv's move caused many to question Russia's operational and strategic dominance because Ukraine was able to keep the Kursk wildcard truly wild and the Russians largely unaware of the looming offensive.[10] In the end, Ukraine's operation in Kursk failed, but it did cause Russia to spend six months (August 2024 – March 2025) diverting attention and resources to address the situation and retake the town.[11]

Further, Alexander Svechin writes that premeditated attritive battles that are not allowed to conclude quickly deplete an opponent's stocks, decreasing their capacity to obtain or retain dominance.[12] Like the surprise attacks mentioned in the previous paragraph, an actor can use these approaches to force an adversary to divert attention and resources from areas of potential or achieved dominance. The goal being to cause the dominant actor to lose proximal dominance, whether that be temporarily or permanently. The combined US-led coalition and Iraqi operation to grind the Islamic State against the millstone of combat in Mosul (October 2016 – July 2017) is an example of this situation.[13] The operation, which lasted nearly as long as World War I's battle of Verdun, depleted the Islamic State's combat power sufficient that they were no longer capable (or willing) of mustering a cohesive military force to threaten Iraqi security in an existential way.[14] In effect, the battle's attritive character broke the Islamic State's metaphorical back, allowing the US-led coalition and Iraqi security force to gain dominance proximal to Mosul, and consequently across the remainder of northern and central Iraq.[15]

9   Olga Tokariuk, "Ukraine's Gamble in Kursk Restores Belief It Can Beat Russia – It Requires a Western Response," *Chatham House*, 19 August 2024, https://www.chathamhouse.org/2024/08/ukraines-gamble-kursk-restores-belief-it-can-beat-russia-it-requires-western-response.
10 Tokariuk, "Ukraine's Gamble in Kursk."
11 Mark Santora, "How Ukraine's Counteroffensive in Russia's Kursk Region Unraveled," *New York Times*, 16 March 2025, https://www.nytimes.com/2025/03/16/world/europe/ukraine-kursk-retreat-russia.html.
12 Alexander Svechin, *Strategy*.
13 Jake Richmond, "Iraqi Forces Begin the Battle of Mosul," *US Department of Defense*, 17 October 2016, https://www.defense.gov/News/News-Stories/Article/Article/975239/iraqi-forces-begin-battle-for-mosul/.
14 Amos Fox, "The Mosul Study Group and the Lessons of the Battle of Mosul," *Association of the United States Army*, Land Warfare Paper 130: 2-7, https://www.ausa.org/sites/default/files/publications/LWP-130-The-Mosul-Study-Group-and-the-Lessons-of-the-Battle-of-Mosul.pdf.
15 Dan Lamothe, Thomas Gibbons-Neff, Laris Karklis, and Tim Meko, "Battle of Mosul: How Iraqi Security Forces Defeated the Islamic State," *Washington Post*, 10 July 2017, https://www.washingtonpost.com/graphics/2017/world/battle-for-mosul/.

A Theory of Dominance

As a theory, dominance can be articulated in the following way. Dominance (D) equals one's resources (Re) plus time (Ti), divided by enemy action (En) plus self-sustainment (Su):

Conceptually, dominance is applied through actualization—or Zones of Proximal Dominance (ZoPD). This is built upon the premise that hard constraints (i.e., resources, manpower and time) curb an actor's realization of dominance and persistence. As a rule, dominance is localized within zones that radiate from a power source. In technical terms, a power source is any formation in contact or that envisions contact with an adversary. For instance, in large-scale combat operations (LSCO), this could be a field army or an army group mustering resources to make headway against an adversary of relatively equal size and strength, akin to Lieutenant General Omar Bradley's First Army during World War II's Operation Cobra (25–31 July 1944; see Map 1).[16] It is important to note that a power source is not bound to just high-level headquarters or large formations. Power sources are found at any echelon where two or more belligerents come into contact or might encounter one another.

Power radiation is proportional to the power source's strength, the strength of the adversary and the ability of both to aptly replenish resources. Further, power radiation is situationally dependent and can be both omnidirectional or unidirectional, but it is most likely focused on adversarial forces. Moreover, ZoPD power radiation derives its level of mobility from its power source.

In ZoPD planning, options are often binary. A broad front results in wide coverage, but it induces fragility across the front because of limited power and resource redundancy. Further, a broad front results in limited operational reach because all (or most) power sources are pushed forward, reducing the ability to resupply, reinforce or generally react in response to adversarial contact. On the other hand, a scaled front with layered power sources increases the ability to offset shock and surprise, and it facilitates operational reach through resource conservation and redundancy.

In summary, applied dominance, or ZoPD is a useful framework to assist in strategy development and planning at both the operational and tactical levels. By the same token, understanding the character of dominance

---

16 Carlo d'Este, *Decisions in Normandy* (New York: Konecky and Konecky, 1994), 400–408.

and ZoPDs can help strategists and planners to frame the operational environment and develop stratagems to disorganize an adversary's grip on dominance. These ideas should be incorporated into MDO to add a layer of practicality to its suppositions about persistence, overmatch and convergence.

Hard Constraints and Frontage Problems

A second flaw with MDO, i.e., its unbalanced focus on persistence and convergence, fails to adequately consider hard constraints and the associated frontage problems.

Disintegrating an adversary's anti-access and area denial (A2AD) system is a central tenet of MDO. "Persistence" is the animating verb used to describe how the US Army and joint force look to accomplish this. MDO posits that persistent reconnaissance, surveillance, mid-range fires and long-range precision fires are key; it also assumes that the Army will have the resources required to meet the demands of the national defense strategy through 2040.[17] However, incorporating recent combat operations and the concept of dominance into the equation makes this assumption look suspect.

Operation Inherent Resolve (OIR) and similar missions rapidly depleted the US military's stores of precision-guided munitions (PGMs).[18] Mosul, Marawi, Raqqa and some engagements in Africa combined to result in a sharp decline in both on-hand and stockpiled PGMs.[19] At Mosul's apex, the US Army and joint force almost ran out of Hellfire missiles and other PGMs, requiring DoD to pursue special funding to help replenish its stocks.[20] Importantly, this occurred within a relatively short amount of time against small, light infantry-type forces, in a small number of locations. If the US Army and joint force were instead engaged with the Russians or Chinese, they would find the opposite—large, heavily armored forces holed up in many urban areas across a vast front. Russia and Ukraine's "shell hunger"

---

17 Carlo d'Este, *Decisions in Normandy*, 49.

18 Marcus Weisgerber, "The US is Raiding its Global Bomb Stockpiles to Fight ISIS," *Defense One*, 26 May 2016.

19 Paul Shinkman, "ISIS War Drains U.S. Bomb Supply," *U.S. News and World Report*, 17 February 2017.

20 Jeff Daniels, "ISIS Fight Shows US Military Can Use Lower-Cost Weapons With Lethal Results," *CNBC*, 19 July 19.

throughout the Russo-Ukrainian War, and each combatants' dependents on external support also makes this argument more tangible.[21]

President of All Russias

In a hypothetical but plausible scenario, the Russian president, nodding to the Romanov Czars, creates the title, "President of All Russias." With historical precedence, he defines "All Russias" in the following manner:[22]

- Muscovy is *Great Russia*;
- Belorussia, or Belarus, is *White Russia*;
- Ukraine is *Little Russia*;
- Crimea (initially annexed by the Romanovs from the Crimean Khanate in 1783; annexed again by Putin in 2014)[23] and southern Ukraine, are *New Russia*; and
- Galacia (parts of modern-day southeastern Poland and portions of western Ukraine) is *Red Russia*.
- Putin doesn't relent in his pursuit of denationalizing Ukraine, eliminating Ukrainian culture, and pulling it back into the Kremlin's yoke. Moreover, Russia launches an assault from Pskov with the battle-hardened 76th Guards Air Assault Division moving toward Riga, Latvia, and toward Tallinn, Estonia.[24] Russia launches this assault as a demonstration intended to lure NATO's attention and resources toward the Baltics—and to freeze the NATO forces that are already there in place.

Simultaneously, Russia launches forces from the 20th Combined Arms Army toward Kharkiv to overcome their failure to take the city in 2022. The 1st Guards Tank Army, mauled early in the war with Ukraine, strikes

---

21 Maria Tsvetkova, Polina Nikolskaya, Anton Zverev and Ryan McNeill, "Russia Building Major New Explosives Facility as Ukraine War Drags On," Reuters, 8 May 2025, https://www.reuters.com/investigations/russia-building-major-new-explosives-facility-ukraine-war-drags-2025-05-08/; David Vergun, "Air Force General Says Ukraine Needs Ammunition," *US Department of Defense*, 28 February 2024, https://www.defense.gov/News/News-Stories/Article/Article/3689717/air-force-general-says-ukraine-needs-ammunition/.

22 Simon Montefiore, *The Romanovs, 1613–1918* (New York: Vintage Books, 2017), 365.

23 Orlando Figes, *The Crimean War, A History* (New York: Metropolitan Books, 2010), 14–16.

24 Bill Sanderson, "Leaked Transcripts Reveal Putin's Secret Ukraine Attack," *NY Post*, 21 September 2014.

out toward Minsk.[25] Concurrently, the Russian Southern Military District's forces, in conjunction with their proxy armies in Donetsk and Luhansk, expeditiously move to seize control of Odesa, something they failed to do in 2014 and 2022, aiming to protect and further reinforce Crimea's northern flank and so to gain control of *New Russia*.

At the same time, the Russian Baltic Fleet sails from Kaliningrad and establishes a Baltic blockade that runs from just north of Szczecin, Poland, to Finland's southern coast to increase its zone of proximal dominance in the Baltic region and to deny free naval passage to Polish ports along the Baltic coast. Further, the Black Sea Fleet launches combat elements from Sevastopol to block the Bosporus Straits, protecting Russia's southern flank and extending its southern zone of proximal dominance.

Assuming that NATO forces are able to react quickly enough to establish a meaningful response (and the US military in particular), the question becomes, does MDO's assumption about sufficient resources (including PGMs) to support its persistent and converging supposition, key to the "penetrate-disintegrate-exploit" framework, hold water when the adversary is defending from multiple urban locations along a 920-mile front, with secure lines of communication to its logistical and manpower base? (See Figure 7.2).

The Battle of Mosul and the PGM crisis from 2015 to today, a vastly smaller problem against a far less numerically significant or resources-rich adversary, suggest that the answer to that question is no.

Given the limiting effect of hard constraints, it is entirely logical to assume that warfighting capabilities such as artillery, rockets, unmanned aerial vehicles, rotary-wing formations, air defense, and other capabilities will diminish as the size of the land force increases. To be sure, the use of main efforts and supporting or shaping efforts already exists to account for this simple resource problem. However, in an LSCO environment with an MDO doctrine built around the idea of the persistence and convergence of enabling-capabilities, the problem of main efforts and supporting efforts illustrates that without significant industrial mobilization, most units will be fighting as an economy of force, while only the lead elements will have access

---

25 David Axe, "Russia's 1st Guards Tank Army Has Won Its First Battle in Two Years – By Advancing a Mile and Capturing a Half-Dozen Buildings, *Forbes*, 30 January 2024, https:// www.forbes.com/sites/davidaxe/2024/01/30/russias-1st-guards-tank-army-has-won-its-first-battle-in-two-years-by-advancing-a-mile-and-capturing-a-half-dozen-buildings/.

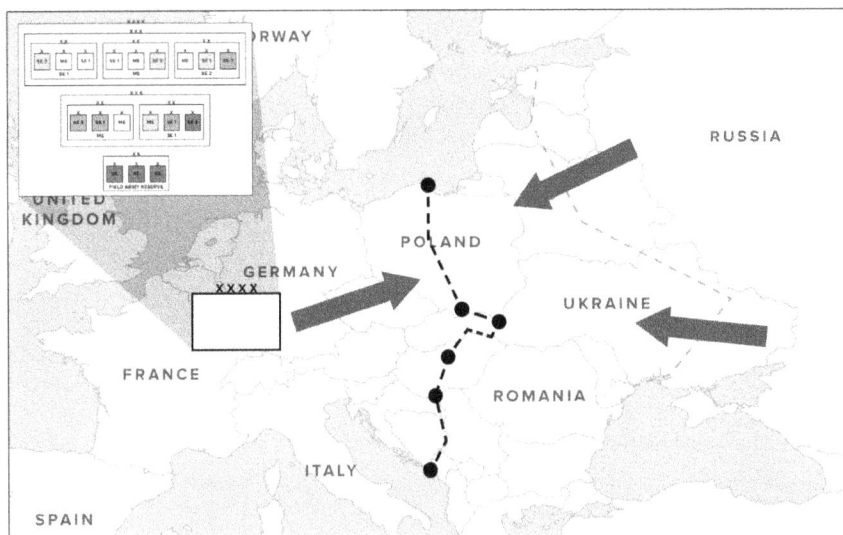

Figure 7.2 Notional Russian Invasion of Europe
Source: Author.

to those combat enabling-capabilities. It is fair to assume that a problem like the one described above results in MDO not passing the feasibility test.

Analysis

MDO's position regarding convergence further compounds the persistence feasibility problem. MDO posits that convergence against a great-power competitor requires, "continuous and rapid integration of multi-domain capabilities to gain cross-domain overmatch at decisive spaces."[26] While MDO does make note of extended time frames, it fails to elaborate on extended physical fronts, i.e., the long distances along which a force will likely find itself operating in LSCO; this is where the problem with persistence and convergence lies.[27]

As the hypothetical scenario illustrates, the US Army and joint force could very well find themselves operating along a highly contested front, with an adversary defending in multiple urban areas spread across hundreds of miles, with secure lines of communication to its rear (i.e., out of contact with US forces). If this becomes the case, the ability to rapidly integrate

26 TRADOC Pamphlet 525-3-1, 20.
27 TRADOC Pamphlet 525-3-1, C-2.

multi-domain capabilities to gain cross-domain overmatch in decisive spaces might not exist because the requirement (i.e., the number of locations, distance between locations and quantity of resources necessary to achieve overmatch) exceeds the capability of the Army and joint force. This question lies beyond the realm of theory and is beginning to surface in defense analysis.[28]

Further, the US Army and joint force's inexperience with operating against this level of applied pressure and immense space will inherently create suboptimization as soldiers of all ranks struggle to work through problems that are outside of the scale of anything that they have previously experienced.

These problems underscore the claim that persistence and convergence in a highly contested environment against a great-power competitor might not be a meaningful solution to the problem. As is currently written, this situation prevents significant problems for the US Army's MDO. Therefore, MDO should be reworked to account for LSCO problems, to include (1) scale (i.e., geographic distances that cover vast stretches of land and sea), (2) multiple decisive spaces stretched across a front, and (3) that those decisive spaces are not adversary tank formations sitting in open terrain, but rather a heavily ensconced adversary in large urban areas.

Expanded Assumptions

Several assumptions, beyond those listed in TRADOC Pamphlet 525-3-1 and FM 3-0, can be drawn from the above examinations of dominance, applied dominance, convergence, persistence and frontage problems. Listed below, these assumptions should be incorporated into the existing MDO to help provide context and logic, and to assist policymakers, strategists, analysts, and military practitioners in better utilizing MDO outside of doctrinal discussions.

1. Self-preservation is every actor's baseline goal.
2. Actors will not intentionally engage other actors in ways that put themselves in existential crises.
3. All international actors operate within an open-system; that system seeks order and will reallocate assets to maintain equilibrium during armed conflict.

---

28 Jen Judson, "Does the US Army Have Enough Weapons to Defend Europe? Exercise Defender 20 Will Reveal All," *Defense News*, 27 December 2019.

4. Actors will kill off elements of their system when sustaining those elements becomes deleterious to the system.

5. Dominance is a matter of perception because incompletion information obscures available resources, intentions and timing.

6. If an actor perceives an adversary as dominant in relation to itself but chooses to engage that belligerent anyway, it will do so in a way that avoids self-destruction, offsets the adversary's strength and seeks tactical parity.

7. If an actor assesses that the cost of direct confrontation with another actor will rapidly exhaust its fixed resources, it will indirectly engage the adversary.

8. Dominance (D) equals one's resources plus time (Ti), divided by enemy action (En) plus self-sustainment (Su):
   $$D = (Re + Ti) \div (En + Su).$$

9. Resource expenditure in an adversarial environment (Rx) is equal to quantity of one's force (Qf) plus one's frontage (Ft) plus the number of points of enemy contact along that front (Pc) plus the duration of enemy contact (Dr) divided by one's on-hand resources (Re) plus an actor's ability to replenish those loss (Rp):
   $$Rx = (Qf + Ft + Pc + Dr) \div (Re + Rp).$$
   Note: This assumption (and equation) should not be law, but instead a model to assist in thinking and understanding resource expenditure. Further, friction or entropy can be factored into the equation to account for the natural tendency of things to not go according to plan.

10. Without economic and industrial mobilization in support of LSCO, battlefield persistence (Pr) is directly linked to resource expenditure in an adversarial context; or,
    $$PR <= Rx.$$
    Therefore, the greater a subordinate unit's contact at one location, the less its higher headquarters can support subordinate unit contact at other locations along its front.

11. Without economic and industrial mobilization in support of LSCO, the ability of a headquarters to impose multiple dilemmas (Md) on an adversary decreases as the number of enemy contact points and duration of that contact increases across its front; or
    $$Md <= (Ft \times Pc \times Dr \div Re + Rp).$$

Conclusion

The purpose of highlighting flaws within MDO is to help illuminate those shortcomings in hopes of achieving positive theoretical and doctrinal change. As J.F.C. Fuller writes:

> Method creates doctrine, and a common doctrine is the cement which holds an army together. Though mud is better than no cement, we want the best cement, and we shall never get it unless we can analyse war scientifically and discover its values.[29]

Illuminating MDOs is done to help create the "best cement" as the Western militaries, the US Army, and the US joint force continue to refine MDO. The US and its allies must incorporate the theory of dominance and ZoPD. Doing so will provide policymakers, strategists, and military practitioners with a useful framework for planning, analysis, and operations, both from a friendly side, but also when thinking about an adversary.

Several derivative assumptions come from analyzing the features and impacts of dominance. Those assumptions should also be added to MDO because they can increase understanding by clarifying the ideas behind dominance, ZoPD, and MDO's challenge-response dynamic among belligerents. In doing so, the assumptions support the practitioner's ability to plan, analyze and develop operations in an MDO environment.

Finally, as the Battle of Mosul (part of a larger, global, counter-terror campaign) illustrates, hard constraints must be accounted for when developing concepts, theories and doctrine. By the time Mosul concluded, the US military was all but out of PGMs. If PGMs are a critical component of MDO, as it is currently written—and the doctrine is written based on best-case scenarios—then the ideas of persistence and convergence are not feasible. Therefore, MDO must rework the role of PGMs, persistence, convergence, and the "penetrate-disintegrate-exploit" model. Further, MDO needs to note that both the best-case and worst-case scenarios are merely planning assumptions; otherwise, the entire theory is of little-to-no utility. To get MDO right, these flaws must be addressed.

---

29 J.F.C. Fuller, *The Foundations of the Science of War* (Leavenworth, KS: Command and General Staff College Press, 1993), 35.

8

Multidomain Operations As Military Strategy

Jeffrey W. Meiser

Introduction

This chapter analyzes multidomain operations (MDO) doctrine as a military strategy. MDO is the US Army conceptualization of "detailed solutions to the specific problems posed by the militaries of post-industrial, information-based states like China and Russia."[1] In other words, MDO is the main concept that drives Army actions to compete with, and if necessary, defeat the most capacious adversaries of the United States. MDO is the US Army's highest-level strategic concept, and it is a presentation of how the Army thinks the joint force, and the US's alliance and coalition partners, should compete, fight, and win on the 21st century battlefield. Furthermore, MDO has spread far and wide, and according to one report "is the dominant intellectual concept within NATO and other technologically advanced militaries."[2] Thus, while MDO contains the word "operations," the top-level, most abstract set of ideas

---

1 US Army Training and Doctrine Command (TRADOC), *TRADOC Pamphlet 525-3-1: The US Army in Multidomain Operations 2028*, 6 December 2018, https://www.army.mil/article/243754/the_u_s_army_in_multi_domain_operations_2028, 5 (subsequently *TRADOC Pamphlet 525-3-1*).
2 Davis Ellison and Tim Sweijs, "Breaking Patterns: Multidomain Operations and Contemporary Warfare," The Hague Centre for Strategic Studies, September 2023, https://hcss.nl/report/breaking-patterns-multi-domain-operations-and-contemporary-warfare/.

contained within MDO suggest a military strategy that attempts to explain the US Army's theory of battlefield success.[3]

For this chapter, we use the "theories of strategy" approach to analyzing multidomain operations.[4] Under this framework, the analyst starts with three questions: What is the challenge the strategy is meant to address? What is the strategy? And what are the likely points of failure? In applying this framework, the analyst defines the challenge as a "theory of the challenge," describing the causes of the challenge and the negative effects of the challenge. The strategy is defined as a "theory of success," and the analyst should assess how well the strategy defines the proposed actions and the effect those actions will have, paying close attention to the presumed causal effect of the proposed actions. Finally, the strategy's likely points of failure are defined as "theories of failure." The emphasis here should be on thinking through alternative causal relationships, especially the possibility that proposed actions will have effects contrary to expectations. Articulating the challenge, strategy, and points of failure as theories (or "causal explanations") gives the analyst a basis for assessing the validity, usefulness, and likely success of a strategy.[5] As a "challenge-based approach" this framework assumes that all strategies are a response to a challenge or problem, and it is impossible to effectively assess a strategy without understanding the challenge on which it is based.

The remainder of the chapter applies this framework by identifying and analyzing the theory of the challenge, theory of success, and theories of failure of the US Army's MDO strategy. To conduct this analysis, MDO is divided into two types of strategy, one for competition and one for armed conflict. This approach is based on the structure of Army MDO doctrine as

---

3 MDO is also rooted in the 2018 and 2022 summaries of the US National Defense Strategy, both of which use phrasing like "all domains" or "across domains." See Summary of the 2018 National Defense Strategy of The United States of America, Office of the Secretary of Defense, 2018, https://www.defense.gov/News/Feature-Stories/story/Article/1656414/what-is-the-national-defense-strategy/, 2; and "2022 National Defense Strategy of the United States of America," Office of the Secretary of Defense, 2022, https://www.defense.gov/News/News-Stories/Article/article/3202438/dod-releases-national-defense-strategy-missile-defense-nuclear-posture-reviews/, 1, 12.

4 Jeffrey W. Meiser, "Bringing Method to the Strategy Madness," *War on the Rocks*, May 2, 2024, https://warontherocks.com/2024/05/bringing-a-method-to-the-strategy-madness/. Others have partially applied this approach, see Ellison and Sweijs, "Breaking Patterns" 17-20.

5 The phrasing "causal explanation" to define theory comes from Stephen Walt, "The Relationship between Theory and Policy in International Relations," *Annual Review of Political Science, 8* (2005), 23–48, https://www.annualreviews.org/content/journals/10.1146/annurev.polisci.7.012003.104904.

seen in *TRADOC Pamphlet 525-3-1: The US Army in Multidomain Operations 2028* (2018), and *FM 3-0 Operations* (2022).

Theory of the Challenge

MDO is explicitly articulated as a response to a multifaceted challenge rooted in both the operational environment and in the strategies of nation-state adversaries. According to Army doctrine, changes in technology and the balance of power have decreased US advantages over time and made it more difficult for the US to deter potential adversaries.[6] Nation-states like Russia and China have taken advantage of these changes to more directly challenge US interests by pursuing "stand-off" or anti-access and area denial (A2AD) strategies.[7] Traditionally, stand-off refers to "the tactic of engaging an enemy element when your weapons can reach him and his return fire cannot."[8] For MDO, stand-off has been elevated to something more like a strategic concept. Under the context of competition, stand-off refers to adversaries' use of subversion to slow down or prevent the US from responding to malign activities. Under the context of armed conflict, stand-off refers to adversaries' use of long-range fires to prevent US expeditionary forces from accessing defended areas and maneuvering effectively in contested terrain.[9] MDO asserts Russia and China are pursuing the same types of strategies and both need to be addressed in competition and in armed conflict. Moreover, MDO identifies Russia as the US Army's pacing threat.[10] Subsequent doctrine describes China as the "pacing challenge" and Russia as the "acute threat"

---

6 *TRADOC Pamphlet 525-3-1*, vi, 6. Though in later Army and Department of Defense documents, there seems to be more confidence in deterrence, see United States Army, *FM 3-0 Operations*. October 2022. Washington, DC: Government Printing Office. https://armypubs. army.mil/epubs/DR_pubs/DR_a/ARN36290-FM_3-0-000-WEB-2.pdf, 1-3 (subsequently *FM 3-0 Operations*); "2022 National Defense Strategy of the United States of America," 1, 8-11.
7 *TRADOC Pamphlet 525-3-1*, v, vi, 11-13; Also referred to as "preclusion" in *FM 3-0 Operations*, 2-9; Robert Rose, "Returning Context to Our Doctrine," *Military Review* (October 2023), 3, https://www.armyupress.army.mil/journals/military-review/online-exclusive/2023-ole/returning-context-to-our-doctrine/; *FM 3-0 Operations*, 1-4.
8 Huba Wass de Czege, *Commentary on "The US Army in Multidomain Operations 2028* (Carlisle: US Army War College Press, 2020), https://press.armywarcollege.edu/monographs/909.
9 See also the discussion of stand-off in Amos C. Fox, *Obstructive Warfare: Applications and Risks for AI in Future Military Operations*, Centre for International Governance Innovation, CIGI Paper No. 307, October 2024, https://www.cigionline.org/static/documents/Fox-Sept2024.pdf. *TRADOC Pamphlet 525-3-1* has a footnote defining stand-off but it is practically incomprehensible (footnote #2 on page vi).
10 *TRADOC Pamphlet 525-3-1*, vii, 6-7.

without clearly prioritizing between them.[11] The following subsections analyze the MDO theory of the challenge for competition and armed conflict.

Competition

From the perspective of MDO, the central organizing challenge in competition is the intent of China and Russia to use subversive and coercive methods to create stand-off or "political separation" between the United States and its allies and partners. Political separation causes "strategic ambiguity reducing the speed of friendly recognition, decision, and reaction."[12] This slower reaction allows (and encourages) China and Russia to achieve their goals without having to engage in armed conflict. Methods of creating separation or "isolation" (a related doctrinal concept) include "using disinformation campaigns and the threat of aggression,"[13] and "creating instability within countries and alliances."[14] This framing of the challenge suggests the causal logic for "peer threats" (i.e., China and Russia) is to undermine and manipulate American allies and partners to create politically distance between them and the United States. The implication of this strategy is that with enough political separation, the US will be isolated, and our allies and partners will react too slowly and without sufficient intensity to deter or prevent hostile actions by China and Russia. What is "enough political separation" to unlock this outcome? That is the key question that all Chinese, Russian, and US, and US-partners' policymakers and strategists wrestle to identify, forestall, and overcome.

A less clearly articulated element of the challenge is the relationship between the hostile actions of adversaries, US allies' perception, and US credibility and deterrence. The concern seems to be twofold. First, malign activities by peer threats can undermine the credibility of US security guarantees.[15] Through disinformation and conventional force deployments and demonstrations, adversaries seek to convince US allies and partners that the American response will be slow and insufficient in the event of a Russian (or Chinese) attack. Second, by acting in the "gray zone" below the threshold of armed conflict, adversaries can make deterrence almost irrelevant. Put another way, adversaries seek to ratchet up the "scope and intensity of

---

11 See General James C. McConville's Forward to *FM 3-0 Operations*.
12 *TRADOC Pamphlet 525-3-1*, p. vi
13 *FM 3-0 Operations*, 2-12.
14 *TRADOC Pamphlet 525-3-1*, p. vi
15 Ibid., 7.

their malign activities conducted below the threshold of armed conflict" having the effect of "diluting the joint force's conventional deterrence."[16] These adversarial efforts have "expanded the battlefield" and "blurred the distinctions between actions 'below armed conflict' and 'conflict,' enabling the achievement of strategic objectives short of what the US traditionally considers 'war.'"[17] This second concern gets most of the attention in Army doctrine, but both pathways are seen to undermine US and NATO deterrence.

MDO doctrine provides concrete examples of the challenge in the cases of Georgia, Ukraine, and Syria.[18] Georgia and Ukraine (prior to 2022) seem most relevant as examples of Russia manipulating each state's internal politics with the goal of keeping each of them out of the West's political, economic, and normative orbit. Furthermore, the Ukraine case (especially in 2014) illustrates Russia's intention and ability to muddy the waters in the information space sufficiently enough to slow the West's reaction time.[19] The Ukraine example in particular seems to be driving much of the concern over how peer threats can achieve strategic goals (e.g., annexing Crimea) without triggering an assertive response by the United States and its allies and partners.[20] It is not clear from Army doctrine where the current threat of Russian competitive actions is most acute, raising the question of where exactly the threat lies. Army doctrinal publications make effective use of passive voice to avoid identifying any specific country that might be particularly vulnerable to Russian subversion and intimidation or whose political separation from the US may be particularly consequential. Presumably, key allies like Germany and France, awash in Russian disinformation, could inhibit NATO's ability to respond to additional Russian aggression in the future.[21] Reconsidering the run-up to the invasion of Ukraine in 2022, it is important to highlight the

---

16 *FM 3-0 Operations*, 1-4.

17 *TRADOC Pamphlet 525-3-1*, 8.

18 Ibid., 7.

19 Rod Thornton, "The Changing Nature of Modern Warfare: Responding to Russian Information Warfare," *The RUSI Journal*, Vol. 160, No. 4 (2015), 40-48, https://doi.org/10.1080/03071847.2015.1079047; Michael Kofman, Katya Migacheva, Brian Nichiporuk, Andrew Radin, Olesya Tkacheva, and Jenny Oberholtzer, *Lessons from Russia's Operations in Crimea and Eastern Ukraine*, RAND Corp., 2017, https://www.rand.org/pubs/research_reports/RR1498.html.

20 Amos Fox, "Russian Hybrid Warfare: A Framework," *Journal of Military Studies* Vol. 10, No. 1 (2021), 6-7, https://sciendo.com/article/10.2478/jms-2021-0004.

21 On Russian disinformation in Germany see Kathrin Wesolowski and Tetyana Klug, "Fact check: Russia's influence on Germany's 2025 election," *DW*, February 18, 2025, https://www.dw.com/en/russian-disinformation-aims-to-manipulate-german-2025-election/a-71664788; *The Economist*, "France Uncovers a Vast Russian Disinformation Campaign in Europe," February 12, 2024, https://www.economist.com/europe/2024/02/12/france-uncovers-a-vast-russian-disinformation-campaign-in-europe.

slow evolution of European states' appreciation of the true Russian threat facing Kyiv and Ukrainian sovereignty. This was such a challenge that the Biden administration had to exert considerable diplomatic effort to create a relatively united opposition front against Russia throughout 2022.[22]

While not stated anywhere in Army doctrine, it is also plausible we are seeing a weakening US commitment to ally and partner security, and this could be the real cause of political separation and diluted deterrence. For example, members of the Trump administration have made pejorative and even threatening statements about NATO allies. Trump has noted his interest in absorbing Greenland, currently part of Denmark, and pointedly refused to rule out using military force to acquire Greenland.[23] Defense Secretary Pete Hegseth called European allies "pathetic," a sentiment seemingly shared by several other senior members of the administration including Vice President J.D. Vance.[24] If the US commitment to European allies is weakening, a central premise of MDO—that it is in the interests of the US to support allies and partners—is severely undermined.

Bringing all these strands together, it is possible to articulate an MDO "theory of the challenge." Given the emphasis in key MDO documents, "political separation" is the best candidate for the challenge, which is caused by adversary's use of subversion (disinformation, political interference, etc.) and coercion (conventional force posture, exercises, etc.). Political separation has the negative effect of isolating the US from its allies and partners and making allied support slower and less dependable. The US loses credibility, adversaries gain the initiative, US deterrence is diluted, and American adversaries achieve strategic goals without having to resort to armed conflict.

---

22 Mark Landler, Steven Erlanger and David E. Sanger, "In Standoff With Putin, Biden Makes Sure European Allies Are With Him," *New York Times*, January 28, 2022, https://www.nytimes.com/2022/01/28/world/europe/biden-putin-ukraine-europe.html/; Humeyra Pamuk and Michelle Nichols, "US Accelerates Ukraine Diplomacy as Europe Slides into Winter," *Reuters*, December 13, 2022, https://www.reuters.com/world/europe/us-accelerates-ukraine-diplomacy-europe-slides-into-winter-2022-12-13/.

23 David E. Sanger and Michael D. Shear, "Trump Floats Using Force to Take Greenland and the Panama Canal," *New York Times*, January, 7, 2025, https://www.nytimes.com/2025/01/07/us/politics/trump-panama-canal-greenland.html.

24 Ellen Francis, Anthony Faiola, Kate Brady, and Annabelle Timsit "For Europeans, Signal Chat Gives Unfiltered view of Trump Team's Disdain," *Washington Post*, Updated March 25, 2025, https://www.washingtonpost.com/world/2025/03/25/trump-administration-europe-signal-chat-leak/.

It is also plausible to see armed conflict as the result of deterrence failure during the competition phase.[25]

In assessing the MDO theory of the challenge, two concerns stand out: (1) the fuzziness of the concepts and the hypothesized relationships and (2) the anachronistic view of US-allied relations. The attempt to use "stand-off" to describe the potential political isolation of the United States under competition and to describe the potential operational separation of elements of the joint force is probably meant to be elegant but the phrase comes off as conceptual stretching.[26] Furthermore, the claim that the heart of the challenge is that US allies will either be persuaded or destabilized by disinformation and subversion is not clearly related to actual events or likely events. In the examples mentioned in Army doctrine—Georgia, Ukraine, and Syria—the problem seems to be the weakness of the US response as much or more than it is allied response. Yes, Georgia and Ukraine were dismembered by Russian subversion, but the US led the way in looking the other way.

The second concern returns us to the issue of what country or countries are most susceptible to peer adversary subversion and coercion. As noted, in this current moment the United States seems more likely to separate itself from allies (especially European allies) than the other way around.[27] NATO members like Germany, France, the UK, and Canada seem more committed than ever to opposing Russian aggression, while the US has already given in to a series of Russian demands.[28] Doctrine written in 2018 and 2022 can hardly be expected to have foreseen the Trump shock of 2025, but MDO must be evaluated with the current situation in mind, and from that perspective the baseline assumptions of MDO are highly questionable. Political separation between the United States and its allies seems increasingly likely to come from within that United States rather than from allies. The real challenge would seem to be the declining political commitment of the United States to stand with allies and partners to oppose Russian (and Chinese) aggression.

---

25 Andrew Feickert, "Defense Primer: Army Multidomain Operations (MDO)," Congressional Research Service, Updated October 1, 2024, https://www.congress.gov/crs-product/IF11409.
26 Ellison and Sweijs discuss the problem of conceptual stretching and MDO in "Breaking Patterns," 16.
27 See for example, David Luhnow and Marcus Walker, "What Does MAGA Have Against Europe?" *Wall Street Journal*, March 28, 2025, https://www.wsj.com/politics/what-does-maga-have-against-europe-96416042.
28 Michael Weiss and James Rushton, "Can Europe Back Ukraine's Fight Alone?" *NewLineMagazine*, March 3, 2025, https://newlinesmag.com/argument/can-europe-back-ukraines-fight-alone/.

Armed Conflict

According to Army doctrine, adversaries create conditions (e.g., fractured alliances, decreased US credibility) during competition to enhance their position in the event of armed conflict. Creating political stand-off, coupled with slowing, and weakening friendly responses gives an adversary an advantage in the transition to armed conflict. This leads to the central challenge in armed conflict: Russia and China use of long-range, mid-range, and short-range fires systems "to separate the Joint Force in time, space, and function." Their goal is to impose high losses on forward positioned forces and to prevent the US and their allies and partners from introducing expeditionary forces. The MDO assumption is that adversaries are focused on surprise, *fait accompli* attacks, and plan to "achieve campaign objectives within days."[29] Most of the concern about the ability of Russia and China to create stand-off stems from the range and effectiveness of their long- and mid-range fires. For both states, these capabilities reside in their ballistic missiles, rocket systems, and anti-aircraft weapon systems. Under the cover of long- and mid-range fires, adversary short-range fires and land forces can destroy friendly forces and seize terrain.[30]

The theory of the challenge during armed conflict starts with the separation of the joint force in time, space, and function at the center. The meaning of separation in time and function, is not well articulated. It presumably refers to the efforts of adversaries to use surprise and speed to complicate the US military's preferred "time-phased and domain-federated operational approaches in armed conflict."[31] Therefore we can infer this means adversaries intend to interfere with the planned "deployment sequence" of US forces, i.e., the deployment or "flow" of forces consistent with the given operational planning.[32] It could also refer to the broader tendency in the American way of war to establish air superiority before subsequently using land forces in close combat to destroy enemy forces.[33] Consistent with this tendency, General James B. Hecker, Commander, US Air Forces in Europe and Allied Air Command recently explained that, "air superiority remains job

---

29 *TRADOC Pamphlet 525-3-1*, 11.

30 Ibid. 12-13.

31 Ibid., vii.

32 *FM 3-0 Operations*, 4-15; *JP 5-0 Joint Planning*, December 2020, IV-38, https://irp.fas.org/doddir/dod/jp5_0.pdf.

33 James B. Hecker, "Air Superiority: A Renewed Vision," *ÆTHER: A Journal of Strategic Airpower and Spacepower*, Vol. 3, No. 2 (Summer 2024), 5-6.

number one" and "it [air superiority] will typically remain our top priority… because it grants us freedom of maneuver to accomplish all other tasks and because attrition rates would otherwise become prohibitive."[34] Therefore, by interrupting the deployment of forces and the phasing between domains (air and land), the adversary could degrade US ability to respond to an attack.

Separation of the joint force in time, space, and function results from adversary use of long- and mid-range fires. Air defense systems prevent the effective combination of US ground and air forces while ballistic missile and artillery prevent expeditionary forces from accessing the battlespace. All of these actions are enabled and enhanced by electronic warfare, cyberspace, and information operations. Once the joint force is "fractured," forward deployed forces can be isolated and destroyed, including "high-value capabilities" like "headquarters, aircraft, and trained combat formations that are difficult to regenerate and essential to achieving US operational and strategic objectives."[35] The *fait accompli* attack succeeds, and political gains consolidated.

The logic of the challenge articulated above distills down to two propositions. First, both sides (US and adversary) are attempting to implement a cohesive joint and combined arms campaign in which both sides wanting to prevent the other from doing the same thing. The theory of the challenge can really be stated as, "the adversary will try and hit us with long-range fires so we can't do the stuff we want to do."

Second, the challenge seems highly focused on the increased range and effectiveness of Russian and Chinese weapon systems and the integration of subversion and coercion with conventional armed conflict. The potential synergy of the stand-off weapons systems with political subversion is hypothesized in doctrine, but the connection is unclear and unevidenced. Here the China case might be considered the pacing threat because it is easier to imagine a situation where China couples long-range missile attacks with political subversion against Taiwan in a modern form of gunboat diplomacy.

Theory of Success

MDO is a response to the challenge caused by adversary strategies of political separation during peacetime and violent separation during armed conflict. This section treats MDO as a strategy with two parts or phases, one during

---

34 Ibid. 6.
35 *TRADOC Pamphlet 525-3-1*, 13.

competition and one during armed conflict. The key issues in making this assessment are: What is the theory of success? Is the theory of success valid? Does it make sense conceptually? Does it seem to have some correspondence to reality? Does the theory of success directly address the theory of the challenge? Does the theory of success describe a coherent set of actions that intervene in the causal process of the challenge and in some way disrupt it?[36]

Competition

Under conditions of competition, MDO is meant to defeat adversaries' efforts at destabilization and coercion, which are the foundation for an advantageous transition by adversaries to armed conflict. Army doctrine suggests that the "conduct of multidomain operations" will cause the joint force to "prevail in competition."[37] The specifics suggest a strategy for successful competition based on "active engagement" by the Army and joint force to counter the coercion and subversion targeting allies and partners. The central role of the Army in competition seems to be putting in place the "posture, capabilities, and readiness" that demonstrate the "capability to prevail in armed conflict."[38] If these elements are in place, the Army can contribute toward deterring adversaries, countering adversary subversion (especially information operations), and preventing successful coercion of partners and allies. Effectiveness in these efforts prevent adversaries from achieving their goal of "winning without fighting."[39] MDO documents also note that effective Army actions and dispositions "expands the competitive space" giving leaders more freedom of action and helps reassure allies of American commitment and strength.[40] Ultimately, Army actions can "seize and sustain the initiative in competition" and "set conditions for a rapid transition to armed conflict."[41] MDO is linked to the sister concept of integrated deterrence which "includes preventing adversaries from increasing the scope and intensity of their malign activities conducted below the threshold of armed conflict."[42]

---

36 For a similar effort attempting a cross-national comparison of MDO, see Ellison and Sweijs, "Breaking Patterns," 17-19, 28-29. They perform a brief cross-national analysis of theories of success in multidomain doctrine.

37 *TRADOC Pamphlet 525-3-1*, vii.

38 Ibid., viii.

39 Ibid., vii.

40 Ibid., vii-viii.

41 Ibid., viii.

42 *FM 3-0 Operations*, 1-4.

The centerpiece of the MDO competition strategy is convincing allies and adversaries that the joint force can win in armed conflict. The joint force can demonstrate this ability through the positioning and readiness of forces and through something called "active engagement." With these pieces in place, the result is deterrence, reassurance, suppression of malign activities, and seizure of the initiative and freedom of action.

Switching to assessment, MDO's theory of success is plausible, but too vague to allow for a confident judgment of the validity of the causal relationships. As noted in the previous section, MDO doctrine had a lot of fuzziness when it comes to defining the competition challenge, so it is no surprise the strategy also lacks clarity. Without having a better sense of which allies and partners might be susceptible and what kind of subversion is most threatening, it is difficult to know what, if anything, the joint force can do about it.

Despite the weaknesses discussed above, there is reason to see some value in the MDO strategy. It is plausible that if US allies and partners see a large, highly capable American military presence postured forward, then they will be reassured of US commitment. The literature on conventional deterrence supports the claim that highly capable, forward deployed, conventional forces are the most effective way to deter a *fait accompli* type attack with limited aims.[43] What is far less certain is how the requirements for conventional deterrence will affect adversary ability to implement successful subversion operations within ally and partner societies. Corruption, election interference, and disinformation are not likely to have a military remedy. However, the forward presence of substantial American forces will create strong inertia against political separation and probably reduce the likelihood of strategic ambiguity. Since MDO does not answer the question of how large the forward presence should be, it is unclear what the basic strategic posture will be (or should be according to MDO). For example, should NATO deploy much larger and more capable forces as far forward as feasible? Should NATO scrap the 1997 NATO-Russia Founding Act, as suggested by Andrea Kendall-Taylor and Michael Kofman?[44] It is also important to note the strong possibility that US posture in Europe will go in the opposite direction and be

---

43 Michael Gerson, "Conventional Deterrence in the Second Nuclear Age," *Parameters*, Vol. 39, No. 3 (August 2009), 32-48, https://press.armywarcollege.edu/parameters/vol39/iss3/8/.
44 Andrea Kendall-Taylor and Michael Kofman, "Putin's Point of No Return How an Unchecked Russia Will Challenge the West," *Foreign Affairs*, Vol. 104, No. 1 (Jan/Feb 2025), 72-87, https://www.foreignaffairs.com/russia/putins-point-no-return.

significantly reduced under the Trump administration.[45] If this were to occur, MDO would have a foundational element eroded, at least in the European theater.

In assessing how well the MDO strategy matches with the MDO framing of the challenge, there does seem to be a plausible intervention point in the theory of the competition challenge. MDO envisions an intervention early in the causal process to prevent political separation. Part of the subversion effort suggested in Army doctrine is the adversarial promotion of a narrative portraying the United States as a weak and fickle ally; MDO's theory of success counters that narrative with a show of strength that reassures allies and partners. As suggested in the analysis of the challenge above, the underdevelopment of the theory of the challenge means there is also an underdevelopment of the theory of success. If the only mechanism of subversion is to portray the US as weak, then a show of strength can remedy that, but if there are other (perhaps stronger) mechanisms of subversion, then the show of strength will not be sufficient. Besides a force posture that shows strength and commitment MDO includes the phrasing "active engagement" which seems to be a place holder for "stuff the Army will do." It is not clear what those activities will be. There is a whole slew of intelligence, counterintelligence, informational operations as well as training, operational plan development, and security force assistance activities, but nothing that would seem to have a significant effect on adversary subversion.[46] In sum, MDO strategy provides an incomplete solution to an underspecified problem.

Armed Conflict

In armed conflict, MDO provides a solution to the challenge posed by an adversary's attempt to fracture the joint force. The US Army, and the joint force more broadly, will conduct operations across domains to penetrate and dis-integrate adversary stand-off weapons systems to enable access and freedom of maneuver for the joint force and the defeat of adversary forces in close battle. The goal is to "fracture the coherence of threat operational approaches by destroying, dislocating, isolating, and disintegrating their interdependent systems and formations, and exploiting the opportunities these disruptions

---

45 Steven Erlanger, "Trump Wants Europe to Defend Itself. Here's What It Would Take." *New York Times*, March 7, 2025, https://www.nytimes.com/2025/03/07/world/europe/europe-self-defense-trump.html.
46 See *FM 3-0 Operations*, chapter 4.

provide to defeat enemy forces in detail."[47] This success will establish favorable terms for continued competition. The three "tenets" of MDO are: "calibrated force posture, multidomain formations, and convergence."[48] The first two seem to apply mostly to the positioning and the capabilities of Army forces, while the last is the key concept for understanding how the Army plans to fight. Convergence explains how the Army will win against the adversaries' stand-off strategy. With the understanding that adversaries will attempt to separate the joint force, convergence calls for the joint force to use "cross-domain synergy and multiple forms of attack" to "overmatch the enemy." According to FM 3-0, "Leaders multiply the effects of lethal force by employing combinations of capabilities through multiple domains to create, accrue, and exploit relative advantages—imposing multiples dilemmas on enemy force and overwhelming their ability to respond effectively."[49] Convergence requires "mission command and disciplined initiative" and "integration" and "synchronization" of capabilities.[50]

The general phrasing used to explain MDO can easily be perceived as simply asserting that the US Army and joint force will just be better at warfighting than the enemy. The more detailed discussion of how to penetrate, dis-integrate, and establish freedom of maneuver, reinforce this perception. For example, "Army forces at echelon employ cross-domain fires to defeat the enemy's long-range systems and begin the neutralization the enemy mid-range systems."[51] Thus, the enemy will try and create stand-off through long- and mid-range fires, but the US Army and joint force will neutralize and destroy those weapons systems to allow for expeditionary force penetration and exploitation. These are really just assertions with no reason to believe that one side or the other will be more likely to achieve their goals. Asserting operational superiority is not a valid causal argument.[52]

Further discussion of convergence gives a bit more clarity on the causal process the Army intends to initiate with multidomain operations doctrine. Yet, MDO never arrives at a fully convincing theory of success especially in the sense of making a convincing case for why the US Army and joint force is going to be more effective at implementing its strategy than an enemy will be at implementing their strategy. The main issue is whether "convergence"

---

47 Ibid., 1-3.
48 *TRADOC Pamphlet 525-3-1*, v, vii, 17.
49 *FM 3-0 Operations*, 1-5.
50 *TRADOC Pamphlet 525-3-1*, vii, 20; *FM 3-0 Operations*, 3-4, 3-5.
51 *TRADOC Pamphlet 525-3-1*, p. ix.
52 Ellison and Sweijs, "Breaking Patterns," 29.

has any causal effect or whether it is a "secret sauce" addition that has no real meaning or value. In unpacking this concept, the key is the claim that by fully integrating capabilities across domains commanders have "multiple forms of attack and...layered options...to impose complexity on the enemy."[53] Doing this "enables the Joint Force to stimulate, see, and strike vulnerabilities in the Chinese and Russian systems and defeat their efforts to create stand-off."[54] Nonetheless, this assumes that the joint force's adversaries do not have the capability to address the challenges, much less the instinct for survival or the training to address these rather common sense approaches. Having multiple methods for attacking enemy vulnerabilities (from different domains) makes it more difficult for the enemy to defend by imposing complexity on enemy decision-making and presenting multiple dilemmas. A multidomain approach has an even more important hypothesized effect in creating "cross-domain synergy," suggesting that it goes beyond simply having different options to achieve "an overall effect greater than the sum of the individual parts."[55] As we get into the details of penetrating, disintegrating, and exploiting enemy systems, there is little reference back to the main tenets of MDO.[56] There are periodic references to "cross-domain fires" and acting across all domains, but not further discussion of convergence or cross-domain synergy.[57]

In sum, it is hard to disagree with Robert Rose's critique that "multidomain operations uses stuff, advantageously, to do stuff."[58] That is, if there is not much method to the MDO madness, success is deeply dependent on what kind of resources and how much are available and whether or not the joint force is better at using those resources and weapons systems than the adversary. In particular, MDO doctrine does not provide a clear sense of what convergence really is in practice and how to create it. More importantly, we have no way of assessing its causal effect. There is a sense that MDO is simply traditional combined arms with cyberspace and space dimensions (domains). According to one graphic, the enemy can be stimulated through cyber-attacks and then seen from space-based capabilities and attacked from the air and from ground-based long-range precision fires.[59] This, apparently,

---

53 *TRADOC Pamphlet 525-3-1*, x.
54 Ibid.
55 Ibid., 20.
56 See Ibid., 32-45.
57 Ibid., 33, 37.
58 Rose, "Returning Context," 3.
59 *TRADOC Pamphlet 525-3-1*, 21.

is convergence. If MDO is simply a dressed-up version of combined arms warfare, this is not necessarily a problem, but it may fall into the trap of equating operational effectiveness with strategy. Many scholars of strategy have noted that simply doing things better is not much of a strategy.[60] All things being equal, operational excellence is preferable, and some would argue the best we can hope for given the challenges of crafting good strategy. However, ideally when the Army creates doctrine for how it plans to fight at the highest level of the organization, it should be more than "we will be excellent" and "we will do combined arms maneuver better and so we will win."

Beyond these conceptual problems, there are also practical issues not addressed by Army doctrine. MDO assumes Army "formations possess the combination of capacity, capability, and endurance which generates the resilience necessary to operate across multiple domains."[61] MDO also assumes "persistence" and "sustainment" of forces.[62] Capacity, capability, endurance, persistence, and sustainment are all necessary prerequisites of creating convergence and cross-domain synergy, the claimed payoff of MDO. As Amos Fox notes, this creates a "feasibility problem;" it is far from certain the United States has the resources necessary to enable the joint force to meet the perquisites of effective MDO, especially the number of precision guided munitions.[63] Of course, it is the constitutional duty of the US Congress to "provide for the common Defence [sic]" and "raise and support Armies."[64] But Congress acts based on budgetary requests by the Department of Defense, presumably based on the Secretary of Defense's analysis of what is required to resource the doctrine of the joint force. Or to put it a bit differently, it is

---

60 Michael E. Porter, "What Is Strategy?" *Harvard Business Review* (November–December 1996), 61–78, https://hbr.org/1996/11/what-is-strategy.

61 TRADOC Pamphlet 525-3-1, 19.

62 Ibid., 18.

63 Amos C. Fox, "Getting Multidomain Operations Right: Two Critical Flaws in the US Army's Multidomain Operations Concept," Land Warfare Paper No. 133, June 2020, Association of the United States Army, https://www.ausa.org/publications/getting-multidomainmain-operations-right-two-critical-flaws-us-armys-multidomainmain, 8; see also Bryan J. Quinn, "Sustaining Multidomain Operations: The Logistical Challenge Facing the Army's Operating Concept," *Military Review* Vol. 103, No. 2 (March-April 2023), 128–38, https://www.armyupress.army.mil/Journals/Military-Review/English-Edition-Archives/March-April-2023/Multidomain-Operations/; Nathan A. Jennings, "Penetrate, Disintegrate, and Exploit: The Israeli Counteroffensive at the Suez Canal, 1973," Modern War Institute Report, October 2024, https://mwi.westpoint.edu/penetrate-disintegrate-and-exploit-the-israeli-counteroffensive-at-the-suez-canal-1973/, 21.

64 US Constitution, Section 8, https://constitution.congress.gov/browse/article-1/section-8/.

the responsibility of the Secretary of Defense to ensure congruence between strategy (articulated as doctrine) and resources.

A related feasibility concern is whether the joint force can actually field the technologies assumed by MDO. Specifically, Ellison and Sweijs argue that while the intelligence, surveillance, and reconnaissance (ISR) capabilities assumed by MDO are mature after decades of development, "the vital 'hinge' between 'sensing and shooting,' the communications systems, are far from sufficient maturity and are highly differentiated multinationally."[65] They note challenges with the US Joint All-Domain Command and Control system (JADC2) as particularly problematic, given the overall challenges with connectivity on the battlefield.[66]

In the end we are left with the sense that MDO could be a productive way forward in thinking about the tactics, techniques, and procedures needed for modern warfare, but at the higher level of thinking about how the Army should plan to achieve (or contribute to achieving) strategic objectives, MDO has major weaknesses. Ideally, the Army would have a strategy for achieving (or contributing to achieving) success in competition and armed conflict, and that strategy (or strategies) would increase the likelihood of success, i.e., would convey some additional advantage beyond operational effectiveness. This might be hoping for too much from doctrine, but top-level Army doctrine like MDO is portrayed as describing the way in which the Army and joint force will defeat an adversary and by taking this posture, fuels these high hopes.

Conclusion: Towards A Theory of Failure

This chapter concludes with a final assessment of MDO by considering theories of failure. Conceptually, a theory of failure is the opposite of a theory of success. To derive theories of failure, an analyst should ask: What if the theory of success is wrong? What if the prosed actions have no effect? What if proposed actions have the opposite causal effect to what we expect? What if the magnitude of effect is too low, or too high?

One short-cut an analyst can take in some contexts is to consider the opponent's strategy to be a theory of failure. If the adversary's theory of success is more compelling than your theory of success, there might be a problem. For MDO we can consider specifically whether adversaries will be

---

65 Ellison and Sweijs, "Breaking Patterns," 27.
66 Ibid., 27, 30.

able to achieve their goals through subversion and coercion or can use these malign activities to produce significant advantages in the context of armed conflict. When compared to the lackluster MDO strategy for competition, the competition strategies of Russia and China look relatively strong. While Army doctrine is not that clear about the exact nature of the threat, it does seem that adversaries have the initiative and an approach that has proven to be at least somewhat successful under some conditions (e.g. in the Russian near-abroad). What is not as clear is whether Russia can continue to effectively use subversion and coercion to make additional gains in the post-Ukraine War environment. If so, MDO provides no credible answers.

A second way to consider failure is to consider weaknesses in the MDO theory of success and contemplate whether there might be modifications that can improve the MDO theory of success. The most important issue is whether the actions proposed to implement MDO will have the causal effect the US Army believes it will have on the contemporary battlefield. A potential theory of failure, in the context of armed conflict, could be stated as: the multidomain capabilities of peer adversaries will cause attempts to implement MDO to fail because adversaries will be able to interrupt some elements of MDO and prevent full integration and convergence.

One way to think through how this theory of failure might cause us to reconsider MDO is through the lens of the 1973 Yom Kippur War. Relying on Nathan Jennings' interpretation of the 1973 War, the multidomain operations of the Israeli Defense Forces were successfully interrupted by the Egyptian Army and had to be "restored" using an armor assault into Egyptian territory to create "an opening in the Egyptian missile shield."[67] In Israel's 1973 operation across the Suez Canal, IDF units specifically targeted anti-air systems to enable the restoration of close air support and convergence across domains.[68]

This brief example suggests the question: should Army doctrine shift to focus specifically on creating "windows of opportunity" to enable MDO? Starting with the assumption that MDO may be mostly impossible, but can be created through focused effort, is different from assuming MDO is the standard operating context for the joint force. The effort then is focused on what it takes to enable MDO and exploit those fleeting moments of convergence. This way of thinking has some consistency with Ukraine's experience in fighting Russian forces. Ukrainian operations have to be

---

67 Jennings, "Penetrate, Disintegrate, and Exploit," 13.
68 Ibid., 19-20.

meticulously planned to account for the jamming of drones (necessary for persistent surveillance and air attack), often requiring efforts to create a temporary corridor of geographical and electromagnetic space to allow for coordination of air and ground forces. According to the *Wall Street Journal* reporting on the Ukrainian assault into Kursk, "Ukrainian electronic-warfare units went into Russian territory ahead of the main mechanized assault forces to jam Russian equipment to stop Russian forces from pinpointing Ukrainian positions or intercepting their communications. That unusual, early deployment—more like that of a reconnaissance unit—created a protective bubble around advancing Ukrainian assault forces."[69] While it is probably not appropriate for US doctrine to reveal if the joint force has secret, prepositioned electronic warfare units in Baltic countries (for example) it would be useful to acknowledge and plan for the need to create temporary windows or bubbles to enable MDO. It is also probably useful to acknowledge it may be necessary to take action in a single domain to open up a window for other domains to enter the fight.

In sum, MDO is well structured in the sense of starting with a theory of the challenge and seeking to respond to that challenge with a theory of success. The articulation of the theory of the challenge, however, is poorly conceptualized in the sense of not clearly articulating the threat under the context of competition, which prevents the development of a clearly targeted strategy. Therefore, it is no surprise that the MDO competition strategy is vague and difficult to analyze. In the context of armed conflict, the challenge is somewhat more specific, but the MDO warfighting strategy is not much clearer than its competition strategy. MDO in armed conflict does not go much beyond the assertion that the joint force will be better at synergy, integration, and convergence than adversaries will be in stand-off and fracturing. A plausible way to create a more powerful theory of success might be to take into account adversary success and consider an approach that assumes the joint force will have to fight to establish the conditions necessary for operations across domains. In other words, perhaps MDO is something to fight for rather than take for granted.

---

69 Isabel Coles, Michael R. Gordon, and Ievgeniia Sivork, "Behind Ukraine's Russia Invasion: Secrecy, Speed and Electronic Jamming," *Wall Street Journal*, Updated Aug. 17, 2024, https://www.wsj.com/world/behind-ukraines-russia-invasion-secrecy-speed-and-electronic-jamming-188fcc22.

9

Attrition, Maneuver, and Lessons from Ukraine for Future Warfighting

Michael Kofman and Franz-Stefan Gady

Due to the scale of the war and technologies employed, Ukraine represents an important case study akin to the 1973 Arab-Israeli War in examining some key Western doctrinal concepts and preconceptions.[1] The 1973 Arab-Israeli War and its perceived lessons had a large impact on Western doctrinal development. Much of Western doctrine for conventional high-intensity warfare to this day is derived from the US Army's AirLand Battle doctrine, which was officially adopted in 1982 and heavily influenced by the Israeli experience of 1973. The doctrine has endured almost mythical status in US military circles to this day given its supposed role in the swift allied victory over the Iraqi military during the 1991 First Gulf War.

Crucially, AirLand Battle emphasized not only long-range precision firepower; but more importantly, it focused on flexible maneuver built around new technologies and mission command to defeat Soviet and Warsaw Pact forces on the Central Front in Europe in the event of war. The doctrine emphasized closer coordination between airpower and land forces in blunting an attack and degrading rear echelons. The maneuver oriented, and mission command component continues to inspire proponents of maneuver

---

1 On the impact of the Arab-Israeli War on US and NATO doctrinal development see, David Johnson and Zach Alessi-Friedlander, "An Alternative History of Airland Battle, Part I", *War on the Rocks*, 4 August 2022, https://warontherocks.com/2022/08/an-alternative-history-of-airland-battle-part-i/.

warfare, a set of principles that aim to break an enemy's will to resist through psychological and physical dislocation rather than just destruction, as the doctrinal ideal worth pursuing in future military campaigns.[2]

Creating a military force largely built around maneuver risks it being incapable of not just dealing with a prolonged military campaign dominated by attrition, but future great power war overall. The risk here is real. There is an established tradition in US and NATO military thinking that seeks to find ways out of attrition via technology and concept development built around an updated maneuver-centric version of the AirLand Battle doctrine. In the US military, this is evident with efforts to transform the individual service branches into organizations capable of conducting multidomain operations or combined arms operations, enabled by artificial intelligence (AI) and other emerging technological capabilities.[3] The US continues to emphasize joint force maneuver, stressing speed and simultaneity, believing that it can achieve significant cognitive and decision advantage in concepts like Joint All Domain Command and Control, or JADC2. There is merit in focusing on Western comparative advantages in force quality and integration, but these tend to be overemphasized to the detriment of other factors that also need be considered in war.

The prevailing idea among future force planners in the United States and NATO of multidomain operations is that emerging technological capabilities such as AI-enabled information superiority or decision dominance—the ability to obtain full situational awareness or a common operating picture while denying this information to an adversary thus enabling a faster decision-making cycle and as a corollary an accelerated operational pace—can compensate for quantitative disadvantages in terms of manpower, equipment, and munitions magazine depths.[4] The idea is that decision dominance can compensate for quantity and firepower, and that is can set the conditions to enable maneuver in the close in battle. A major premise of Multidomain Operations, or MDO, is that decision dominance or information superiority can contribute to severely degrading an adversary's combat power during a deep battle campaign fought with

2 See, Franz-Stefan Gady, Manoeuvre Versus Attrition in US Military Operations, Survival, 63:4, 27 July 2021, 131-148. DOI: 10.1080/00396338.2021.1956195.
3 For example, Congressional Research Service, 'Defense Primer: Army Multidomain Operations', 21 November 2022, https://crsreports.congress.gov/product/pdf/IF/IF11409.
4 Strategic Comments, 'The US Army's Multidomain Operations Doctrine,' Volume 28, Comment 29, Institute for International Strategic Studies, December 2022, https://www.iiss.org/publications/strategic-comments/2022/the-us-armys-multidomain-operations-doctrine.

long-range precision missiles and drones, supported by offensive operations in cyberspace and the electromagnetic spectrum.

Such an approach is envisioned to reduce attrition in terms of men and materiel and stands as the antithesis to a more direct attritional (or destruction-based) approach of warfighting. Maneuver has long been seen as the preferred approach to battles of annihilation, enveloping and displacing enemy forces, and avoiding the carnage of World War I. The key is to avoid a direct symmetric confrontation with your opponent in a close battle, because both sides would suffer huge losses in such an attritional contest without the means available to force a quick decision.[5] This view of war is naturally focused on the initial period, which is seen as intense, and decisive.

To be clear, maneuver and attrition are not discrete forms of warfare, or stratagems. War exists on a continuum, featuring maneuver, attrition, or positional approaches throughout. However, there is an important question on doctrinal emphasis, and different military communities' belief systems as to which proves more decisive versus which one should be avoided. The challenge is that wars between major powers are rarely resolved in the decisive period, and historically it has proven difficult to deal a knockout blow in one or two operations. Nonetheless, most military establishments orient their operational concepts towards that initial period of war, with much less thought given as to what follows. Consequently, there are prolonged attritional phases, and see-saw battles that nations find themselves unprepared for as the war eventually takes on a more positional character with prepared defenses, and technologies deployed that make maneuver much more difficult or costly. As the current Russia-Ukraine war exemplifies, the contest then evolves into both sides seeking to break out of the dynamic that they themselves established, via new technologies, tactics, and more importantly new forms of force employment. Yet the Russia-Ukraine war, and other wars, teach us that maneuver is unlikely to prove as decisive as its proponents' hope, and that it is essential to plan for high levels of attrition and protraction.

Maneuver Woes

The major issue with an approach based on maneuver as envisioned is that while it promises more immediate military success on its own, and lower

---

5   B.H. Liddel Hart, *Strategy* (Pentagon Press; Reprint Edition: 1 March 2012).

loss rates of men and materiel, it often fails to deliver.[6] Too much emphasis is placed on the cognitive dimensions of war, and precision strike as a key enabler of maneuver. Success in the initial period of war also depends on successful offensive actions in global domains, like space and cyberspace, which are essential for the complex operations that modern warfare entails, they too will serve as the initial battlegrounds for the contending parties.

This may prove effective if the context of the war is primarily an air or sea battle, and there is a vast asymmetry between opponents, as in Israel's strike campaign against Iran. The element of surprise, leadership decapitation, and on ground special forces strike teams may have proven decisive early on. Even when highly successful from an operational point of view, bombardment campaigns have a poor track record of attaining significant political war aims. Therefore, their application to wars between major powers is limited, since these tend to involve more expansive war aims, and often the war escalates even when geographically confined. In such cases the evidence is unconvincing that an effective deep battle campaign can compensate for decisively defeating an opponent in close battle, or that cognitive displacement can be easily achieved against a prepared opponent. This should be unsurprising as there often seems to be a false dichotomy between attrition versus a maneuver approach with the former in a doctrinal idea associated with a direct and the latter with a more indirect method of warfighting. A maneuver approach in this context is seen as a smart or more decisive way of fighting, presuming that a military force can swiftly achieve its objectives with a reduced need for firepower and also suffer less casualties.[7] Conversely an attritional approach is associated with an uninspired or slower way of fighting with a military force only incrementally able to achieve its objectives while taking heavy casualties.[8]

Presently, definitional issues remain between attrition, positional, and maneuver warfare, which this chapter is unlikely to resolve. As Christopher Denzel wrote in *USNI Proceedings*, many of the supposed drivers contained within maneuver warfare are in practice features of attritional or positional

---

6 For example, see discussions on German and allied offensives in 1918 and the Wehmacht's Operation Barbarossa in 1941 in chapters 11 and 12 in Cathal J. Nolan, "The Allure of Battle," *Oxford University Press*, New York, 2017.

7 Amos Fox, "Manoeuvre is Dead? Understanding the Conditions and Components of Warfighting," *RUSI Journal* Vol. 166, no. 6-7: 10-18. DOI: 10.1080/03071847.2022.2058601.

8 Amos Fox, "On Attrition: An Ontology of Warfare," *Military Review* (September-October 2024): 51-61.

approaches.[9] Those elements that focus on capacity and capability, and which can be measured or assessed, are most closely associated with attrition and positional forms of fighting, rather than the maneuverist school of thought that seeks to break an opponent's will to fight or to impose cognitive dilemmas. Unsurprisingly, the ideas in deep battle first belong best under attrition, destroying an opponents' capacity. However, a variation of this discussion, which emerged in 2023 in looking at the Ukrainian offensive, promulgated the theory that deep battle could achieve decisive effects without the need for close battle success if fires could be moved close enough to enemy logistics.[10]

This became its own spinoff theory of victory through long range fire control, which as will be discussed below, experience in Ukraine aptly disproves as a serviceable substitute for an unsuccessful ground offensive. Long-range precision strike inflicts attrition and can drive an opponent to reorganize in a manner that is less efficient, but it has not been demonstrated to confer fire control over an opponent's rear areas. In short, long range fire control and defeat through deep battle, when imagined as causal of victory on their own, are simply the latest iteration of wish-casting in thinking about war. This may be particularly the case due to the domain, whereas, for example, at sea it is likely easier to pursue sea denial and sea control over areas. In one sense, the maneuver versus attrition discussion is not just about warfighting approaches, but about the narratives surrounding outcomes. If a swift victory can be achieved it must be because of a maneuver-based approach, while defeat or slow progress is the result of an attritional or positional style of warfighting. Throughout military history, however, maneuver-based approaches were not necessarily more decisive or relatively less cost intensive in terms of men and materiel than their attritional or positional counterparts.[11]

---

9 https://www.usni.org/magazines/proceedings/2023/november/maneuver-warfare-just-operational-art.

10 Jan Kalberg, "Ukraine – Victory is Closer Than You Think," *CEPA*, 23 August 2023 https://cepa.org/article/ukraine-victory-is-closer-than-you-think/.

11 For example, Compare the maneuver-based approaches of the 1862 and 1863 campaigns of Army of North Virginia versus the attritional- and positional-based approaches of Army of the Potomac's 1864 campaign; compare the maneuver-based opening campaigns of the French, German, and Austro-Hungarian Armies of 1914 to the attritional and positional-based Allies' 100 Days Offensive of 1918; Compare the German *Wehrmacht's* maneuver-based campaigns of Operation Barbarossa in 1941 to the Western Allies' attritional-, positional-based Normandy campaign of 1944.

The Russia-Ukraine war has also brought to light discussions over the need for air superiority in reference to failures in maneuver warfare. Airpower packages mobility and firepower on the battlefield, it may be the best, when available, but it is hardly the only way to deliver firepower. Artillery-heavy armies, like those of Russia and Ukraine, had been bringing intense levels of firepower to the front line. Artillery had been the main form of fire support long before the development of airpower. If airpower is necessary for maneuver warfare to work, then in practice the necessary factor is firepower, and the attrition it inflicts. Fire and maneuver are essential elements of modern combat operations, but demanding air superiority is another way of saying maneuver requires a decisive, and sustained, fires advantage at operational depths. In contemporary cases, airpower has been used less to enable maneuver via close air support, and more to destroy an opponent's air defenses, then pulverize them such that the force was significantly degraded. There are very few recent cases of airpower being used, in conjunction with ground forces, to overcome a prepared defense. In short, if the contemporary conception of maneuver warfare is often one that requires a decisive fires advantage to succeed then it is largely borrowing attritional methods to attain its outcomes. It is also then easier to explain either Ukrainian or Russian successes in this war when neither side had air superiority, because they were able to inflict attrition, and attain fires superiority without having to rely on airpower.

Ukraine As A Case

Although the Russian-Ukrainian War began with an intense maneuver phase, for much of the conflict it has been an attritional contest dominated by tube, rocket artillery, and increasingly strike drones. What the course of the war points to is the underappreciated importance of attrition to enable maneuver, with emphasis on firepower advantage over combined arms integration. Positional fighting, taking advantage of terrain, and the geometry of the battlefield has been a key feature of the war.[12] Whether in defensive operations, or to more up fires to displace enemy forces, both sides have skillfully fought for position in order to attain tactical aims.

Maneuver warfare has consistently struggled to succeed against a prepared defense, especially when the opponent enjoys a favorable position.

---

12 Amos C. Fox, "A Solution Looking for a Problem: Illuminating Misconceptions in Maneuver Warfare Doctrine," *Armor Magazine* Vol. 129, no. 4 (2017): 18-22.

This is not an unexpected finding. Position and attrition have proven more deterministic of outcomes. In the West, these results have at times been hand waived away because neither side in the war could effectively scale combined arms operations, or joint force employment. Others emphasize the lack of air superiority, which is problematic if that is a prerequisite since it cannot be presumed in any near peer conflict. If air superiority can be established decisively early on then of course it can set critical conditions for success, but such an outcome would inherently imply a significant asymmetry between the forces involved at the outset. But this misses the more important, and historically consistent lesson of the Russia-Ukraine war, that these advantages in force employment are often insufficient to provide a decisive advantage absent other factors which are more reliably predictors of outcomes. This includes attaining fires superiority, having sufficient magazine depth, establishing a positive localized correlation of forces, and having key enablers in sufficient numbers. Force employment is critical in the ability of a military to scale more complex types of operations, particularly in the early period of war. However, by itself it may often prove insufficient of an advantage to overcome a lack of superiority or deficit in these other categories.

Ukraine's summer 2023 offensive is a useful example, because it was in part conceived with Western doctrinal precepts, and wargaming support.[13] The operation employed Western equipment, and long-range precision strike systems, both air and land based. It should have been an example of what a motivated force, with formations trained in the West, using Western equipment, and to an extent concepts, could achieve in a set-piece battle. Instead, the summer offensive began with a failed breaching effort against a prepared defensive line. The initial attack failed due to a combination of planning, force employment issues, shortage of enablers, and most importantly a lack of clear advantage relative to a well-prepared defense.[14] This was not just a function of lacking air superiority, or specific strike capabilities. Ukrainian force structure, and doctrine, was not designed around attainment of air superiority or the need for substantial air delivered fires. Neither was Russia's for that matter, and many of the challenges posed

---

13 Adam Entous, "The Partnership: The Secret History of the War in Ukraine," *The New York Times*, 29 March 2025, https://www.nytimes.com/interactive/2025/03/29/world/europe/us-ukraine-military-war-wiesbaden.html.
14 "In Ukraine, A War of Incremental Gains as Counteroffensive Stalls," *Washington Post*, December 4, 2023, available at: https://www.washingtonpost.com/world/2023/12/04/ukraine-counteroffensive-stalled-russia-war-defenses/.

by Russia's defenses did not have obvious airpower solutions. The resulting battle was one largely prosecuted by dismounted infantry, using small group tactics, with artillery seeking to establish a fires advantage.[15] Ukraine thus preserved equipment, but at the price of infantry, and ammunition provided by the West. The deep battle was hardly decisive, and fire control via long-range precision strike did not exist in practice. The persistent ISR, magazine depth, and other requirements to establish something akin to fire control at range was simply not possible to achieve and maintain. This would have required both air superiority well behind the forward line of troops (FLOT), persistent ISR, and the munitions depth to consistently deny the rear to the adversary.

Even in very close ranges, pervasive ISR and strike capabilities yielded intermittent denials of mobility, rather than fire control. Much of the battle was fought within the same 10-15 kilometer range over the course of the offensive. Most battles in Ukraine have played out with both sides being able to range each other's ground lines of communication, command and control, and forward logistics, with the lines often separated by a few kilometers from each other. With rare exceptions, this did not allow either side to 'control' the engagement via fires but did instead lend itself to prolonged attritional warfare. When nearly enveloped, with lines of communications disrupted by enemy indirect fires, both sides have been able to continue fighting and hold on for weeks or months in battles like Bakhmut, Robotyne, or Avdiivka. Presently, drones offer a degree of fire control often at 15-20 kilometers behind the FLOT, with persistent ISR presence now often at greater depths of 50 kilometers or more. While this effectively denies mobility, a prepared defense is still manageable, and neither side has been able to achieve an operationally significant breakthrough even when enjoying a degree of fire control over the opponent upwards of 20 kilometers in depth. This points to both forces' ability to adapt, and continue operating under such conditions, with low force density still able to deny maneuver. To put it simply, prepared defense backed by mass precision has significantly lowered the forces required to deny maneuver in the close in fight and as a consequence many of the deep battle or precision-strike based approaches have reduced efficacy. The assaulting force cannot mass without suffering losses and cannot reach the enemy line with sufficient combat power intact, and therefore it cannot complete its requisite combat tasks.

---

15 See Jack Walting and Nick Reynolds, *Stormbreak: Fighting Through Russia's Defenses in Ukraine's 2023 Offensive* (London: Royal United Services Institute, 2024).

Consequently, while maneuver is not dead, it is going to be much harder to achieve in the future and with greater attendant costs. This raises questions about the continued trend in Western defense communities to invest in complex concepts, and forms of force employment, over mass, and munitions depth. When such investments fail to deliver, a force can face munitions hunger, and can struggle, especially if it is brittle. Maneuver faces several constraints in the Ukrainian battlespace making it an unreliable instrument at best for Ukraine to escape high rates of attrition. Attrition, primarily reliant on ground-based fires that include tube and rocket artillery, has been and likely will continue to be the key enabler to maneuver in Ukraine. Attrition is more reliable, and easily attainable. The force quality required to execute combined arms operations is often difficult to maintain and reconstitute later in a conventional war. Hence there is a challenge in that the "cult of the maneuver" can mislead militaries into underinvesting in the right areas, or building sufficient hedge into their force.

Ukraine's experience illustrates that much of the current discourse does not center on the character of war between major or great powers, but more accurately on the early maneuver phase of such a war. This sort of thinking is deeply reminiscent of the conceptual preferences found among Western European armies ahead of World War I. Moreover, modern armies do not have the mobilization capacity or reserves available to them and will prove far more difficult to regenerate. The current approach leads to armies that are brittle, defense industrial capacity geared towards peacetime and difficult to mobilize, and little thinking about the natural attritional phases that are likely to be found in a war. Much the same can be said for the interaction between force quality and complex concepts that demand a great deal to go right in force employment. These make little room for the reality that synchronization of effort at scale is hard, the better part of the force is likely to be lost early on in a war due to attrition, and the reconstituted components will struggle to execute something that heavily depends on integration and simultaneity.

Looking at Russian and Ukrainian losses in the early part of the war, it seems clear that many European armies would have been off the battlefield within weeks unable to recover from loss of manpower, materiel, and ammunition. However, any force would have to ask whether it had the depth and reconstitution capacity to replace those kinds of losses, and what it might look like several months into a war with this rate of attrition. What sort of operations might it be able to execute at that point? They too might be attrition driven and sequenced.

Furthermore, Ukraine's principal problem was not an inability to conduct combined arms maneuver. The matter lay in Ukraine's inability to scale offensive action, a deficit of capable assault units in the force, low availability of enablers to support a breach, and insufficient firepower concentration to create a decisive fires advantage.[16] The Ukrainian offensive was an attempt at breakthrough against a well-prepared defense without having set the conditions via attrition, or established a decisive advantage in firepower. While it is true the new brigades trained by Western countries struggled to coordinate combat arms, this was ancillary to, rather than the chief reason for why the offensive proved unsuccessful.[17] Therefore the incorrect interpretation is that Ukrainian forces could not succeed because they could not fight like a Western military, or that fighting like a Western military doctrinally requires air superiority, and absent that condition there is no way to succeed. Instead, Ukraine made progress by pursuing attrition, fighting for position and for relative fires advantage, which reduced overall losses, and made Russia pay a high price to defend. On the basis of just tactical performance, Ukraine's offensive did not go particularly poorly, and the attrition ratio was not highly unfavorable to the attacking side. Conversely, the maneuver centered concept of operations proved unworkable and unachievable given the conditions and the forces involved.

Lessons

In drawing lessons from Ukraine some Western military thinkers and commentators, steeped in Western military doctrine, were quick to identify an indirect approach by the Ukrainian armed forces, referred to as a strategy of "corrosion" that aims to erode "the Russian physical, moral and intellectual capacity to fight" by "attacking the weakest physical support systems of an army in the field—communications networks, logistic supply routes, rear areas, artillery and senior commanders in their command posts".[18] According

---

16 Michael Kofman and Rob Lee, "Perseverance and Adaptation: Ukraine's Counteroffensive at Three Months," *War on the Rocks*, September 4, 2023, available at: https://warontherocks.com/2023/09/perseverance-and-adaptation-ukraines-counteroffensive-at-three-months/.

17 "Franz-Stefan Gady and Michael Kofman on What Ukraine Must Do to Break Through Russian Defenses," *Economist*, July 28, 2023, available at: https://www.economist.com/by-invitation/2023/07/28/franz-stefan-gady-and-michael-kofman-on-what-ukraine-must-do-to-break-through-russian-defences.

18 Mick Ryan, 'The Strategy of Corrosion', *Twitter* thread, 1 August 2022, https://twitter.com/WarintheFuture/status/1553973368809918464?s=20&t=Qq4fAIBG_ZZQyM4_NIqdbA.

to these thinkers, this is primarily done via two means: long-range precision fires and maneuvers, exemplified in the successful Ukrainian offensives in Kharkiv and Kherson over the summer months and the fall of 2022. However, in both instances attrition in the close battle set the conditions for success. Indeed, the Kherson offensive, while seen by commentators as exemplifying an indirect approach, was a months-long, firepower-intense grind through Russian layered defenses that included extensive minefields with Ukrainian forces taking heavy casualties.

The Kherson operation was overall a success, in that it displaced Russian forces, but a maneuver failure. The initial attack failed, and after replacing the commander a month later, the second assault made modest progress. Still, Russian forces were able to withdraw in detail, and with their equipment intact, despite a significant long-range fires advantage on the Ukrainian side. Were it not for the Dnipro River, and the fact that Russian forces had to supply themselves across one bridge and a ferry network, the Russian airborne might have continued holding that terrain for much longer. The geometry of the battlefield was deeply unfavorable to Russian forces, yet the operation lasted more than two months until the Russian withdrawal. The net result was a success, but less so for advocates of long-range fire control, given this should have been the easiest battlefield case for long-range fires to establish fire control and deny a force the ability to retrograde.

The Kharkiv offensive took advantage of a thinly manned Russian line, with many units woefully understaffed, and key forward positions held by irregulars. The force was depleted and variegated, a patchwork lacking established defensive lines. Essentially, prior attrition leading up to the offensive resulted in Russian forces having a structural manpower deficit and large gaps in the lines. The low force to terrain ratio, lack of prepared defenses, and lack of cohesion was a leading factor in why Kharkiv proved more successful than the Ukrainian armed forces expected. Russian units detected the Ukrainian buildup but could not act upon it. The Kharkiv breakthrough was also preceded by intensive supporting fires at the beginning of a speedy advance.[19] Yet Kharkiv looked to be *sui generis*, a case where the conditions affecting the Russian military, which in fairness were imposed by Ukraine, were more deterministic of outcomes than how the attack was conducted. The results were then routinely misinterpreted, leading to outsized expectations

---

19 For details on Kharkiv and Kherson offensives see, Franz-Stefan Gady & Michael Kofman, "Ukraine Strategy of Attrition," *Survival*, 65:2, 7-22, 28 March 2023 https://www.tandfonline.com/doi/full/10.1080/00396338.2023.2193092.

for what the summer offensive could achieve, given that Ukrainian forces would face a much stronger defend than in Kherson and with a high density of forces relative to the terrain holding the lines. Russian units were also not in an unfavorable geography, having fortified the high ground with cities as key anchor points, and with ground lines of communication that would prove difficult to sever.

Another misinterpretation followed the lack of fast progress in the close battle during Ukraine's military operations in 2023, proponents of Ukraine's indirect approach or strategy of corrosion doubled down on the importance of the deep battle focused on the destruction or disruption of Russian command and control and logistical centers deep behind the frontline with a special focus on Russian military assets on the Crimea peninsula.[20] The main idea being that success in the deep battle will ultimately indirectly erode Russian combat power on the frontline and reduce the need for a close fight as Russian forces would eventually be forced to withdraw from their defensive positions due to a collapse in logistics and an inability to deploy reserves. The more extreme of these viewpoints argue that just by destroying the Kerch strait bridge Ukraine can cause Russian lines to collapse across the south and put in jeopardy Russian forces in Crimea.[21] Russia maintained the peninsula for four years before building the bridge. The bridge had been previously struck severely degrading rail or road traffic, and this did not lead to noticeable supply effects on Russian forces. Ukraine's occupied south represents a sizable ground corridor for resupply of Russian forces there (this is what the 2023 offensive was meant to sever). Destroying the bridge would have hampered Russian resupply via that corridor, but hardly a magic bullet, as seen during periods when its operation was degraded.

In the case above, expectations for what can be accomplished with long-range missiles proved grossly beyond what was possible. Systematically dismantling the Russian military's support in rear areas, and its logistical base in Crimea proved beyond the means of these capabilities in the volumes available. There is also a general conceptual contradiction in this thinking: If it is indeed possible to dismantle Russian supply lines and reserves with a deep battle campaign alone there would be no need for a Ukrainian

---

20 Jake Epstein and Chris Panella, 'No Russian ammo depots would be safe in Ukraine if the US provided longer-range HIMARS missiles and better drones, retired US general says,' *Business Insider*, 6 July 2023, https://www.businessinsider.com/atacms-gray-eagle-drones-kyiv-hit-russian-targets-inside-ukraine-2023-7.

21 Ben Hodges, et. al., "Putin's Weak Link to Crimea," *Foreign Affairs*, 5 December 2023. Available at: https://www.foreignaffairs.com/united-states/putins-weak-link-crimea.

counteroffensive in the first place. The deep battle sets the conditions to be exploited in the close battle, but in this specific case neither proved successful. What operations in Ukraine have consistently illustrated is that long range strikes are not a substitute for a major ground offensive, nor are they likely to succeed in the absence of one. Missiles and drones do not take, or hold, terrain. Ukraine may well conduct an expanded strike campaign against Russian forces in Crimea, or the links to the peninsula. This will undoubtedly create burdens for the Russian military, but it is not likely to collapse Russian forces without a close battle component, or a major ground offensive.

Maneuver Deniers

This section lays out the difficulties of maneuver in the close fight by identifying a select number of so-called maneuver deniers. The principal challenge for the close in fight is the combination of traditional prepared defense backed by mass precision and pervasive presence of ISR. Mass precision is a combination of high-fidelity sensors, strike munitions, and networks that enable effective integration in real time. Low-cost, expendable capabilities can effectively find and fix the target, plus engage it. Whereas previously in precision strike munitions range or performance exceeded one's ability to find and fix the target, today this is no longer an issue. Scaled production such that strike munitions exceed the number of troops and equipment on the battlefield turns precision into a mass, or economy of scale effort, which makes a difference in kind from previous improvements to said capabilities. This is not a revolution, but it is an evolution in the precision-strike regime.

Mass precision-guided munitions in the form of different types of remotely operated drones usually fitted with high-explosive anti-tank munitions have emerged as one of the principal maneuver deniers on the battlefield in Ukraine. Strike drones are often paired with munitions dropping (bomber) drones, heavy-multirotor drones, and various reconnaissance drones. In addition, forces employ loitering munitions, and various types of cheap one-way attack drones against positions, fortifications, and vehicles. Most importantly, a panoply of drones are now available for battlefield surveillance during the day or night, at 0-30 kilometer, and more complex platforms able to cover 30-100 kilometer. The density of ISR is such that massing for an attack within 15 kilometers is now exceedingly difficult, and non-line of sight munitions can begin engaging a force long before it makes contact with the FLOT. Traditional prepared defenses canalize the attack, making it difficult to switch from column march to an assault formation. Enablers are easily destroyed, often stranding the attacking vehicles in minefields and rendering them vulnerable.

Ukraine's relative weakness in long-range precision fires can be traced back to three major factors. First, Ukraine lacks the necessary magazine depth of precision-guided munitions, and is throttled in the few means of deliver available. A systematic long-range strike campaign would require not hundreds but thousands of missiles over the course of a year. Ukraine was only, however, supplied with several hundred air-launched SCALP EG/ Storm Shadow land-attack cruise missiles by the UK and France in 2023 the majority of which has likely already been used up with limited capacity in Europe for resupply in the short to medium term.[22] Also, Germany remains reluctant to supply several hundred air-launched land attack cruise missiles which would help replenish Ukraine's long-range strike arsenal. Ukraine was initially also limited to its Su-24 tactical bombers as a means of delivery, and that constrained salvo size considerably, with often no more than 12 missiles being fired in a sortie. When stacked against intercept rates, heavily defended targets would likely see a smaller number of Air-Launched Cruises Missiles (ALCMs) getting through.

Additionally, the US supplied ground-launched ballistic missiles, variants of the Army Tactical Missile System (ATACMS) in relatively low quantities.[23] The Ground Launched Small Diameter Bomb (GLSDB), a weapon system that can be fired from High Mobility Artillery Rocket Systems (HIMARS) launchers with an estimated maximum range of around 165 kilometers, have thus not worked.[24] This has left the family of Guided Multiple Launch Rocket Systems (GMRLS) missiles for HIMARS as the workforce of Ukraine's precision strike from 30-85km. To a much lesser extent the Ukrainian air force has been able to employ stand-off precision guided munitions such as the SDB, JDAM, Hammer, family of guided aerial bombs.

---

22 Fabian Hoffman, "If Germany Doesn't Deliver Quickly, Ukraine Will Run Out of Cruise Missiles," *Der Spiegel,* August 11, 2023, available at: https://www.spiegel.de/politik/deutschland/marschflugkoerper-fuer-die-ukraine-warum-deutschland-liefern-muss-a-3c8bd74d-b8f2-4eb5-9ea0-7c5d5e4b07da.

23 John Hardie, 'Ukraine Now Has ATACMS-But Will Soon Need More,' *Long War Journal,* October 19, 2023, available at: https://www.longwarjournal.org/archives/2023/10/ukraine-now-has-atacms-but-will-soon-need-more.php.

24 *Boeing Backgrounder*, 'Ground Launched Small Diameter Bomb,' October 2015, https://www.boeing.co.kr/resources/ko_KR/Seoul-International/2015/GLSDB.pdf.

Other means for long-range strike, such as Ukrainian-made drones, have been available in larger quantities for the past year with Ukraine, however constraints remain in scaling up production and the effectiveness of their strike potential.

Second, to conduct a systematic deep battle campaign Ukraine would not only need a larger arsenal of long-range precision-guided munitions, but also require persistent intelligence, surveillance and reconnaissance (ISR) capabilities (a mix of satellites and drones to help identify and track Russian targets). Based on field research, Ukraine has made substantial progress in this area with the deployment of new indigenous ISR drones and the more effective fusion and dissemination of data. However, while Ukrainian ISR is relatively persistent for up to 20-30 kilometers behind the main Russian frontline, ISR becomes spottier in the rear where larger platforms would need to be deployed given that neither Ukraine, nor Russia have control of the skies. This makes any form of dynamic targeting, that is hitting a moving target, difficult, while stationary targets are dispersed, camouflaged and protected by electronic countermeasures. Even assuming, hypothetically, that such coordinates are being passed in relatively close to real time from Western sources, the time necessary to complete the kill chain, and engage the target, is still likely to be significant. For high value assets that displace immediately after firing, like air defense, this is often insufficient. Hence Ukraine is often reliant on its own tactical ISR to identify these targets even if their location is being acquired from Western sources. Commercially available satellite sources tend not to provide real time ISR sufficient for dynamic targeting, or the other types of information needed to properly route a strike package towards fixed targets.

Part of the challenge observed is that a prepared opponent can disperse logistics, command and control, while hardening key nodes. Counters are also available and become increasingly effective over time. Capabilities have their greatest impact when first introduced onto the battlefield at scale. The challenge for Western militaries in the case of Russia, and other adversaries, is that this form of warfare has already been demonstrated and applied. The Russia-Ukraine War illustrates that in a matter of months an adversary can adapt quickly, alter how they organize sustainment or C2, and continue fighting. Subsequently, it becomes much more difficult to affect the rear with precision fires, requiring more munitions than many Western militaries doctrinally anticipate. Destroying bridges and disrupting ground lines of communication can be effective in preventing the enemy force from reacting in a timely and cohesive manner. However, this is most useful once a breach has succeeded, and in a maneuver war, any near-peer adversary can do much the same to Western forces. In the close fight, a prepared defense, backed by mass precision, can effectively deny maneuver to the attacking force with a fair low density of forces at the initial contact line. Units that breakthrough can find themselves isolated and unable to exploit those breaches. Temporary gaps in ISR coverage can then quickly close, with precision munitions denying sustainment or follow on forces from advancing.

Surprise is still possible and can be effective. During the Kursk 2024 operation Ukrainian forces were able to achieve tactical surprise, breaching two lines of defense. However, Russian ISR coverage was quite minimal, and the lines were held by conscripts or lower quality supporting forces. Ukrainian forces were able to leverage mobility and quickly bypass defending formations via roads. However, this breach quickly ran out of momentum, and by the third day leading formations found themselves isolated without secure lines of communication. Kursk illustrates that with the right pieces in place maneuver is possible even on a static battlefield, but this operation could not have been replicated against the main Russian lines where the force density and ISR coverage was far higher. Most of the advance, beyond the initial breach, was not a combined arms operation but a series of thunder runs by highly mobile air assault units. Mobility can yield a major temporally advantage, but as Kursk illustrated once those forces run into organized defense the momentum is quickly halted. Similarly in 2024 Russian forces achieved tactical surprise at Kharkiv, advancing dismounted infantry units through tree lines and catching the initial Ukrainian defenders by surprise. The Armed Forces of Ukraine (AFU) rushed units into meeting engagements, stabilizing the situation within a week. Neither offensive resulted in operational success, nor significantly changed the dynamic of the war. They illustrated that at the tactical level maneuver was still possible, but this was heavily context dependent, and at a relatively small scale.

Doctrinal Implications

As Russia's war of aggression against Ukraine continues along in its third year, the fighting along the frontlines has steadily become positional in character. The lack of substantial territorial gains and decisive breakthrough by either side has led to a sense of deadlock. Yet neither side accepts this as a durable or stable situation, seeking to break out from this equation to attain offensive goals. Consequently, this period of the war is not uncommon from that witnessed in other conventional wars, which saw periods of positional fighting, and efforts to establish a decisive advantage, perhaps restoring the ability to succeed in maneuver warfare.

The war in Ukraine shows that high rates of attrition of personnel and materiel remain a main characteristic of high-intensity warfighting as they have been throughout military history, regardless of the warfighting approach or method chosen. Western force planners tend to ignore this salient feature of warfare at their own peril. Attrition, consequently, needs to be factored into any deliberations on future military doctrine, and expectations of success in combat operations.

Attrition on the modern battlefield can be both a denier and enabler of maneuver possibilities. It is therefore up to a military force and its commanders to exploit attrition to its own advantage. In the case of Ukraine, this is best done by focusing on establishing a favorable loss ratio vis-à-vis Russia in the close battle and in the deep battle. This may seem like a straightforward suggestion, but the way to establish this advantage is best done via quantitative and qualitative fires superiority, not maneuver, or the pursuit of cognitive advantage.

The effective exploitation of attrition in the close fight may be the best chance to enable maneuver on the modern battlefield. The focus in Western military thinking about future warfare has by and large focused on the deep battle as a key maneuver enabler. This includes ongoing conceptual work on the more effective use of precision-guided munitions, and the integration, synchronization, and convergence of kinetic and non-kinetic effects across multiple warfighting domains. Our field research, however, suggests that in the case of Ukraine the deep battle cannot substitute for the close battle, especially when poorly synchronized. Even with more effective coordination mechanisms in place, practice has shown that establishing fire control beyond tactical ranges is much more challenging than theory would lead one to believe. As such, some rebalancing in doctrinal discussions in the West may be in order. The debate will undoubtedly continue: did it not work for Ukraine because they couldn't execute well, or is this another example that the practice will come up short of expectations?

The deep battle is no end in itself, but a means to an end, that is to effectively conduct and ultimately win the close fight. In Western militaries targeting remains heavily dependent on space, and exquisite sensors, as opposed to numerous expendable ones. This remains a limiting factor since adversaries like Russia and China will seek to deny space-based sensing, contest it, or compromise it in the initial period of war. Here Ukraine is not a useful guide because of the peculiar context of the conflict, which may not be representative of how the next conventional war will unfold.

Consequently, future military doctrine may do well to emphasize that multidomain or joint all-domain operations as envisioned by the US and NATO can create windows of opportunity for military success, but they are no magic bullet, especially since they are premised on Western militaries retaining decision dominance—an informational advantage acquired when a force obtains "full situational awareness" or a "common operating picture," including knowledge about the composition and structure of adversary forces, faster than the adversary. The problem with the "window of opportunity" thesis is that this speaks primarily to an operation, and even if successful the initial operation is unlikely to prove decisive. Therefore, even if proven right, Western doctrinal thinking needs to look considerably beyond this initial warfighting phase.

Multidomain doctrine, however, does not discuss what actions a force might take if decision dominance cannot be achieved. The lack of operational guidance in this area could cause paralysis at critical moments and could cause Western militaries to lose the operational initiative in a future fight. Especially, given ongoing trends toward dispersing forces and spreading them out over larger geographical areas to avoid detection.

Ukraine shows that western militaries should be cautious in pursuing multidomain concepts built around the deep battle and need to especially rebalance thinking that leans on resource-heavy tech-driven efforts to attempt to establish effective fire control beyond tactical ranges. Instead, examining in more detail how Western militaries can sustain the close battle by embracing a new destruction-maneuver approach could prove beneficial. Such an approach entails more strongly emphasizing the symbiotic relationship between attrition and maneuver, with the former setting the conditions for the successful exploitation of the latter. As the conflict in Ukraine has shown, this is first and foremost accomplished by prevailing and sustaining a favorable attrition ratio of personnel and materiel in the close fight through the application of fire power. This also means balancing Western forces from the current over-emphasis on capabilities that are *few and complex* to the *many but simple*. Having a suitable number of cheap expendable strike systems, and ISR systems, would help enable this approach and reduce the presently observed brittleness. That said, it would also require deliberate changes in force structure allocation, which will prove a sticking point. To employ these technologies Western maneuver formations need drone companies, or battalions, along with supporting electronic warfare units. This means that the forces need to undergo expansion, or make trade-offs in force structure, getting rid of something to add something else.

More effectively fighting the close battle has direct consequences for future acquisitions and force structure. On the fires side, it includes a stronger emphasis on tube and rocket artillery. It also entails the integration of large quantities of loitering munitions and short-range strikes drones. Large quantities of man-portable anti-tank guided munitions are also required.

On the enablers side, it requires the acquisition of large quantities of ground-based electronic warfare capabilities including short-range jammers that are either man-portable or mountable on individual vehicles. Mobile, short- and medium -range air defenses are another priority. Additionally, it requires a larger variety of breaching equipment. This is one of the areas where the introduction of scalable robotic systems needs to be emphasized.

More effectively fighting the close battle also demands a doctrinal emphasis on establishing an advantage in firepower and being able to then effectively scale offensive operations to exploit the opponent's degraded capacity to resist. For many Western militaries scaling is a bigger issue than combined arms integration, since they regularly train to do the latter on a small scale and the former not at all. However, small scale force employment is not likely to survive levels of attrition seen in a sustained conventional war, whereas a force that can't scale is one that ultimately depends on supporting a force that can. Discussions around accelerated kill-chains and long-range precision strikes have led to a deemphasis on the importance of the effective synchronization and integration of firepower and movement in close combat. A special emphasis here needs to be put on timing and tempo. That is the ability of a dispersed force to quickly aggregate individual capabilities for an attack, execute the assault, and then quickly disperse, while retaining the capability to exploit a breach and fight off counterattacks. This in turn requires effective means to exercise command and control at various echelons of the force.

In conclusion, the ongoing Russian war of aggression in Ukraine has exposed the limitations of maneuver-centric doctrines with multidomain operations and the enduring importance of attrition in high-intensity conflict. While Western military thinking has long favored maneuver, enabled by technological superiority and information dominance, Ukraine's experience demonstrates that attrition—achieved through firepower, positional advantage, and sustained close battle—remains a primary determinant of battlefield outcomes. Efforts to substitute deep battle and long-range precision strikes for success in the close fight have consistently fallen short, as neither side in Ukraine has been able to achieve decisive breakthroughs without first degrading the opponent through attrition. The conflict underscores the need for Western militaries to rebalance doctrine and force structure, emphasizing mass, munitions depth, and scalable offensive operations over complex, brittle concepts that depend on fleeting technological advantages. Ultimately, future warfighting will require a symbiotic approach, where attrition sets the conditions for maneuver, and operational success depends on the ability to sustain and exploit favorable loss ratios in protracted military contests.

PART 4

INTERNATIONAL AND ALLIANCE PERSPECTIVES

10

The Expanding Role of Ground Forces in Multidomain Operations across East Asia and the Indo-Pacific

Andre Luiz Viana Cruz de Carvalho and Sandro Texieira Moita[1]

Introduction

The growing centrality of the Indo-Pacific and East Asia to international security imposes unique strategic challenges and reflects a sustained focus on the region by the United States, underscored by its strategic "Pivot to Asia."[2] As then-Secretary of State Hillary Clinton outlined in her influential article, "America's Pacific Century," the United States recognized the Indo-Pacific as critical to its long-term interests and security and pledged to shift its diplomatic, economic, and military resources toward ensuring stability in the region.[3] This pivot has led to a recalibration of American defense priorities and greater operational demands in East Asia and the Indo-Pacific.

---

1 The authors would like to thank Dr. Andrea Resende for her insightful suggestions and for doing an in-depth and thorough review of this chapter.
2 Hillary Clinton, "America's Pacific Century," *Foreign Policy*, Oct. 11, 2011, https://foreignpolicy.com/2011/10/11/americas-pacific-century/.
3 Clinton, America's Pacific Century, 2011.

Multidomain Operations (MDO) have emerged as a response to a threat environment that transcends the limits of any single operational domain, in a region marked by intense territorial disputes and competition among great powers such as the United States and China, and more asymmetric threats such as North Korea (DPRK). In East Asia, the Taiwan Strait and Korean Peninsula demand operational flexibility and rapid responses in contexts with a high risk of conflict. Meanwhile, in the Indo-Pacific, it has become essential to secure maritime communication lines and strategic positions[4] and integrate ground, naval, and air forces to sustain operations in contested areas like the South China Sea (SCS).

MDO reflects a paradigm shift towards the integration of warfighting domains to overcome complex obstacles, such as anti-access and area denial (A2AD) networks. A2AD strategies are widely used by China,[5] which deploys an integrated arsenal of anti-air defenses, long-range missiles, and electronic warfare capabilities to limit access and movement of its adversary forces - namely the United States and its allies. Such networks, particularly in East Asia's coastal regions and along the Indo-Pacific sea routes, represent a major challenge for conventional operations. Overcoming these access barriers demands a more integrated posture and continuous adaptation by the ground forces to coordinate with the other branches.

The ground forces' role must now be expanded beyond the traditional functions of logistical sustainment and rear area protection. In the context of MDO, ground forces proactively protect critical infrastructure, deter adversary advances, and ensure freedom of action for allied forces. In East Asia, the ground forces' geographic proximity to disputed areas allows them to occupy and secure strategic terrain, enhancing deterrence and strengthening military presence in sensitive regions. In the Indo-Pacific— where operations on small islands and the protection of logistical routes are essential—ground forces work directly with air and naval forces, using new capabilities to respond to threats and ensure flexibility in the maritime environment.

To meet these challenges, ground forces have adapted their doctrines and incorporated technologies that enable them to operate in a coordinated

---

4 Andrea Luiza Fontes Resende de Souza, 'United States Overseas Basing System in the Indo-pacific: Sea Power and Competition for Strategic Access.' (PhD diss., Pontifical Catholic University of Minas Gerais, 2022).

5 "China's Anti-Access Area Denial," *Missile Defense Advocacy Alliance*, accessed October 27, 2024, https://missiledefenseadvocacy.org/missile-threat-and-proliferation/todays-missile-threat/china/china-anti-access-area-denial.

manner with other domains. This involves developing new tactics for rapid mobility, decentralized operations, and command-and-control systems that facilitate effective communication even in contested environments. Disruptive technologies such as mobile missile launch systems, surveillance drones, autonomous vehicles, and electronic warfare capabilities are also increasingly integrated into the repertoire of ground forces. These innovations not only expand their reach and impact but also enable them to confront denial strategies to support joint and combined operations in the region.

In this sense, this chapter highlights the critical role of ground forces in the complex security environment of East Asia and the Indo-Pacific. It explores how evolving doctrinal concepts and technological advances are reshaping land power to meet emerging multidomain operational challenges, particularly in the context of A2AD strategies. The chapter is organized into three main sections: first, we analyze the regional security dynamics, the geopolitical context and their impact on force posture; second, we discuss doctrinal and technological innovations driving ground force modernization; finally, we assess the case of US and allied forces on the practical application of these concepts.

Geopolitical Context and Regional Challenges

The Asian continent has long held a position of geopolitical importance. Historically, East Asia was recognized as a critical geostrategic location well before the concept of the broader Indo-Pacific region took shape. Today, the Indo-Pacific region is a center of global economic activity,[6] characterized by heightened military tensions. In this context, the US pivoted to Asia and the development of the AirSea Battle[7] strategic concept helped to justify the region's prominence.. There is an argument of a "pivot before the pivot", signaling the region's enduring importance on the country's strategic horizons with the maintenance of prepositioned forces in Japan and South Korea during the post-Cold War.[8]

Ground forces are critical to MDO in the complex, contested Indo-Pacific, as they secure strategic land masses and enable rapid, integrated responses

6 Ashok Kapur. *Geopolitics and the Indo-Pacific Region* (London & New York: Routledge, 2019).
7 US Department of Defense, Air-Sea Battle: Service Collaboration to Address Anti-Access & Area Denial Challenges (Washington, D.C.: US Department of Defense, May 2013), ii.
8 Robert D. Blackwill and Richard Fontaine, *Lost Decade: the US Pivot to Asia and the Rise of Chinese Power* (Oxford: Oxford University Press, 2024).

alongside joint forces.[9] Ground forces must adapt to an evolving balance of power shaped by omnipresent weapons and omniscient surveillance.[10] However the exact strategic demands and operational requirements placed on and facing the ground forces vary, depending on the specific flashpoint; a tailored approach must be adopted to support air and maritime operations and reinforce presence and counteract adversarial tactics (see Table 10.1).[11]

Taiwan

Taiwan's strategic importance stems from its geographic position. Geographically, Taiwan sits along the first island chain, serving as a "natural barrier" that restricts China's access to the broader Pacific Ocean and offers a potential outpost for power projection into critical areas such as the South and East China Seas.[12] Control of Taiwan would grant any regional power substantial influence over these waterways, impacting local security, global trade, and energy flows.[13] Symbolically, Taiwan embodies competing ideological models and national identities in East Asia, especially since it represents a profound ideological and political challenge to the Chinese Communist Party, which views any assertion of Taiwanese independence as a fundamental threat to its sovereignty and national unity.[14] Ground forces supporting Taiwan would not only navigate complex logistics but also contribute to broader defense and deterrence strategies.[15]

---

9   Michael E. Lynch and Brennan Deveraux, 'Landpower, Homeland Defense, and Defending Forward in US Indo-Pacific Command', *Journal of Indo-Pacific Affairs* Vol. 7, No. 4 (2024).

10 Collin Fox, 'The Porcupine in No Man's Sea: Arming Taiwan for Sea Denial,' *Center for International Maritime Security (CIMSEC)*, Aug. 4, 2021, https://cimsec.org/the-porcupine-in-no-mans-sea-arming-taiwan-for-sea-denial/.

11 Robert B. Brown, 'The Indo-Asia Pacific and the Multidomain Battle Concept', *The Military Review* Vol. 97, No. 5 (2017).

12 Xiaobing Li, *The History of Taiwan* (New York: Bloomsbury Publishing USA, 2019).

13 Mark F. Cancian, Matthew Cancian, and Eric Heginbotham, 'The First Battle of the Next War: Wargaming a Chinese Invasion of Taiwan' (Washington, DC: Center for Strategic and International Studies, 2023).

14 Kerry Brown and Kalley Wu Tzu Hui, *The Trouble with Taiwan: History, the United States and a Rising China* (New York: Bloomsbury Publishing, 2019).

15 Cancian, Cancian, and Heginbotham, 'The First Battle of the Next War,' 2023; Xiaobing Li, *The Cold War in East Asia* (London & New York: Routledge, 2017).

Flashpoint	Strategic Significance	Role of Ground Forces
Taiwan	Key position along the First Island Chain; critical for controlling access to the Pacific and regional influence	Support deterrence and defense strategies; manage complex logistics; enable integration with joint forces
Korean Peninsula	Long-standing conflict with North Korea; nuclear threat and need for constant readiness	Maintain forward presence near DMZ; enable rapid joint responses; reinforce South Korean defense
Yellow Sea	Overlapping territorial claims; frequent naval incidents; proximity to North Korea	Ensure cross-domain coordination; support coastal defense; respond to missile threats and asymmetric tactics
Senkaku Islands	Disputed territory with China; near major shipping lanes; important to US-Japan alliance	Enable rapid response via Okinawa; support logistics and infrastructure; signal allied commitment
South China Sea	Strategic waterway for global trade; subject to multi-claimant disputes and Chinese militarization	Secure infrastructure and sustainment nodes; reinforce freedom of navigation operations through integrated support

Table 10.1 Role of Ground Forces Across Flash Points
Source: Authors.

Korea

The Korean Peninsula's sustained geopolitical tensions are distinct from sovereignty disputes over Taiwan. The region's volatility stems from enduring North-South Korea tensions and the Democratic People's Republic of Korea's (DPRK) persistent nuclear ambitions.[16] This environment demands a strong and continuous US military presence with ground forces playing a crucial role in deterrence.[17] Positioned near the fortified demilitarized zone (DMZ), these forces bolster South Korea's defenses and enable rapid, joint responses to potential North Korean aggression.

---

16 Ankit Panda, *Kim Jong Un and the Bomb: Survival and Deterrence in North Korea* (Oxford: Oxford University Press, 2020).
17 Hal Brandsand Zack Cooper. "Dilemmas of Deterrence: The United States' Smart New Strategy Has Six Daunting Trade-offs." 2024.

The Yellow Sea

The Yellow Sea further complicates the regional security landscape. This semi-enclosed waterway, bordered by China and the Korean Peninsula, is marked by overlapping territorial claims and frequent maritime disputes, particularly between North and South Korea, including naval skirmishes and fishing rights conflicts.[18]

For ground forces, the Yellow Sea's proximity necessitates multidomain readiness, as operations may require swift responses to support naval engagements or secure coastal areas critical to South Korean defense. Additionally, North Korea's advancing missile capabilities[19] and asymmetric tactics elevate the strategic need for MDO in the area. The peninsula's challenging geography—including mountainous terrain and densely populated urban centers—adds to the complexity, hindering mobility and logistics, which underscores the need for precise cross-domain coordination.

The Senkaku Islands

The Senkaku Islands are also critical, representing a broader struggle for regional dominance and control over vital maritime routes. Strategically located near key shipping lanes and resource-rich waters, the Senkakus carry significant economic and security implications, consistently heightening tensions. The US-Japan alliance is crucial in managing these risks.[20]

The American military presence in Okinawa—including ground, air, and naval forces—acts as a strong deterrent to potential aggression.[21] Okinawa's proximity to the Senkakus enhances rapid response capabilities, strengthening Japan's ability to assert its territorial claims and project power.[22] US ground forces are essential in the multidomain strategy, signaling a commitment to allied security and critical support. While the

---

18 Terence Roehrig, 'North Korea: Gray Zone Actions in the Yellow Sea', In *Maritime Gray Zone Operations* ed. Andrew S. Erickson (London & New York: Routledge, 2022): 127-147.
19 Defense Intelligence Agency, 'North Korea Military Power: A Growing Regional and Global Threat' (Washington, D.C.: US Government Publishing Office, 2021).
20 Misato Matsuoka, *Hegemony and the US–Japan Alliance* (City Name Here: Routledge, 2018); Takashi Inoguchi and G. John Ikenberry, eds. *The US-Japan Security Alliance: Regional Multilateralism* (London: Springer, 2011).
21 'About USFJ', US Forces Japan, accessed Jan. 14, 2025, https://www.usfj.mil/About-USFJ/.
22 Ryo Hinata-Yamaguchi, 'Developments and Transformations in Japan's Defense Planning and Readiness', in *Handbook of Japanese Security*, ed. Leszek Buszynski (Amsterdam: Amsterdam University Press, 2024), 113.

Senkaku dispute primarily involves maritime and air domains, ground forces contribute indirectly by securing infrastructure, supporting logistics, and ensuring operational continuity.

The South China Sea

The South China Sea (SCS) poses a complex challenge, where maritime disputes and freedom of navigation issues highlight the need for effective power projection across vast waters. Unlike the localized Sino-Japanese tensions over the Senkaku Islands, the SCS has overlapping claims from multiple nations—including Vietnam, the Philippines, Malaysia, and China. China's A2AD strategy is now a challenge to US ground forces since they are located in strategic positions across the SCS.[23] This environment underscores the critical role of multidomain integration in the Indo-Pacific.[24]

Beyond its geography, the SCS is vital for global trade and energy supply, making its stability crucial to regional and international economies.[25] For the United States and its allies, maintaining freedom of navigation and countering unilateral militarization are key to preserving a rules-based order. While primarily a maritime theater, ground forces contribute by securing infrastructure, supporting logistics, and ensuring operational continuity— reinforcing a cohesive, multidomain approach similar to strategies in the Senkakus and along the First Island Chain.

These flashpoints, with their idiosyncrasies, highlight the critical role of ground forces in MDO within the region. Modern warfare's complexity demands that ground forces not only adapt to varied operational needs but also integrate effectively with other domains. Meeting these challenges requires ongoing doctrinal and technological advancements to boost readiness, resilience, and effectiveness.

Doctrinal and Technological Evolution

In the Indo-Pacific, sophisticated denial strategies, emerging technologies, and the growing complexity of multidomain threats have constrained traditional

---

23 Alex Vershinin, 'The Challenge of Dis-Integrating A2AD Zone', *Joint Force Quarterly*, Vol. 97, No. 2 (2020): 13-19.
24 Robert D. Kaplan, *Asia's Cauldron: The South China Sea and the End of a Stable Pacific* (New York: Random House Trade Paperbacks, 2015).
25 Sigfrido Burgos Cáceres, *China's Strategic Interests in the South China Sea: Power and Resources* (London & New York: Routledge, 2013).

US force projection, creating contested battle spaces where maneuvering conventional forces is ever more difficult. The US Army is adapting its doctrinal approach to maintain both (a) operational effectiveness in such highly dynamic threat environments and (b) relevance in large-scale combat operations. The US Army has successfully adjusted to emerging threats in the past, for example, when it shifted from AirLand Battle, designed for high-intensity conflicts against peer adversaries, to Full-Spectrum Operations, aimed at counterinsurgency and irregular warfare.[26] The US Army must adapt again as China develops new capabilities to disrupt traditional US dominance in land, air, and sea.[27]

Historical Developments

The US Army's doctrinal evolution reflects geopolitical shifts, external strategic changes, and technological advancements.[28] The AirLand Battle doctrine, introduced in 1982, responded to the Soviet threat in Central Europe by prioritizing high-speed, coordinated air-land operations. AirLand Battle aimed to counter Soviet numerical superiority through deep strikes that disrupted enemy echelons before they reached the front lines. However, as the US military transitioned into the post-Cold War era,[29] characterized by 1990s peacekeeping missions and 2000s counterinsurgency (COIN) campaigns in Iraq and Afghanistan, it demanded a more flexible approach.[30]

This shift led to Full-Spectrum Operations (FSO), a doctrine that expanded beyond conventional warfare and promoted decentralized structures, intelligence-driven targeting, and civil-military integration. However, FSO's reduced emphasis on large-scale combat left ground forces less prepared for high-intensity conflicts.

By the early 2010s China and Russia were advancing their military capabilities, particularly with A2AD systems.[31] As great power competition

---

26 Benjamin Jensen, *Forging the Sword: Doctrinal Change in the US Army* (Stanford, CA: Stanford University Press, 2016).

27 Bill Benson, 'Unified Land Operations', *Military Review* (2012): 47-57.

28 Jensen, *Forging the Sword*, 2016; Ben Buley, *The New American Way of War: Military Culture and the Political Utility of Force* (New York: Routledge, 2007).

29 Antulio J. Echevarria II, 'Rediscovering US Military Strategy: A Role for Doctrine', *Journal of Strategic Studies*, Vol. 39, No. 2 (2016): 231-245.

30 Buley, *The New American Way of War*, 2007.

31 Sam J. Tangredi, *Anti-Access Warfare: Countering A2AD Strategies* (Annapolis: Naval Institute Press, 2013); Taylor M. Fravel, *Active Defense: China's Military Strategy since 1949* (Princeton: Princeton University Press, 2020).

re-emerged, the US Army recognized the need to refocus on large-scale operations, leading to the 2011 Unified Land Operations (ULO) doctrine. ULO aimed to balance offense, defense, and stability operations while enhancing adaptability in complex, multidomain environments. It emphasized joint and coalition warfare, highlighting the importance of interoperability with allied forces. However, as adversaries continued expanding their capabilities, it became evident that an even more integrated approach was necessary to meet 21st century strategic challenges.

Multidomain Operations Today

MDO, formally outlined in the Army Training and Doctrine Command (TRADOC) Pamphlet 525-3-1 (2018) and Field Manual 3-0 Operations (2022), represents the most significant transformation of US military doctrine since AirLand Battle. Designed to counter near-peer adversaries, MDO seeks to integrate all warfighting domains into a unified operational framework.[32] A core objective of MDO is to penetrate and disintegrate A2AD networks, create multiple dilemmas for the enemy, force them to respond to threats from multiple vectors simultaneously, and continuously emphasize convergence.

Warfare evolves through technological advancements, doctrinal shifts, and structural transformations. Just as doctrine guides military responses to strategic challenges, new weapons systems and technologies emerge to enable its implementation, ensuring effectiveness in modern conflicts. This interplay shapes military adaptation, aligning operational capabilities with evolving threats and strategic demands.[33] Ground forces play a critical role in securing key territories and enabling joint operations through advanced capabilities such as long-range precision fires, next generation combat vehicles, tactical airlifting, as well as networked sensors and electronic warfare.[34]

---

32 Moita, Sandro Teixeira. "The Permanence of Land Power at the Center of Military Clashes to Conquer Territories," *Coleção Meira Mattos* 16, special issue (July 2022): v–vii. http://www.ebrevistas.eb.mil.br/RMM/article/view/11383/9122.

33 Christopher Bellamy, *The Evolution of Modern Land Warfare: Theory and Practice* (London & New York: Routledge, 2015); Andrew Feickert, 'US Ground Forces in the Indo-Pacific: Background and Issues for Congress', *Current Politics and Economics of South, Southeastern, and Central Asia*, Vol. 31, No. 4 (2022): 369-420.

34 US Department of the Army, *Army Modernization Strategy 2021* (Washington, D.C.: US Department of the Army, 2021), 1, https://dml.armywarcollege.edu/wp-content/uploads/2023/01/Army-Modernization-Strategy-2021.pdf.

Incorporating Technological Advances into MDO

MDO requires seamless operations across warfighting domains to achieve operational dominance, positioning land forces as decisive actors rather than mere support elements.[35] Advanced technologies enable ground forces to penetrate contested environments, disrupt adversary defenses, and deliver critical intelligence, surveillance, and targeting data that enhance overall operational effectiveness. Given the dynamic interplay between doctrine and technology, key advancements such as long-range precision fires (LRPFs), the Future Vertical Lift (FVL), tactical airlift platforms, and networked sensors are particularly well-suited to support MDO frameworks in the Indo-Pacific and East Asia. [36]

Long-range precision fires have become a cornerstone of MDO, enabling ground forces to strike high-value targets at extended ranges while remaining outside the reach of enemy defenses. Advances in hypersonic missiles and long-range artillery provide the capability to engage strategic targets deep within contested areas, disrupting command nodes, logistics hubs, and integrated air defense systems before direct contact with enemy forces.[37] According to the Army Modernization Strategy, these capabilities are essential for countering A2AD networks.[38] One of the most notable advancements in this field is the Precision Strike Missile (PrSM),[39] a next-generation system designed to engage targets beyond 500 kilometers with enhanced accuracy and survivability. As a replacement for the Army Tactical Missile System (ATACMS), PrSM represents a significant leap in long-range strike capability, allowing US forces to project power from dispersed positions while minimizing exposure to enemy counterfire.

The Joint All-Domain Command and Control (JADC2) initiative is central to the US military's push for seamless integration of sensor networks

---

35 Bryan J. Quinn, 'Sustaining Multidomain Operations: The Logistical Challenge Facing the Army's Operating Concept', *Military Review*, Vol. 103, No. 2 (March-April 2023): 128–139, https://www.armyupress.army.mil/Portals/7/military-review/Archives/English/MA-23/Multidomain-Operations/Sustaining-Multidomain-Operations.pdf.

36 Even though tanks and armor in general play an important role in the Korean Peninsula, the feasible and comprehensive use of armor in the region is still questioned to some extent.

37 Stephen Lanza and Daniel S. Roper, 'Fires for Effect: 10 Questions about Army Long-Range Precision Fires in the Joint Fight', *Association of the United States Army*, Aug. 30, 2021, https://www.ausa.org/publications/fires-effect-10-questions-about-army-long-range-precision-fires-joint-fight.

38 US Department of the Army, *Army Modernization Strategy*, 2021.

39 Operational Test and Evaluation Director, 'Precision Strike Missile (PrSM)', in *FY2020 Annual Report* (Washington, D.C.: US Department of Defense, 2021), 105–106, https://www.dote.osd.mil/Portals/97/pub/reports/FY2020/army/2020prsm.pdf.

---

and data-driven decision-making across all warfighting domains. JADC2 facilitates real-time data sharing by connecting diverse military assets into a unified operational framework, boosting situational awareness and response times in contested environments. For ground forces, this capability transforms operations within MDO by providing instant access to reconnaissance assets, sensor grids, and targeting data, even under electronic warfare conditions. It enhances survivability and effectiveness by enabling rapid, intelligence-driven actions that accelerate the observe, orient, decide, act (OODA) loop. This advanced integration is particularly crucial in the Indo-Pacific because of the vast distances involved and contestation by near-peer threats, something that makes it a rather unique region.

In contested environments where movement is constrained, tactical airlifting is vital for ensuring force and strategic mobility, as well as logistical and theater sustainment. Ground forces need operational agility in MDO, including rapid deployment, resupply, and reinforcement across dispersed areas. Aircraft like the C-130J Super Hercules, CH-47 Chinook, and V-22 Osprey provide critical support by transporting troops over extended ranges to facilitate ground maneuvers. Future advancements in tactical air mobility—such as the FVL program, advanced tiltrotor aircraft, and unmanned aerial platforms—promise even greater operational flexibility. The FVL initiative's next-generation rotorcraft will offer enhanced speed, range, and survivability, even against robust A2AD systems, further solidifying the Army's strategic relevance in the region's complex battlespace.

In the next section, we analyze how doctrinal and technological advancements shape US ground force postures in East Asia and the Indo-Pacific, emphasizing terrain, proximity, force projection, and operational flexibility in a complex maritime domain.

Ground Forces in East Asia and the Indo-Pacific

A US Army War College wargame conducted in 2014 affirmed the continuing relevance and predominance of land forces in the operational environment of the Indo-Pacific.[40] The final report concluded that even though the Indo-Pacific region is dominated by water, it is not an exclusively naval domain,

---

40 Pacific Partners Wargame Analysis, Strategic Wargaming Series, US Army War College Center for Strategic Leadership & Development, 24-25 Sep. 2014, Carlisle Barracks, PA, 5. https://csl.armywarcollege.edu/DSE/StrategicWargamingDivision/publications/WG_6-14_Pacific_Partners_Wargame_Report.pdf.

especially because the army dominates the military services in US regional allies, and [cooperation with these partners] can yield political and military dividends.[41]

China's growing military modernization and assertive posture in the Indo-Pacific have been key factors shaping US Army strategic planning.[42] Official documents such as the Army Modernization Strategy (2021) and FM 3-0 Operations highlight China as the United States' primary "pacing threat", stressing the importance of countering its efforts to create regional spheres of influence through A2AD strategies.[43] The Joint Concept for Competing[44] elaborates on China's preference for deterrence strategies that present the United States with a fait accompli, consolidating its regional sphere of influence before Washington can effectively respond.

The operational challenges posed by China in the Indo-Pacific have prompted the US Army to prioritize targeted modernization efforts aimed at ensuring power projection, geographic flexibility, and rapid response capabilities. This is reflected in the establishment of Multidomain Task Forces (MDTFs), with three specifically assigned to the Indo-Pacific theater.[45] The 1st MDTF, aligned with the US Army Pacific (USARPAC), serves as a critical component in addressing China's regional ambitions. Similarly, the 3rd MDTF and the 4th MDTF are dedicated to Indo-Pacific operations and reinforce the Army's strategic presence in the region.

A2AD systems provide defenders with distinct advantages over air and maritime operations, particularly when deployed on controlled landmasses. Biddle and Oelrich highlight that A2AD is most effective in complex terrestrial environments, where natural concealment and reduced detection profiles enhance defensive capabilities.[46] However, the effectiveness of these systems diminishes as they extend farther from their support networks, creating zones with varying degrees of influence. Thus, the strategic distribution of A2AD across East Asia's landmasses fundamentally shapes the contribution

41 Ibid, 15.

42 Zenel Garcia, *China's Military Modernization, Japan's Normalization and the South China Sea Territorial Disputes* (Cham: Springer International Publishing, 2019).

43 US Army, 'Army Modernization Strategy: Investing in the Future' (Washington, DC: Department of the Army, 2021).

44 Joint Chiefs of Staff, 'Joint Concept for Competing' (Washington, DC: US Department of Defense, 2023).

45 Congressional Research Service, 'The Army's Multidomain Task Forces (MDTFs)' (Washington, DC: CRS Reports, 2022).

46 Stephen Biddle and Ivan Oelrich. 'Future Warfare in the Western Pacific: Chinese Anti-Access/Area Denial, US AirSea Battle, and Command of the Commons in East Asia', *International Security*, Vol. 41, No. 1 (2016): 7-48.

of ground forces to MDO.[47] Proximity to key landmasses—such as mainland China and allied territories like Taiwan and Japan—creates fortified zones of influence, while more distant areas become contested spaces where neither power achieves full control over air and naval movement.[48]

Recognizing A2AD as both a denial tool and a power-balancing mechanism is crucial, as ground forces must adapt to fortified regions and shifting influence zones. This adaptability goes beyond tactical and operational considerations, shaping regional deterrence. While A2AD strategies limit force projection, ground forces play a vital role in reinforcing allied commitments, controlling critical terrain, and enabling rapid crisis response. The TRADOC Pamphlet 525-3-1 emphasizes the importance of disrupting China's strategic timelines through sustained multidomain competition, integrating land forces with naval and air assets to deny adversaries the ability to establish uncontested control over critical maritime corridors.[49]

Ground Forces and Deterrence in East Asia

The role of ground forces in deterrence within East Asia is a crucial component of regional security. While naval and air assets are typically associated with power projection, land forces provide persistent presence, territorial control, and rapid crisis response capabilities, reinforcing the credibility of US and allied defense commitments.[50] In the region, US and allied ground forces serve as tripwires against potential aggression, enhancing joint and coalition operations, contributing to the broader deterrence framework through force posture, training, and logistical infrastructure.[51] East Asia's challenging geography—which includes mountainous terrain, swamps, waterways, and densely populated urban areas—limit the scope of tank armored operations but do not diminish their strategic importance.

The Korean Peninsula remains one of the most heavily militarized regions in the world, with over 28,000 American troops stationed in South

---

47 Hassan M. Kamara, 'Countering A2AD in the Indo-Pacific: A Potential Change for the Army and the Joint Force', *Joint Force Quarterly*, Vol. 97 (2020): 97-102.
48 Biddle and Oelrich, 'Future Warfare in the Western Pacific', 2016.
49 US Army Training and Doctrine Command, 'The US Army in Multidomain Operations 2028' (Fort Eustis, VA: US Army Army Training and Doctrine Command, 2018).
50 Feickert, 'US Ground Forces in the Indo-Pacific', 2022.
51 Kaname Kuniyuki, 'Multidomain Operations: US Army and the Indo-Pacific', 2019.

Korea as a central pillar of deterrence against North Korea.[52] The Eighth Army, operating under US Forces Korea (USFK), is fully integrated with Republic of Korea (ROK) forces, ensuring readiness through combined exercises like Foal Eagle and Ulchi Freedom Shield.[53] The presence of US and allied ground forces not only reinforces the defensive posture against conventional threats—namely the DPRK—but also demonstrates the ability to rapidly escalate towards an enhanced troop presence if deterrence fails. This force posture is further enhanced by prepositioned equipment—especially under the Eighth Army—and rapid deployment capabilities, ensuring that reinforcements can arrive swiftly to support combat operations if necessary.

On the Korean Peninsula, where the DPRK employs massed land warfare tactics, platforms like the M1A2 Abrams and K2 Black Panther provide robust frontline deterrence.[54] In a conflict scenario, these armored units would be essential for breaching defensive lines, securing critical terrain, and ensuring the survivability of mechanized formations in combined arms operations.[55]

Japan, particularly Okinawa, plays an equally strategic role in the US deterrence posture. The island hosts a significant US military presence, including the III Marine Expeditionary Force (III MEF) and elements of Army and Air Force units that support regional contingencies. Japan's Self-Defense Forces (JSDF) have developed naval and air capabilities, but there is a growing recognition of the importance of ground forces in defending key islands along the First Island Chain. The Japan Ground Self-Defense Force (JGSDF) has been expanding its amphibious capabilities, establishing the Amphibious Rapid Deployment Brigade (ARDB) as a response to potential island seizures, showcasing the evolving role of ground forces in Japan's

---

52 'US Security Cooperation with Korea', US Department of State, Jan. 20, 2025, https://www.state.gov/u-s-security-cooperation-with-korea/.

53 'Exercise Foal Eagle', Seventh Air Force, accessed Feb. 22, 2025, https://www.7af.pacaf.af.mil/About-Us/Fact-Sheets/Display/Article/408383/exercise-foal-eagle/; 'UFS 24 Successfully Concludes', United States Forces Korea, Aug. 29, 2024, https://www.usfk.mil/Media/Press-Products/Press-Releases/Article/3889608/ufs-24-successfully-concludes/; 'Full Spectrum of Eighth Army Capabilities Tested During Ulchi Freedom Shield 2024', Eighth Army Public Affairs Office, Sep. 5, 2024, https://www.army.mil/article/279423/full_spectrum_of_eighth_army_capabilities_tested_during_ulchi_freedom_shield_2024.

54 Andre L.V.C. Carvalho and João Paulo Nogueira, 'Tanques na Península Coreana: uma avaliação das plataformas K2 e M2020', *Panorâmico*, Vol. 3, No. 9, Sep./Dec. 2024, https://ompv.eceme.eb.mil.br/images/publicacoes/panoramico/panoramico-vol3-n09-set-dez2024/Tanques_na_peninsula_coreana-_uma_avaliaco_das_plataformas_K2_e_M2020_Andre_Luiz_Viana_Cruz_de_Carvalho_e_Joo_Paulo_Ribeiro_Nogueira.pdf.

55 Ibid.

deterrence architecture.[56] This is relevant given China's increased military activities in the East China Sea and its territorial claims over the Senkaku Islands.[57]

Japan's evolving defense posture underscores the growing importance of armor within its archipelagic defense strategy,[58] particularly against potential Chinese incursions near the Senkaku Islands and the Ryukyu chain. While no formal US armored units are stationed in Japan, coalition operations could leverage Japan's Type 10 Main Battle Tank (MBT), designed specifically for the country's geographic and urban conditions. Additionally, the Type 16 Maneuver Combat Vehicle (MCV) offers highly mobile firepower for rapid deployment, enhancing operational flexibility across the archipelago.[59]

Taiwan, positioned at the center of the First Island Chain, presents one of the most complex deterrence challenges in the region. Unlike South Korea and Japan, Taiwan lacks a formal US troop presence but relies on US arms sales, military training, and the strategically ambiguous US defense commitments. US Army and Marine advisors play a crucial role in Taiwan's defense preparations, particularly in areas such as asymmetric warfare, coastal defense, and urban operations.[60] Taiwanese ground forces are being restructured to maximize survivability and resilience against a potential Chinese invasion,[61] incorporating lessons from A2AD strategies and multidomain coordination. The First Island Chain, which includes Taiwan, Okinawa, and the Philippines, forms a critical geographic barrier, limiting China's ability to project force into the broader Pacific. Ensuring that ground forces can rapidly reinforce and sustain operations within this contested space is essential for maintaining deterrence.

Taiwan employs armor in its urban and coastal defense strategies, using CM-11 Brave Tiger and M1A2T Abrams tanks to reinforce critical choke

---

56 Jeffrey W. Hornung, Scott Savitz, Jonathan Balk, Samantha McBirney, Liam McLane, and Victoria M. Smith, *Preparing Japan's Multidomain Defense Force for the Future Battlespace Using Emerging Technologies* (Santa Monica, CA: RAND Corporation, 2021).

57 Hughes, Christopher W. *Japan as a Global Military Power: New Capabilities, Alliance Integration, Bilateralism-Plus*. Cambridge University Press, 2022.

58 Japan Ministry of Defense, *Defense of Japan 2024* (Tokyo: Japan Ministry of Defense, 2024), 274, https://www.mod.go.jp/en/publ/w_paper/wp2024/DOJ2024_EN_Full.pdf.

59 Hinata-Yamaguchi, 'Developments and Transformations in Japan's Defense Planning and Readiness', 113.; Christopher W. Hughes, *Japan as a Global Military Power: New Capabilities, Alliance Integration, Bilateralism-Plus* (Cambridge, UK: Cambridge University Press, 2022).

60 Ian Easton, 'The Chinese invasion threat: Taiwan's defense and American strategy in Asia', Project 2049 Institute, 2017.

61 Ian Easton, Mark Stokes, Cortez A. Cooper, and Arthur Chan, *Transformation of Taiwan's Reserve Force* (Santa Monica, CA: RAND Corporation, 2017).

points against potential amphibious assaults. Given Taiwan's restrictive geography, these armored assets are strategically pre-positioned to slow and disrupt invasion forces, contributing to a layered defense approach.[62]

The future relevance of armored forces within MDO in East Asia will hinge on their adaptability to maintain lethality, survivability, and strategic flexibility in a complex battlespace. Ground forces in the region serve not merely as a defensive measure but as an active component of deterrence. They reinforce allied commitments, secure strategic terrain, and enhance regional resilience against coercion. Their capacity to hold territory, control critical logistics hubs, and integrate with air and naval forces makes them indispensable within the broader security architecture. By maintaining a credible armored presence, ground forces contribute significantly to deterring adversarial actions and sustaining stability across East Asia and the Indo-Pacific.

Operational Flexibility in the Maritime Domain

While traditionally viewed as a naval and air-dominated theater, the Indo-Pacific's vast maritime expanse requires highly mobile and adaptable ground forces to ensure operational effectiveness across dispersed environments. Given the logistical complexities, contested access, and the necessity of rapid deployment, ground forces must integrate into joint and multinational frameworks to enhance deterrence, sustainment, and crisis response capabilities. This is particularly relevant for US Army and Marine Corps forces, as well as allied armies in Japan, Australia, and the Philippines, all of which play a pivotal role in the protection of critical maritime terrain, securing strategic locations to enable expeditionary operations.

A core requirement for ground forces in this theater is strategic mobility and expeditionary capabilities, allowing them to project power and respond flexibly to emerging threats. The MDTFs, designed to integrate long-range precision fires, electronic warfare, and joint interoperability, exemplify how land forces are evolving to meet the operational demands of the Indo-Pacific.[63] This evolution is exemplified by the successful deployment of the

---

62 Caitlin Campbell, 'Taiwan: Defense and Military Issues' (Washington, DC: CRS Reports, 2024). https://crsreports.congress.gov/product/pdf/IF/IF12481; Tommy Jamison, 'Taiwan's Theory of the Fight', *War on the Rocks*, Feb. 21, 2024, https://warontherocks.com/2024/02/taiwans-theory-of-the-fight/.
63 Lynch and Deveraux, 'Landpower, Homeland Defense, and Defending Forward in US Indo-Pacific Command', 87.

Precision Strike Missile (PrSM) during the Valiant Shield 2024 exercise, where the 3rd Multidomain Task Force launched two PrSMs from an Autonomous Multidomain Launcher (AML) to engage a moving maritime target off the coast of Palau. This landmark event marked the first use of both the PrSM and AML outside the continental United States, demonstrating the Army's capability to project power across vast oceanic distances and effectively contribute to maritime domain operations.[64]

The integration of long-range precision fires, such as the PrSM, into the US Army's arsenal addresses the unique challenges of the Indo-Pacific's dispersed geography and adversarial A2AD strategies. With a range exceeding 400 kilometers, the PrSM enables ground forces to target enemy vessels and critical infrastructure from extended distances. This capability enhances deterrence, complements naval and air operations, and introduces multiple threat vectors, complicating adversaries' strategic calculations while bolstering the joint force's freedom of maneuver in contested environments.

Ground forces are also essential in amphibious operations and securing logistics hubs. While the US Marine Corps is the primary amphibious force, the Army has increasingly focused on joint integration to support contested littoral environments. Through the Pacific Pathways initiative, the Army deploys rotational forces for joint exercises across the region, enhancing shore-to-shore maneuvering, reinforcement of key terrain, and sustained island chain operations. Pre-positioned assets, such as Army Prepositioned Stocks-3 (APS-3) afloat, enable rapid deployment without a heavy logistical footprint, crucial for maintaining stability across maritime chokepoints.

Interoperability with the Marines and allied forces is a cornerstone of Army operations in the region. US Army watercraft, including maneuver support vessels (MSVs) and logistics support vessels (LSVs), enhance personnel and equipment transport across maritime spaces, augmenting naval capabilities. Collaboration with allied militaries in Japan, Australia, and the Philippine allows for combined amphibious training, coastal

---

64 'US Army Conducts First Anti-Ship Ballistic Missile SINKEX Using PrSM," *Naval News*, Jun. 23, 2024, https://www.navalnews.com/naval-news/2024/06/u-s-army-conducts-first-anti-ship-ballistic-missile-sinkex-using-prsm/; Stephen Page, '3d MDTF Demonstrates Ability to Operate in the Indo-Pacific', US Army, Jun. 21, 2024, https://www.army.mil/article/277487/3d_mdtf_demonstrates_ability_to_operate_in_the_indo_pacific; Daniel Lopez, 'Multidomain Transformation in the Indo-Pacific', US Army, Oct. 16, 2024, https://www.army.mil/article/280588/multi_domain_transformation_in_the_indo_pacific; Jennifer Hlad, 'Army Tests Next-Gen Long-Range Fires Capability in Pacific', *Defense One*, Jun. 27, 2024, https://www.defenseone.com/technology/2024/06/army-tests-next-gen-long-range-fires-capability-pacific/397724/.

defense, and rapid island reinforcement missions. These efforts support a distributed, networked operational model where ground forces must remain light, mobile, and technologically integrated to operate in highly contested maritime regions.

Ultimately, ground forces in the Indo-Pacific operate within a cohesive joint force structure, extending their role beyond traditional land warfare. As competition intensifies, ground forces must continue adapting to expeditionary warfare, refining their ability to maneuver across dispersed terrain, and reinforcing joint force lethality in contested maritime environments. Their adaptability and integration into broader strategies make them a crucial component of the region's defense and stability.

Combat Effectiveness

Unlike conventional theaters where armies can mass forces and sustain prolonged engagements,[65] the Indo-Pacific's dispersed geography, contested logistics, and adversary A2AD strategies demand a more adaptive force posture for ground forces to be effective in combat.[66] Furthermore, given that a brigade combat team may be operating in the Korean Peninsula's mountainous terrain, Taiwan's urban and coastal defenses, or the Philippine archipelago's island chains, teams must tailor their capabilities to maximize operational effectiveness in whichever environment they find themselves in.

A critical metric in this context is combat power per unit of mass[67] which measures how effectively a force generates combat effects relative to its size and logistical footprint. In high-intensity theaters like Korea and Taiwan, armored and mechanized formations provide robust firepower and protection, essential for decisive engagements in direct combat scenarios. Platforms such as the M1A2 Abrams and K2 Black Panther are optimized for open battlefields and combined arms maneuver warfare, delivering shock effect and survivability. However, in dense urban environments and complex coastal defenses, these heavy units must integrate with lighter, more mobile assets—including infantry fighting vehicles, anti-tank missile teams, and loitering munitions—to maintain flexibility in confined battlespaces.

---

65 Bellamy, *The Evolution of Modern Land Warfare*, 2015.
66 Tangredi, *Anti-Access Warfare*, 2013.
67 Richard E. Simpkin, *Race to the Swift: Thoughts on Twenty-First Century Warfare* (Sterling: Brassey's, 1985).

In the maritime theaters of the Indo-Pacific, mobility is prioritized over raw combat mass. Forces must operate across vast distances and maintain combat effectiveness with minimal fixed infrastructure. This necessitates highly expeditionary formations, leveraging prepositioned stocks, rapid deployment airlift, and maneuver support vessels (MSVs) to enable dynamic operations in contested maritime spaces. The US Army's MDTFs exemplify this approach, combining long-range precision fires (PrSM), electronic warfare, and cyber capabilities to allow land-based forces to strike naval assets and disrupt A2AD networks from dispersed island positions.[68] The integration of FVL platforms and unmanned logistics systems further enhances mobility, allowing rapid repositioning to reinforce key terrain, sustain forward-deployed forces, and respond flexibly to emerging threats.

Achieving optimal combat effectiveness in the Indo-Pacific requires a well-balanced integration of firepower, protection, and mobility. Firepower shapes the battlespace before direct engagement, with systems like the PrSM enabling ground units to strike maritime and land targets from extended ranges. Loitering munitions, precision-guided artillery, and networked targeting systems enhance lethality, allowing dispersed units to coordinate strikes efficiently. However, firepower alone is insufficient without robust protection measures, crucial for survivability in sensor-driven and precision-strike environments. Active Protection Systems (APS) for armored formations, hardened logistics networks, and integrated air defense capabilities ensure resilience against enemy countermeasures while maintaining operational readiness. Finally, mobility remains a decisive factor in the Indo-Pacific's maritime geography, where success depends on the ability to rapidly deploy, reposition, and sustain forces across multiple islands and key terrain. The use of MSVs, airlift platforms, and joint naval transport assets provides the strategic mobility needed to reinforce forward positions, support expeditionary operations, and exploit maneuver opportunities across island chains and contested corridors.

By maintaining a balance between firepower, protection, and mobility, ground forces can effectively adapt to the evolving operational challenges of the Indo-Pacific. This integrated approach ensures not only deterrence and survivability but also sustained combat effectiveness in one of the world's most complex and contested security environments. This model can be summarized in Table 10-2.

---

68 Vershinin, 'The Challenge of Dis-Integrating A2AD Zone', 2020.

Combat Element	Description	Key Capabilities/Systems
Firepower	Ground forces must generate kinetic effects to shape the battlespace before direct engagement, ensuring deterrence and disruption of adversary forces.	• LRPFs; • Loitering munitions and precision-guided artillery; • Networked targeting systems for JADC2.
Protection	Survivability in contested environments is enhanced through active and passive defense systems, distributed force postures, and resilient logistics.	• APS for armored units; • Hardened logistics networks and APS-3 afloat; • Integrated air and missile defense for force survivability.
Mobility	Success in a dispersed maritime environment requires rapid force deployment, sustainment, and maneuver between key locations.	• MSVs for amphibious logistics; • FVL platforms for rapid air mobility; • Sea-land integration for transport and sustainment assets.

Table 10.2 Balance of Firepower, Protection, and Mobility
Source: Authors.

Conclusion

The evolving security landscape in East Asia and the Indo-Pacific underscores the enduring importance of ground forces within MDO. While the region is often perceived as a primarily maritime and aerial theater, the integration of land forces remains crucial for ensuring deterrence, securing key terrain, and countering adversarial A2AD strategies. As highlighted in US military doctrine and strategic planning, China's rapid military modernization and assertive regional posture necessitate continuous adaptation and innovation within ground force capabilities. The establishment of MDTFs, the refinement of joint interoperability, and the expansion of long-range precision fires illustrate the Army's commitment to maintaining operational flexibility in contested environments. Moreover, the combination of conventional deterrence and non-kinetic coercion requires ground forces to prepare not only for high-intensity conflict but also for sustained competition, ensuring presence, resilience, and strategic signaling. The ability of ground forces to balance firepower, protection, and mobility is what will enable them to operate effectively in both the high-intensity battlefields of East Asia and the expeditionary, maritime-dominated Indo-Pacific. By integrating these capabilities, land forces will remain a decisive element in regional deterrence, ensuring strategic advantage and operational flexibility in an evolving security landscape. Ultimately, the ground force's ability to integrate with naval and air assets, reinforce allied commitments, and project power across the Indo-Pacific will be instrumental in sustaining a favorable balance of power in the years to come.

11

Multidomain Operations with Japanese Characteristics

Rintaro Inoue and Yuka Koshino

The increasing use of the new domains of space, cyber, electromagnetic spectrum and even cognitive warfare by major military powers such as the United States, China, and Russia since the 2010s has prompted the Japan Self-Defense Force (JSDF) to rethink its strategies, operational concepts, and capabilities to adapt to this evolving security landscape.

In 2010, the Japan Ministry of Defense (JMOD) released the 2010 National Defense Program Guidelines (NDPG), a document setting out Japan's defense strategy and capability development plans for the next ten years and stated plans to enhance JSDF's capabilities in space and cyberspace domains for the first time. The 2018 NDPG further introduced Cross-Domain Operations (CDO), a new operational concept that mirrors the US military's Multidomain Operation (MDO), designed to tackle the growing threat posed by the Chinese People's Liberation Army (PLA), although it does not explicitly mention that in the official document. According to the document, CDO seeks to "organically fuse capabilities in all domains to generate synergy and amplify the overall strength" to mitigate the limitations of individual domains and effectively offset any inferiority in a particular domain by combined effects of operations in multiple domains.[1] Building

---

1 'National Defense Program Guidelines for FY 2019 and beyond', Ministry of Defense, Japan, Dec. 18, 2018, last modified, Dec. 18, 2018, https://warp.da.ndl.go.jp/info:ndljp/pid/11591426/www.mod.go.jp/j/approach/agenda/guideline/2019/pdf/20181218_e.pdf.

on these advancements, the 2022 National Defense Strategy refined the CDO concept by emphasizing the creation of "asymmetrical advantage."[2]

However, despite the introduction of CDO in 2018, there remains a notable absence of official documents outlining how the JSDF will fight under CDO. Additionally, much of the prior research has focused on the JSDF's equipment rather than its operational concepts. In the event of a Chinese invasion, the JSDF will face the formidable challenge of countering the PLA, necessitating the need to maximize its military effectiveness. To address this challenge, it is crucial for Japanese and allied strategic communities to analyze how the JSDF currently plans to fight, assess whether this operational concept is adequate, and explore alternative approaches that might enhance its ability to achieve success in such a scenario.

Against this background, this chapter undertakes a comprehensive examination of the JSDF's CDO concept by drawing on strategic documents, as well as retired and current SDF senior officers' scholarly contributions to prominent defense and security journals in Japan. In the following sections, this research aims to distill the essence of CDO and shed light on its development, evolution, and implications for Japan's military modernization.

The Rise of China and the Shift in Japan's Threat Perception

There are several factors behind the development and implementation of the CDO concept in the 2010s. One is a fundamental shift in Japan's threat perception and its approaches to defense planning, driven by the Chinese rapid military development and its growing assertiveness in the East China Sea. Prior to 2010, the Japanese government planned its defense under the long-held lack of threat perception, rooted in the Cold War *détente* period, that the likelihood of a large-scale war between major powers was low and that a full-scale invasion of Japan, posing an existential threat, was unlikely.[3] Thus, under the "Basic Defense Force Concept," which was introduced in the 1976 NDPG, the JSDF's defense capabilities and capacities were limited to "repel limited and small-scale aggression" without careful consideration of the specific threats and potential aggressor's warfighting capabilities.

---

2 'National Defense Strategy', Ministry of Defense, Japan, Dec. 16, 2022, 12, last modified Dec. 16, 2022, https://www.mod.go.jp/j/policy/agenda/guideline/strategy/pdf/strategy_en.pdf.
3 Japan, Security Council, 'National Defense Program Guidelines for FY 2011 and beyond', Dec. 17, 2010, 4, https://japan.kantei.go.jp/kakugikettei/2010/ndpg_e.pdf.

Needless to say, the JSDF developed warfighting concepts to deal with threats posed to Japan in each era. During the Cold War, it developed a concept to prevent a potential Soviet amphibious landing and advancement in Hokkaido.[4] As North Korea's ballistic missile threats grew in the late 1990s, the JSDF began developing a ballistic missile defense system with space-based assets to intercept North Korean missile threats in cooperation with the US from 2003.[5] However, invasion was deemed unlikely and these capabilities were developed for specific missions, not broader JSDF-wide operations to counter Japan's existential threats.[6]

In the 2010s, several incidents and developments contributed to drastically changing Japan's threat perception amid the rapid military rise of China. One is the collision between the Japanese Coast Guard and a Chinese trawler in 2010 near the Senkaku Islands (China claims the island as Diaoyu Islands), and the growing Chinese naval presence since then. The Chinese Coast Guard's ships began to regularly operate around the Senkaku Islands in 2012.[7] In 2012, the JSDF reportedly began studying contingency operations to recapture Japan's remote islands together with the US Forces Japan.[8]

Second is the Chinese leadership change in November 2012. Conservative leader Xi Jinping came into power in China with strong ambitions to develop a "world-class military," and began to rapidly expand maritime, air, and missile capabilities including nuclear arsenals.[9]

---

4 Narushige Michishita, 'Summary of Sessions', International Forum on War History, 2nd Forum, National Institute for Defense Studies, Mar. 2004, 255-258, last modified Mar. 2004, https://ndlsearch.ndl.go.jp/books/R100000039-I1283024.
5 Masato Toki, 'Japan's Evolving Security Policies: Along Came North Korea's Threats', Nuclear Threat Initiative, Jun. 3, 2009, https://www.nti.org/analysis/articles/japans-evolving-security-policies/; 'Dandō misairu bōei shisutemu no seibitō nitsuite [On the Development of Ballistic Missile Defense System, and others], Security Council, Japan, Dec. 19, 2003, 9, https://www.kantei.go.jp/jp/singi/anzenhosyoukaigi/Missile.pdf.
6 'National Defense Program Outline', The World and Japan Database, Oct. 29, 1976, last modified Feb. 1, 1999, https://worldjpn.net/documents/texts/docs/19761029.O1E.html.
7 Japan, Ministry of Foreign Affairs, 'Trends in China Coast Guard and Other Vessels in the Water Surrounding the Senkaku Islands, and Japan's response', Dec. 1, 2024, https://www.mofa.go.jp/region/page23e_000021.html.
8 'Senkaku yūji, jūninen ni kenkyūan Nichi Chū kinpaku uke Nichi Bei, dakkan o sōtei kyōdō sakusen keikaku no soan ni [United States and Japan to prepare a draft study to prepare for a Senkaku contingency, assuming recapturing operations, to use for joint operational planning, driven by rising tension between U.S. and Japan]', Asahi Shimbun, Jan. 24, 2016.
9 'Secure a Decisive Victory in Building a Moderately Prosperous Society in All Respects and Strive for the Great Success of Socialism with Chinese Characteristics for a New Era— Delivered at the 19th National Congress of the Communist Party of China,' Xinhua News Agency (China), Oct. 18, 2017, https://interpret.csis.org/translations/secure-a-decisive-victory-in-building-a-moderately-prosperous-society-in-all-respects-and-strive-for-the-great-success-of-socialism-with-chinese-characteristics-for-a-new-era-delivered-at-the/.

Furthermore, since 2015, Beijing began to focus on space and cyber as a critical warfighting domain, and to seek superiority in them. At the 19th Party Congress of the Chinese Communist Party in 2017, Xi also guided the PLA to adopt artificial intelligence, and emphasized the need to "accelerate intelligentization, and improve joint operations capabilities and all-domain combat capabilities based on network information systems."[10]

Tokyo was particularly concerned about PLA's rapid missile developments, blue water capabilities, including the development of aircraft carriers, and the increasing capability developments and use of new domains of space, a cyber and electromagnetic spectrum that are supporting the Chinese's anti-access/area-denial capability. If China were to attempt an invasion and escalate into a war, it was clear that the JSDF would be forced to fight with a high degree of intensity and at a disadvantage.

Around the same time, Xi's rhetoric regarding Taiwan intensified with Beijing increasingly asserting its intention to bring Taiwan under the control of the People's Republic of China. The shift was particularly notable after Tsai Ing-Weng assumed office in 2016. The PLA has significantly altered its behavior, with the PLA Navy regularly encircling Taiwan and expanding its presence around the Nansei Islands to advance into the Pacific Ocean.[11] In Japan, concerns about China's intentions have grown with a former US Indo-Pacific Commander's warning that China could invade Taiwan by 2027 sparking serious debate. Senior political leaders began to openly begin to acknowledge that "Taiwan contingency is a Japanese contingency."[12] The large-scale military exercises conducted by China around Taiwan in August 2022, which saw five ballistic missiles land in Japan's exclusive economic zone, only served to reinforce these concerns among the Japanese public.[13]

The unfolding developments around Japan and in the region contributed a greater sense of urgency for the JSDF to create an operational concept to prepare for a broader East Asia contingency, and to "fundamentally reinforce

10 Yuka Koshino, 'New Domains of Chinese Military Modernization: Security Implications for Japan', *NBR Special Report*, No.103 (2022), 55, https://www.nbr.org/wp-content/uploads/pdfs/publications/sr103_meetingchinasemergingcapabilities_dec2022.pdf.

11 'China's Activities in East China Sea, Pacific Ocean, and Sea of Japan', Ministry of Defense, Japan, Oct. 2024, last modified Oct. 2024, https://www.mod.go.jp/en/d_act/sec_env/pdf/ch_d-act_a.pdf.

12 'Taiwan contingency also one for Japan, Japan-U.S. alliance: ex-Japan PM Abe', *Kyodo*, Dec. 1, 2021, https://english.kyodonews.net/news/2021/12/b38433927c1e-taiwan-contingency-also-one-for-japan-japan-us-alliance-abe.html.

13 Ryo Nemoto and Rieko Miki, '5 Chinese Missiles Land in Japan's EEZ: Defense Chief', *Nikkei Asia*, Aug. 4, 2022, https://asia.nikkei.com/Politics/International-relations/Taiwan-tensions/5-Chinese-missiles-land-in-Japan-s-EEZ-defense-chief.

its defense capabilities" based on the opponent's capabilities and new ways of warfare. The ongoing conflict in Ukraine, which began in 2022, has also catalyzed the JSDF's strategic reassessment. The heavy reliance on drones, space-based assets, cyber warfare, electromagnetic spectrum operations, and cognitive warfare in the Ukraine conflict has underscored the need for the JSDF to adapt its warfighting concept to address these emerging threats.[14]

Developing a JSDF Strategy to Defend the Nansei Islands

Amid the growing threats from the PLA, the JSDF developed its military strategy to enhance the defense of the Nansei Islands partially influenced by insights from the US military.[15] The JSDF's strategy has evolved in two distinct phases. The first phase began in 2010 with the introduction of the "Dynamic Defense Force," which was later enhanced and rebranded as the "Dynamic Joint Defense Force" in 2013. This emphasized the traditional approach previously employed by the JSDF, wherein manned assets in the maritime and air domains use short-range weapons to establish superiority within their respective domains. To augment this approach, the JMOD proposed stationing of ground-launched anti-ship missiles on remote islands and the deployment of several divisions from across Japan as reinforcements. This strategy aimed to either prevent a PLA invasion or reclaim the islands if seized by the PLA.[16] To implement this strategy, various initiatives were undertaken, including modernizing fighter jets, introducing amphibious vehicles, establishing the Amphibious Rapid Deployment Brigade, and promoting jointness across services.[17]

However, as China rapidly enhanced both the capability and capacity of its military, it became increasingly apparent that this strategy was insufficient to ensure the defense of the Nansei Islands. Maintaining maritime and air superiority in the waters surrounding the Nansei Islands was becoming increasingly challenging. It was evident that sending reinforcements to

---

14 Ministry of Defense, Japan, 'National Defense Strategy', 8-9.

15 *JMSDF Command and Staff College Review*, Vol. 3 No. 1 extra issue (2013); 'On Joint Operational Access Concept (JOAC)', JMSDF Command and Staff College, Dec. 21, 2011, last modified Dec. 21, 2011, https://www.mod.go.jp/msdf/navcol/index.html?c=topics&id=008.

16 'National Defense Program Guidelines for FY 2014 and beyond', Ministry of Defense, Japan, Dec. 17, 2013, 17, last modified Dec. 17, 2013, https://warp.da.ndl.go.jp/info:ndljp/pid/11591426/www.mod.go.jp/j/approach/agenda/guideline/2014/pdf/20131217_e2.pdf.

17 'Medium Term Defense Program (FY 2014-FY 2018)', Ministry of Defense, Japan, Dec. 17, 2013, 4-6, last modified Dec. 17, 2013, https://warp.da.ndl.go.jp/info:ndljp/pid/11591426/www.mod.go.jp/j/approach/agenda/guideline/2014/pdf/Defense_Program.pdf.

the Nansei Islands in scenarios where "a war in the East China Sea... could likely turn the surrounding waters into a 'no man's sea' devoid of any living humans," would be extremely dangerous.[18] Consequently, there was a growing need to explore a new strategy that would enable the defense of the Nansei Islands that complements this approach.[19]

Thus, the government responded by introducing the "Multidomain Defense Force" strategy in 2018. Building on the previous approach, this new strategy continued to emphasize the deployment of missiles in the Nansei Islands, but with two significant enhancements. Firstly, the adoption of stand-off missiles enabled the JSDF to engage invading forces from beyond their weapon engagement zones, even in scenarios where Japan had lost maritime and air superiority near the island chain.[20] Secondly, the strategy placed unprecedented emphasis on space, cyber, and electromagnetic domains as a foundation for network-centric warfare.[21] However, despite this upgrade, the prioritization of these domains was not necessarily based on detailed operational research, and there was no clear plan for how to develop and employ these capabilities in practice.[22]

In 2022, the Multidomain Defense Force strategy underwent a substantial update, shifting its focus towards stand-off defense capabilities and decision superiority to disrupt and defeat invasion much "earlier and at places further afield."[23] Specifically, this renewed strategy centered on stand-off defense capabilities and Integrated Air and Missile Defense (IAMD) while combining manned and unmanned assets, space, cyber, and electromagnetic capabilities, as well as command, control, and intelligence functions to achieve asymmetric superiority. To support these efforts, the strategy also prioritizes mobile deployment and resilience capabilities to ensure sustained

18 Tomohisa Takei, 'Kaijō Jieitai no genkyō to shōrai [Current Status and Future of JMSDF]', *Ships of the World*, Jan. 2022, 117-123.

19 Yasushi Sugimoto, '"Kekkyoku nan nin shinu no ka" shimyure-shon ni Abe shi kikikan ["How many people will die in the end?" Abe concerned by simulation result]', *Sankei Shimbun*, Sep. 29, 2023, https://www.sankei.com/article/20230929-T3JBQU3ZLVJYXKCUU44VQJFPLM/.

20 Ministry of Defense, Japan, 'National Defense Program Guidelines for FY 2019 and Beyond', 12.

21 Ibid., 19-21.

22 Yasushi Sugimoto, '"Kono mama ja Chūgoku ni katenai" kuniku no usaden [We can't beat China if we don't do something about it." Usaden as a desperate measure]', *Sankei Shimbun*, Sep. 27, 2023, https://www.sankei.com/article/20230927-HASC5DBRVFIWZLSQYMQGOR 5Q2Q/?959896.

23 Ministry of Defense, Japan, 'National Defense Strategy', 12.

effectiveness in the face of evolving threats.[24] In essence, the updated strategy is built around three key pillars:

1. JSDF's firepower will primarily be delivered by stand-off and air defense missiles
2. the force multiplication of these capabilities through unmanned assets, advanced network systems, and enhanced decision-making capabilities
3. the sustainment of these efforts through ammunition stockpiling and other defense infrastructures

Although JMOD does not explicitly define "earlier" in the aforementioned phrase, it likely refers to the aim of gathering information at earlier stages and accelerating the observe, orient, decide and act (OODA) loop to achieve decision superiority. The National Defense Strategy highlights the use of high time-resolution satellite constellations and unmanned assets for intelligence, surveillance, and reconnaissance (ISR), with real-time data transmission via resilient networks and processing enabled by artificial intelligence.[25] The emphasis on decision superiority reflects a growing awareness in Japan's defense community that the rapid and complex nature of modern warfare continues to amplify uncertainty and friction on the battlefield.[26]

Stand-off defense capabilities offer the JSDF several operational advantages to the JSDF, enabling them to address long-standing challenges in Japan's defense strategy. One key benefit is the ability to apply firepower from locations far behind the front lines, thereby reducing the vulnerability of JSDF assets to enemy counterattacks.

When the Japanese government decided to introduce stand-off defense capabilities in 2018, its primary objective was "to effectively intercept attack against Japan while ensuring the safety of its personnel."[27] Given that China significantly outnumbers the JSDF's personnel and assets, it was essential for Japan to adopt a strategy that minimizes attrition. This recognition led to lifting the restrictions on Japan's long-range strike capabilities, including

---

24 Ibid.
25 Ministry of Defense, Japan, 'National Defense Strategy', 27.
26 Ibid., 26.
27 'Medium Term Defense Program (FY 2019 - FY 2023)', Ministry of Defense, Japan, Dec. 18, 2018, 12, last modified, Dec. 18, 2018, https://warp.da.ndl.go.jp/info:ndljp/pid/11591426/www.mod.go.jp/j/approach/agenda/guideline/2019/pdf/chuki_seibi31-35_e.pdf.

counterstrike capabilities, which the Japanese government has self-imposed due to political factors.

A second key advantage is the reduction of logistical challenges associated with deploying troops to the Nansei Islands. Under the "Dynamic Joint Defense Force" strategy, troop deployments required movement from distant regions of Japan, such as Hokkaido, where ground forces are heavily concentrated, to the Nansei Islands.[28] However, this plan was unrealistic, as its implementation would require transport vessels to navigate waters where Japan's air and maritime superiority is contested, significantly increasing the risk of substantial personnel losses. In contrast, the use of cruise missiles with a range of 1,000 km, launched from Kyushu, or hypersonic missiles with a range of 3,000 km from Hokkaido, to destroy or disrupt enemy forces near the Nansei Islands would reduce the need to deploy ground troops to these islands. Moreover, since approximately 70 percent of the JSDF's ammunition stockpiles are in Hokkaido, it serves as a more stable and sustainable base for projecting firepower compared to the Nansei Islands, which have very limited storage facilities.[29]

JSDF's Cross-Domain Operation and Key Features

To implement the Multidomain Defense Force strategy, the JSDF has been developing the CDO, with a primary focus on enhancing the effectiveness of firepower. The term first appeared in 2015 US-Japan Guideline, a strategic document that defines the division of role in the Alliance and has gradually evolved and currently being developed into a doctrine.[30] The 2022 National Defense Strategy defines CDO as a operational concept that aims to "overcome inferiority in individual domains" by "organically fuse capabilities in all domains including space, cyber, electromagnetic spectrum, land, sea, and airspace to generate synergy and amplify the overall strength."[31]

---

28 'Defense of Japan 2014', Ministry of Defense, Japan, 2014, 164, last modified 2014, https://warp.da.ndl.go.jp/info:ndljp/pid/11591426/www.mod.go.jp/e/publ/w_paper/pdf/2014/DOJ2014_2-5-1_web_1031.pdf.
29 Ryo Nemoto, 'Japan to expand fuel and ammo storage on islands near Taiwan: defense chief', *Nikkei Asia*, Sep. 6, 2022, https://asia.nikkei.com/Editor-s-Picks/Interview/Japan-to-expand-fuel-and-ammo-storage-on-islands-near-Taiwan-defense-chief.
30 'Meguro chūtonchi sōsetsu 29 shūnen ando kyōiku kunren kenkyū honbu sōritsu 6 shūnen kinen, honbuchō purezente-shon [29th Anniversary of the Meguro Garrison and 6th Anniversary of TERCOM, Commanding General's Presentation]', JGSDF TERCOM YouTube Channel, Oct. 11, 2024, last modified, Oct. 11, 2024, https://www.youtube.com/watch?v=_pflwzl2Ebs&t=785s.
31 Ministry of Defense, Japan, 'National Defense Strategy', 25.

When aligned with the approach of the Multidomain Defense Force, CDO can be interpreted as a concept in which long-range attacks and decision superiority are achieved by enhancing kinetic and non-kinetic firepower in each domain through force multipliers such as fast, resilient, and flexible kill chains. Simultaneously, this approach involves protecting the satellite systems, cyberspace, and communication systems that serves as the basis of the kill chain from enemy attacks while attempting to disrupt the enemy's kill chain.[32] Implementing this approach, for example, would allow the JSDF to respond more effectively even in situations where it lacks air superiority in contested regions. Instead of limiting its means to fighter jets and air defense missiles, the JSDF could counterstrike enemy air bases supported by long-range kill chains enabled by space and electromagnetic domains, thereby reducing potential airborne threats and recovering air superiority.[33]

While the CDO is a joint operational concept of the JSDF, there is a noticeable variation in how each service embraces them. Among the three branches, JGSDF has been the most proactive in promoting CDO.[34] According to the National Defense Strategy, the JGSDF's interpretation of CDO emphasizes "disrupting invading forces from a distance by reinforcing surface-launched stand-off defense capability."[35] This suggests that the JGSDF aims for operations in which ISR at long ranges is carried out by unmanned or space assets, with the information transmitted through networks to enable stand-off strikes.[36] The JGSDF is also working on gaining decision superiority, building an AI-enhanced command and control network, and simultaneously participating in the US Army-led Project Convergence.[37]

JMSDF embraces the concept of CDO while also advocating for a doctrine suited to maritime operations, namely "distributed maneuver operations." In the context of CDO, the JMSDF believes that this concept will

---

32 'Defense of Japan 2020', Ministry of Defense, Japan, 2020, 12, last modified 2020, https://www.mod.go.jp/en/publ/w_paper/wp2020/DOJ2020_EN_Full.pdf; 'Defense of Japan 2024', Ministry of Defense, Japan, 2024, 281, last modified, 2024, https://www.mod.go.jp/en/publ/w_paper/wp2024/DOJ2024_EN_Full.pdf.

33 Ministry of Defense, Japan, 'National Defense Strategy', 14.

34 'Rikujō Jieitai no torikumi [JGSDF's initiatives]', Ground Self Defense Force, Japan, last accessed, Dec. 22, 2024, https://www.mod.go.jp/gsdf/about/structure/2023.pdf#page=2.00.

35 Ministry of Defense, Japan, 'National Defense Strategy', 30.

36 'Reiwa 4 nendo seisaku hyōkasho [FY 2022 Policy Evaluation Report]', Ministry of Defense, Japan, Dec. 2022, 75, last modified Dec. 2022, https://www.soumu.go.jp/main_content/000854078.pdf.

37 'Rikujō bakuryōchō no gaikoku shutchō nitsuite [Foreign Business Trip of the Chief of Staff of the Ground Staff]', Ground Self Defense Force, Japan, Mar. 1, 2024, last modified, Mar. 1, 2024, https://www.mod.go.jp/gsdf/news/press/2024/pdf/20240301.pdf.

become essential to deny enemy actions in contested waters thus focusing on securing underwater superiority through enhanced submarine forces and stand-off capabilities.[38] Regarding distributed maneuver operations, which resemble the US Navy's distributed maritime operations, the JMSDF has been improving its fleet's air defense capabilities, electronic warfare capabilities, and other aspects to realize this approach.[39]

The JASDF is the most reluctant to use the CDO phrase, with little to no mention of CDO in its strategic documents or press releases. However, JASDF is concentrating its effort on IAMD, which coordinates various air defense assets through advanced networks, a warfighting concept akin to the CDO.[40] Furthermore, JASDF is developing space operations capabilities to support JSDF's joint effort in CDO.[41]

One of the reasons JGSDF has been proactive in adopting CDO is the need to address significant changes in its strategic posture, which were more pronounced than those faced by the other services. In the early 2010s, the JGSDF's posture was deemed inadequate for combat operations in the Nansei Islands, unlike the JMSDF and JASDF, which directly affected the budget allocation between the services.[42] However, with the adoption of the CDO concept and the subsequent development of defense capabilities aligned with it, the JGSDF began acquiring long-range ISR and targeting capabilities through satellites and unmanned aerial vehicles (UAVs), allowing them to strike distant targets with stand-off missiles. As a result, the JGSDF can now contribute to the defense of the Nansei Islands through a much more advanced approach than in the past.

---

38 'Defense Buildup Plan', Ministry of Defense, Japan, Dec. 16, 2022, 22, last modified Mar.14, 2023, https://www.mod.go.jp/j/policy/agenda/guideline/plan/pdf/program_en.pdf; Plans and Programs Division, Maritime Staff Office/ Strategic Studies Department, MSDF Command and Staff College, "Henkasuru senryaku kankyō to kaijō Jieitai 'senryaku san bunsho' o kaiyō ryōiki de gugensuru [Changing Strategic Environment and the JMSDF: Embodying the 'Three Documents' in the Maritime Domain]," *JMSDF Command and Staff College Review*, Vol. 13 No. 2 (2024), 5-17; Omachi Katsushi, "Aratana jidai no shi-pawa- toshite no kaijō Jieitai [Maritime Self-Defense Force as Sea Power in a New Era]," *JMSDF Command and Staff College Review*, Vol. 11 No. 1 (2021), 12-39.
39 Ministry of Defense, Japan, 'Defense Buildup Plan', 22.
40 'Defending Japan's peace from the skies, Evolution in JASDF', Air Self Defense Force, Japan, last accessed Dec. 22, 2024, https://www.mod.go.jp/asdf/en/roles/role04/page04/.
41 Ministry of Defense, Japan, 'Defense Buildup Plan', 26.
42 Satoru Fuse, 'Yosanmen kara miru" kokka anzen hoshō senryaku" no jikkōsei [Effectiveness of the National Security Strategy from a Budgetary Perspective]', *Kokusai Anzenhosho [Journal of International Security]*, Vol. 42, No. 4 (2015), 16-31.

Key Features

Like other military operations, the CDO is structured around the seven core functions: command and control, information, intelligence, fires, movement and maneuver, protection, and sustainment[43] However, the significance and characteristics of each function vary depending on the specific operation. To better understand CDO's focus on maximizing firepower effectiveness, this chapter concentrates on four key functions: command and control, information and intelligence, kinetic fires, and non-kinetic fires.

Command and Control

JSDF's command-and-control is an indispensable function to execute CDO. The JMOD established a Joint Operation Command (JJOC) in 2024, though its structure and operations remain unclear. Since the characteristics of warfighting in each domain differ substantially, tactical-level command and control will continue to be allocated to each Joint Task Force (JTF) established in the command centers of ground, maritime, and air components.[44] The JJOC is expected to coordinate the JTFs to execute effects at the campaign or operational level, encompassing functions such as ISR, kinetic and non-kinetic fires, protection, and sustainment. It is also aimed at improving operational coordination with the US forces. In July 2024, the Pentagon announced plans to upgrade US Forces Japan (or USFJ) to a joint force headquarters, expand USFJ's missions and operational responsibilities, and task it to serve as the counterpart for JJOC.[45]

The command-and-control system of the JSDF, including the JJOC, faces several challenges in implementing CDO due to limitations in the joint military decision-making process and the capacity of staff officers. The JGSDF has updated its manual to initiate targeting alongside mission

43 'Joint Operations, Joint Publication 3-0', U.S. Joint Chiefs of Staff, Oct. 22, 2018, III-1-35, last modified, Oct. 22, 2018, https://irp.fas.org/doddir/dod/jp3_0.pdf; JGSDF TERCOM YouTube Channel, 'Meguro chūtonchi sōsetsu 29 shūnen ando kyōiku kunren kenkyū honbu sōritsu 6 shūnen kinen, honbuchō purezente-shon [29th Anniversary of the Meguro Garrison and 6th Anniversary of TERCOM, Commanding General's Presentation]'.
44 Katsuki Takada, 'Jōsetsu tōgō shireibu nitsuite [On Permanent Joint Operation Command]', Anzen hoshō o kangaeru [Contemplating Security], Sep. 1, 2023, last modified, Sep. 1, 2023, http://anpokon.or.jp/pdf/kaishi_820.pdf.
45 C. Todd Lopez, 'U.S Intends to Reconstitute U.S. Forces Japan as Joint Forces Headquarters', Jul. 28, 2024, U.S. Department of Defense, last modified, Jul. 28, 2024, https://www.defense.gov/News/News-Stories/Article/Article/3852213/us-intends-to-reconstitute-us-forces-japan-as-joint-forces-headquarters/.

analysis, enabling more detailed planning to locate the adversary's center of gravity amid expanding domains in the same time frame.[46] This approach synchronizes kinetic and non-kinetic fires with battle rhythms, aligns them with force objectives, and coordinates with US forces and Japanese space, cyber, and electromagnetic units.[47] While already implemented at the army corps level with plans to expand to divisions and brigades, this approach doubles the analytical and planning workload for staff officers, posing challenges amid personnel shortages.[48] Similar challenges are likely across other services as they seek to expedite command and control. For example, the JASDF's Command and Staff College is exploring a modified version of the US Air Force's air tasking cycle to fit its defensive operations but its implementation is likely hindered by personnel shortages.[49] Meanwhile, the JJOC's military decision-making process is likely still underdeveloped.

Japan plans to address the issue of lack of an effective system for rapid decision-making by introducing artificial intelligence (AI)-supported command and control.[50] However, the JSDF currently lacks the digitized data required to teach AI.[51] Moreover, the advanced networks and information-sharing clouds needed for real-time command and control are still under development.

Additional challenges include the JSDF's limited expertise in operational art. Although the JSDF began incorporating operational art into its curriculum at Command and Staff Colleges around 2015, this was more than 30 years after the concept gained prominence in the US military.[52] The Commanding General of Training, Evaluation, Education, Research and Development Command (TERCOM), an organization established in 2018 to serve a role similar to that of the US Army's Training and Doctrine Command

---

46 JGSDF TERCOM YouTube Channel, 'Meguro chūtonchi sōsetsu 29 shūnen ando kyōiku kunren kenkyū honbu sōritsu 6 shūnen kinen, honbuchō purezente-shon [29th Anniversary of the Meguro Garrison and 6th Anniversary of TERCOM, Commanding General's Presentation]'.
47 Ibid.
48 Ibid.
49 Osamu Yanagida, 'The U.S. Military's Air Tasking Cycle and its Challenges: From the viewpoint of establishing an "Operation Cycle"', *Air Power Studies*, Vol. 5 (2018), 138-158.
50 Koji Yamazaki, 'Tōgō un'yō no genjō to kadai [Current Status and Issues of Joint Operations]', National Defense Academy of Japan Alumni Association, Mar. 2, 2024, last modified, Mar. 2, 2024, https://www.bodaidsk.com/news_topics/docs/20240302koen_tougou-unyou.pdf.
51 'Bōeishō AI katsuyō suishin kihon hōshin [Basic Policy for the Promotion of AI Application by the Ministry of Defense]', Ministry of Defense, Japan, Jul. 2, 2024, 22, last modified Jul. 2, 2024, https://www.mod.go.jp/j/press/news/2024/07/02a_03.pdf.
52 Hiroyuki Terada, 'Kaijō Jieitai kanbu gakkō no kyōiku kaikaku [Education Reform at the JMSDF Command and Staff College]', *JMSDF Command and Staff College Review*, Vol. 8 No. 2 (2019), 6-17.

(TRADOC) and Army Futures Command (AFC), has also noted that teaching operational art to JSDF officers remains one of the most significant challenges in the effort to operationalize the CDO.[53] Until these deficiencies are resolved, JSDF's capacity to effectively conduct CDO will remain significantly constrained.

Information and Intelligence

To achieve decision superiority, the JSDF must continuously collect information from geographically distant areas, provide it to relevant organizations in near real time, and analyze it effectively. Recognizing the need for improvement, the JSDF plans to significantly enhance its information and intelligence capabilities, which have historically received less emphasis. Key initiatives include strengthening its ISR assets significantly. Beginning in 2025, the JSDF plans to launch a satellite constellation for target observation, aiming for full operational capability by 2027.[54] Furthermore, introducing numerous unmanned assets will enable sustained ISR operations within enemy's weapon range, ensuring continuous and comprehensive information-gathering capabilities.[55] The JSDF is also updating its common cloud-based platform for all services, implementing AI-driven automated information collection and analysis systems, and adopting predictive services for future intelligence estimates.[56] Additionally, the JSDF aims to elevate the status of its intelligence units, ensuring that intelligence operations can be conducted independently from operations divisions. This change is intended to establish a robust intelligence collection system with minimum influence from the end-users.[57]

---

53 JGSDF TERCOM YouTube Channel, 'Meguro chūtonchi sōsetsu 29 shūnen ando kyōiku kunren kenkyū honbu sōritsu 6 shūnen kinen, honbuchō purezente-shon [29th Anniversary of the Meguro Garrison and 6th Anniversary of TERCOM, Commanding General's Presentation]'.
54 'Progress and Budget in Fundamental Reinforcement of Defense Capabilities, Overview of FY 2025 Budget Request', Ministry of Defense, Japan, Nov. 26, 2024, 7, last modified Nov. 26, 2024, https://www.mod.go.jp/en/d_act/d_budget/pdf/20241126a.pdf.
55 Ministry of Defense, Japan, 'Defense Buildup Plan', 9.
56 Ministry of Defense, Japan, 'Defense of Japan 2024', 318.
57 Daizo Udagawa and Kentaro Hirose, 'Kaijō Jieitai Reiwa 7 nendo gaisan yōkyū no jūten [Draft on the Budget of JMSDF For FY 2025]', *Ships of the World*, January 2025, 156-163.

Kinetic Fires

Kinetic and non-kinetic fires are critical pillars of the CDO. The former is being emphasized to neutralize distant threats quickly and efficiently. To this end, the JSDF is actively enhancing the range, speed, and diversification of launch platforms for stand-off missiles. Historically, the JSDF only possessed ground-launched anti-ship missiles with a range of approximately 200 km. However, since 2018, the acquisition of missiles such as the JASSM-ER (900 km range) and, more recently, Tomahawk missiles (1500 km range) and hypersonic glide vehicles (3000 km range) have significantly extended JSDF's fire projection.[58] Additionally, multiple hypersonic weapon systems are under development to minimize enemy reaction times, with live-fire testing already conducted for the glide projectile.[59] Diversification of launch platforms aims to increase the element of surprise by attacking the enemy from unexpected angles and timing. For example, the JSDF is equipping submarines with cruise missiles and developing a system similar to the Rapid Dragon, which fires large numbers of missiles from transport aircraft.[60] Furthermore, the JSDF is focusing on leveraging unmanned aerial vehicles for precise targeting to improve fire accuracy.[61] These developments underscore the JSDF's commitment to enhancing its kinetic firepower as a core element of CDO.

Non-Kinetic Fires

The advancement of non-kinetic fires, particularly effective in disrupting enemy networks and weakening their decision-making capabilities, has gained significant momentum under CDO. The JSDF has been reorganizing its structures, increasing personnel, and upgrading command positions for space, cyber, and electromagnetic domains. Space Operations Squadron was established in 2020 to oversee space domain awareness and has expanded

---

58 Shun Kawaguchi and Yusuke Kaite, 'Chijō misairu, 3 dankai haibi" hangeki nōryoku" shatei 1000~3000 kiro seifu kentō [Ground-launched missiles: Government considering 3-stage deployment of ground-based missiles with "counterstrike capability" range of 1,000 to 3,000 km.]', *Mainichi Shimbun*, Nov. 11, 2022, https://mainichi.jp/articles/20221125/ddn/001/010/004000c.
59 'Tōsho bōeiyō kōsoku kakkūdan no jizen hassha shiken [Pre-fire test of Hyper Velocity Gliding Projectile for island defense]', Acquisition, Technology & Logistics Agency Official YouTube Channel, Jul,4, 2024, last modified Jul. 4, 2024, https://www.youtube.com/watch?v=dgaGUb1GAEo.
60 Ministry of Defense, Japan, 'Defense Buildup Plan', 7.
61 Ibid., 9.

capacity to 200 personnel. Plans are underway further to grow this unit into a Space Operations Command manned with 670 personnel.[62] There are also initiatives to increase the personnel of cyber units—initially numbering 890 in late 2022—to a force of 4,000 by 2027.[63] Furthermore, the government plans to expand the role of the Cyber Defense Unit, which currently exclusively focuses on defend in JSDF cyber systems, to assume responsibilities for active cyber defense in the future.[64] In the electromagnetic spectrum, efforts are progressing across all services of the JSDF. JGSDF, for instance, established the Electronic Warfare Unit in 2022 under the Ground Component Command and has been restructured to facilitate joint operations.[65]

The MOD also plans to significantly strengthen its cognitive domain capabilities. They plan to create a new command hub within the ministry's administrative structure to oversee operations related to information warfare, including the cognitive domain.[66] These efforts reflect a comprehensive approach to enhancing the JSDF's non-kinetic capabilities as a critical component of CDO.

Despite the JSDF's substantial efforts to enhance non-kinetic fire capabilities, these remain largely defensive.[67] JSDF has begun deploying stand-off electronic warfare aircraft and stand-in electronic warfare assets to disrupt and jam Chinese military networks.[68] However, initiatives in other domains remain limited, particularly in offensive non-kinetic operations.

For instance, the JSDF has not pursued research into offensive non-kinetic methods in the space domain, such as satellite dazzling or other counterspace measures. In the cyber domain, while legislative efforts are underway to authorize active cyber defense, the timeline for implementation remains uncertain. Moreover, it appears unlikely that the JSDF will be granted authority with a wide range of cyber offensive tools since the scope

---

62 'Kūji ni" uchū sakusendan" senmon butai ōhaba kaihen de shinsetsu e ["Space Operations Command" to be established in the JASDF as part of a major reorganization of specialized units.]', *Mainichi Shimbun*, Oct. 3, 2024, https://mainichi.jp/articles/20241003/ddm/012/010/045000c.

63 Ministry of Defense, Japan, 'Defense Buildup Plan', 11-12.

64 Ministry of Defense, Japan, 'National Defense Strategy', 26.

65 'Reika butai shōkai [Introduction of the subordinate units]', Ground Component Command, Japan, last accessed Dec. 22, 2024, https://sec.mod.go.jp/gsdf/gcc/hq/220-unites.html.

66 Ministry of Defense, Japan, 'Defense of Japan 2024', 318-319.

67 Ministry of Defense, Japan, 'Defense of Japan 2020', 12.

68 Ministry of Defense, Japan, 'Defense Buildup Plan', 12.

of discussion in the government's expert group continues to be limited.[69] These limitations underscore the challenges the JSDF faces in advancing its non-kinetic fire capabilities beyond a defensive posture.

Cognitive warfare efforts within the JSDF remain largely concentrated on safeguarding Japan's decision-making capability, with proactive initiatives confined to the realm of strategic communication. At the operational and tactical levels, attempts to influence or disrupt adversary cognition are also defensive and still in their early stages. For example, in 2024, JGSDF conducted its first exercise employing balloon decoys, an initial step toward deceiving enemy sensors and decision-makers.[70] However, this initiative is still in its infancy and far from being fully operationalized. These lack in offensive capabilities raise concerns about their effectiveness in undermining enemy decision-making processes.

Complementing CDO

Given the disadvantage in kinetic fire capacity and the restrained approach toward non-kinetic capabilities, it seems unlikely that the JSDF could reclaim superiority over the PLA through CDO alone. Against this backdrop, the JGSDF's TERCOM published a document titled *JGSDF 2040* in 2024, which outlines future operational concepts for the JSDF beyond 2040.[71]

JGSDF 2040 draws on various studies and papers previously released by TERCOM, presenting insights into the evolution of JSDF's operational concepts. One of the most influential contributions referenced is a 2023 paper titled "Temporal Warfare." This paper critiques the suitability of maneuver-based doctrines such as CDO for resource-constrained nations like Japan. It argues that maneuver warfare inherently favors resource-rich powers like China or the United States. Instead, the paper calls for a shift in focus from

69 'Saiba- anzen hoshō bun'ya de no taiō nōryoku no kōjō ni muketa yūshikisha kaigi [Expert Group on Enhancing Response Capabilities in Cyber Security]', Cabinet Secretariat, Japan, Jun. 7, 2024, last modified Nov. 29, 2024, https://www.cas.go.jp/jp/seisaku/cyber_anzen_hosyo/index.html.

70 Reiwa 6 nendo ho-ku chū SAM butai sōgō kunren [FY 2024 Comprehensive Training of Hawk and Medium SAM Units]', 1st Antiaircraft Artillery Brigade, Ground Self Defense Force, Japan, last accessed, Dec. 22, 2024, https://www.mod.go.jp/gsdf/nae/katudou/1aab/katudou_syoukai/butai_kunren/R6_1AAB-sougou-boukuu.html.

71 'Rikujō Jieitai 2040 [JGSDF 2040]', TERCOM, Ground Self Defense Force, Japan, Sep. 18, 2024, last modified, Sep. 18, 2024, https://www.mod.go.jp/gsdf/tercom/riku2040.html.

spatial and domain-based strategies to those that emphasize the temporal dimension of warfare.[72]

By leveraging time as a critical factor, temporal warfare advocates for approaches that disrupt the enemy's plans, delay their operational momentum, and exploit time-dependent vulnerabilities, thereby offsetting resource disadvantages. This concept challenges traditional approaches prioritizing operational tempo, such as accelerating the OODA loop. While faster decision cycles have been regarded as advantageous in modern warfare, such methods strain resource-limited forces like the JSDF significantly. Instead, temporal warfare seeks to overcome material inferiority by focusing on the "pre-operational" period—leveraging actions that occur before what is conventionally recognized as the start of an operation.[73]Temporal warfare aims to create a situation in which the enemy loses the ability to achieve the end state when they realize it or are forced to divert or abandon the operation. This goal is achieved through two main methods: (1) disorientation, which affects the enemy commander's intentions through cognitive warfare, such as confusion, and (2) dislocation, which forces the enemy to waste forces or change targets.[74] By implementing these efforts before the exchange of kinetic fire, the JSDF could reduce the size of the enemy's force that they would need to confront using CDO, reducing the disadvantage in capacity. However, it remains uncertain whether the JSDF will adopt this approach or, if implemented, whether it will enable them to regain an advantage over Chinese forces remains uncertain. JGSDF's organizational structure, which is tailored to traditional land warfare, may face challenges in achieving timely decision-making.

Conclusion

In conclusion, the JSDF's implementation of CDO marks a significant milestone in Japan's defense strategy development, despite its nascent stage. When viewed through the lens of Japan's historical constraints on capability development and the evolution of the JSDF's operational concept, this development represents a profound shift. For the first time, the JSDF is

---

72 Anna Ozaki, 'Jikūsen jikan no" hayasa" o kachime to shita kidōsen no aratana tatakaikata [Temporal warfare - A new way of fighting maneuver warfare using the "quickness" of time as a winning factor]', TERCOM, Ground Self Defense Force, Japan, Apr. 21, 2023, last modified, Apr. 21, 2023, https://www.mod.go.jp/gsdf/tercom/img/file2184.pdf.
73 Ibid.
74 Ibid.

developing a warfighting concept and defense capabilities tailored specifically to address the mounting threats posed by China's military modernization and to prepare for a potential major conflict. As the risk of Chinese aggression toward Taiwan and the Nansei Islands increases, the JSDF must continue to enhance its military effectiveness. This requires not only the procurement of advanced equipment but also the continuous refinement and dissemination of its operational concepts through the development of a joint CDO doctrine. By doing so, the JSDF can ensure its forces are better prepared to meet future challenges.

12

Warfighting Concepts and the State of MDO Development
Does the Emperor Have Any Clothes?

Davis Ellison and Tim Sweijs[1]

Introduction

Defense planners will recognize China's "core military concept" of multidomain precision warfare. The concept leverages a Command, Control, Communications, Computers, Intelligence, Surveillance and Reconnaissance (C4ISR) network -enhanced by big data and artificial intelligence- to quickly identify and exploit vulnerabilities in operational systems through coordinated, cross-domain strikes.[2] This mirrors the multidomain operations (MDO) concept gaining traction in militaries on both sides of the Atlantic.

---

1 This chapter is adapted from previous work developed by Davis Ellison and Tim Sweijs for The Hague Centre for Strategic Studies and *War on the Rocks*. See Davis Ellison and Tim Sweijs, 'Breaking Patterns: Multidomain Operations and Contemporary Warfare' (The Hague, Netherlands: The Hague Centre for Strategic Studies, September 2023), https://hcss.nl/report/breaking-patterns-multidomain-operations-and-contemporary-warfare/; Empty Promises? A Year Inside the World of Multidomain Operations'. *War on the Rocks* (blog), 22 January 2024. https://warontherocks.com/2024/01/empty-promises-a-year-inside-the-world-of-multidomain-operations/.

2 'Military and Security Developments Involving the People's Republic of China', Annual Report to Congress (US Department of Defense, 2023), https://media.defense.gov/2023/Oct/19/2003323409/-1/-1/1/2023-MILITARY-AND-SECURITY-DEVELOPMENTS-INVOLVING-THE-PEOPLES-REPUBLIC-OF-CHINA.PDF.

Yet, while MDO is driving force modernization,[3] its effectiveness in winning wars remains uncertain.[4]

New warfighting concepts frequently exhibit excessive optimism and a lack of coordination with other services and allies, compounded by the absence of a clear theory of success. A warfighting concept is a description in general terms of the application of military art and science within a defined set of parameters. For many contemporary concepts, what has stood out is a mix of ideas, visions, and terms that often have little to do with one another.[5] For some, such as in the US Army headquarters, MDO is simply the latest iteration of a Revolution in Military Affairs (RMA), envisioning a technologically driven approach focused on networked missiles and satellites for decisive destruction.[6] Others, including the British armed forces and NATO, see it as a call for enhanced whole-of-government integration, promising universal deterrence and military victory. Neither perspective, however, fully captures the complexities and limitations of MDO.

Recent battlefield outcomes expose a glaring gap between MDOs promises and reality.[7] Ukrainian forces discarded NATO's maneuver-centric concepts to breach Russian defenses through attrition, not maneuver. Similarly, Israel's multidomain concept—predicated on total surveillance and rapid precision strikes,[8] failed to anticipate Hamas's incursion or avoid grinding urban warfare in Gaza.[9] These cases challenge core MDO assumptions,[10] namely, the ability to see everything, move quickly, and strike anywhere to rapidly resolve a conflict with minimal civilian harm. Promises of seamless multidomain dominance increasingly ring hollow.

---

3 Chiara Libiseller, '"Hybrid Warfare" as an Academic Fashion', *Journal of Strategic Studies* 0, no. 0 (2023): 1–23.

4 James N. Mattis, 'USJFCOM Commander's Guidance for Effects-Based Operations', *Parameters* 38, no. 3 (2008): 18–25.

5 Ellison and Sweijs, 'Breaking Patterns: Multidomain Operations and Contemporary Warfare'.

6 Colin Gray and Wiliamson Murray, *Strategy for Chaos: Revolutions in Military Affairs and the Evidence of History* (London: Frank Cass Publishers, 2002).

7 Jamie Dettmer, 'Ukraine's Forces Say NATO Trained Them for Wrong Fight', *Politico*, 22 September 2023, https://www.politico.eu/article/ukraine-war-army-nato-trained-them-wrong-fight/.

8 Haleigh Bartos and John Chin, 'What Went Wrong? Three Hypotheses on Israel's Massive Intelligence Failure - Modern War Institute', Modern War Institute, 31 October 2023, https://mwi.westpoint.edu/what-went-wrong-three-hypotheses-on-israels-massive-intelligence-failure/.

9 Amos Fox, 'Urban Warfare, Sieges, and Israel's Looming Invasion of Gaza', *War on the Rocks*, 27 October 2023, https://warontherocks.com/2023/10/urban-warfare-sieges-and-israels-looming-invasion-of-gaza/.

10 Yaakov Lappin, 'The IDF's Momentum Plan Aims to Create a New Type of War Machine', *Begin-Sadat Center for Strategic Studies* (blog), 22 March 2020, https://besacenter.org/idf-momentum-plan/.

This chapter evaluates whether MDO can fulfill its promise as a viable warfighting concept. Drawing on desk research across eight cases (Denmark, France, Germany, Israel, NATO, Taiwan, UK, US), interviews with practitioners, field visits, and insights derived from the Symposium Future War: Rethinking Fire and Manoeuvre-a forum for transatlantic defense experts—we develop a framework to assess MDO's strengths and weaknesses.[11] This chapter offers a framework for evaluating the viability of MDO across national contexts. We first define warfighting concepts and identify success factors through historical analysis of Western doctrines (e.g., AirLand Battle, Network-Centric Warfare) and military innovation theory. Central to our analysis are two pillars: 1) theory of success, or the causal logic explaining how a concept achieves strategic objectives and 2) defeat mechanisms, or processes that systematically degrade an adversary's capacity to fight. Applying this framework, we categorize current MDO approaches and assess their implementation. Ultimately, we address the chapter's core question: does MDO, in its current form, have substantive value—or is it merely a rhetorical 'emperor's new clothes'?

The Evolution of Warfighting Concepts

Why do some warfighting concepts succeed while others fail? And do discarded concepts vanish entirely, or do their core ideas persist in institutional memory, only to resurface later in new forms? History suggests two primary pathways for conceptual failure: battlefield failure or a sudden shift in the security environment which renders existing concepts obsolete.

Military defeat creates significant evolutionary pressure to revise conceptual approaches. Clearly, current approaches to warfare are facing challenges, prompting a reevaluation of core assumptions. This reevaluation occurs both from a functional perspective:— (1) how to improve our armed forces' military effectiveness, and (2) institutional factors, like preventing leaders from repeating past mistakes.

The origins of MDO lie in the US military's defeat in Vietnam in the 1970s. The post-Vietnam adaptation of the US Army is an indicative case here, one that is particularly relevant for the evolution of military thinking into what would become MDO. As described by Peter Mansoor, "In the

---

11 The webpage of the Symposium 'Future War: Rethinking Fire and Manoeuvre' with individual papers and interviews with the key contributors can be accessed at https://hcss.nl/hcss-nato-hq-sact-symposium/.

post-Vietnam period, army leaders not only relegated counterinsurgency doctrine to the ash heap of history but also adjusted the force structure to eliminate the types of capabilities needed to pursue such operations."[12] As the US Army abandoned counterinsurgency and sought to leave the experience of Vietnam behind it, it refocused its conceptual efforts in the European theatre on countering Soviet and Warsaw Pact conventional force advantages. This, also informed by the Israeli experience in the 1973 Yom Kippur War which was studied closely by US strategists, led to the doctrine of Active Defense, which then evolved into AirLand Battle. AirLand Battle featured methods and capabilities (such as newly developed precision strike weapons underpinned by early versions of reconnaissance strike complexes) designed to defeat both the first and follow-on echelons of Red Army armored forces.[13]

AirLand Battle would become enshrined in NATO doctrine by 1986.[14] Central to this new approach was a new generation of conventional weapons and the proliferation and enhancement of digitized Command and Control (C2) systems, a collective development that formed the basis of what would be classified as the Revolution in Military Affairs (RMA).[15] The AirLand Battle concept was an important driver of these new capabilities. Additionally, the concept revived the Corps-level echelon as the principle fighting formation that could generate sufficient mass, effectively distribute airpower, coordinate theatre-level logistics, and achieve campaign objectives at scale.[16] Importantly, the AirLand Battle concept supported itself with a simple logic that argued that new weapons in larger formations, which were already being fielded and therefore both sufficiently mature and present in sufficiently large numbers, would offset Soviet conventional advantages by striking rear-area targets (e.g. field headquarters, supply lines, and depots) to disrupt the Warsaw Pact's ability to sustain a prolonged fight in Central Europe.[17] Focused solely on a specific threat within a defined theatre, the

---

12 Peter R. Mansoor and Williamson Murray, *The Culture of Military Organizations* (Cambridge, UK: Cambridge University Press, 2019), 301.

13 Mansoor and Murray, 303.

14 'Deep Battle: Showing How Its Done', *Field Artillery Journal*, February 1986, 22–23.

15 Michael E. O'Hanlon, *The Science of War: Defense Budgeting, Military Technology, Logistics, and Combat Outcomes* (Princeton, NJ: Princeton University Press, 2009), 171–87.

16 Douglas W. Skinner, 'AirLand Battle Doctrine', Professional Paper (Arlington, VA: Center for Naval Analyses, September 1988), 12, https://apps.dtic.mil/sti/pdfs/ADA202888.pdf.

17 Hugo Wass de Czege, 'Commentary on "The US Army in Multidomain Operations 2028"' (Carlisle Barracks, PA: US Army War College, April 2020), https://press.armywarcollege.edu/monographs/909/.

clearly articulated concept provided a testable theory of success. This underpinned experimentation, exercises, and substantial force development efforts, including major procurement programs.[18]

The RMA and its attendant post-Cold War concepts of "full-spectrum dominance," "network-centric warfare," and "effects-based operations" took hold in military thinking as a result. The first wave was predominantly led by post-Vietnam airpower theorists who stressed the revolutionary potential of precision guided munitions.[19] This trend went on to be institutionalized in the US Office of the Secretary of Defense's Command and Control Research Program (now an external organization called the International Command and Control Institute).[20] The active study and development of new concepts became increasingly institutionalized through such programs, efforts which expanded to other states such as the United Kingdom and other European states.[21] The proliferation of new concepts, supporting sub-concepts, and government-funded research underpinning such work led to the rapid increase of new terms and ideas within and between the armed services, ministries, and allied states. The stage was set for a new era of jargon-centric military confusion.

Several consistent features emerge from the late-Cold War and post-Cold War eras of military conceptual thinking. First, there was a pronounced technological focus on longer-range precision fires, which were considered ideal for neutralizing hardened or mobile enemy forces without requiring a significant ground commitment. Second, the integration of modern computing into military operations introduced the concept of connectivity, enabling enhanced coordination and information sharing across units. Third, there was an emphasis on rapid decision-making and execution, deemed essential for achieving decisive effects early in a conflict and avoiding prolonged engagements. The military successes of the 1991 Persian Gulf War and the 1999 Kosovo air campaign appeared to validate these principles as

---

18 Edward C. Keefer, *Harold Brown: Offsetting the Soviet Military Challenge, 1977-1981*, Secretaries of Defense Historical Series (Washington, D.C.: Office of the Secretary of Defense Historical Office, 2017).

19 Mansoor and Murray, *The Culture of Military Organizations*, 445–46.

20 'Command and Control Research Portal', Command and Control Research Portal, accessed 5 July 2023, https://internationalc2institute.org.

21 'Ministry of Defence | Fact Sheets | DCDC - Background', The National Archives, accessed 10 July 2023, https://webarchive.nationalarchives.gov.uk/ukgwa/20080205182025/http://www.mod.uk/DefenceInternet/FactSheets/DcdcBackground.htm; Yotam Feldman, 'Dr. Naveh, Or, How I Learned to Stop Worrying and Walk Through Walls', *Haaretz*, 25 October 2007, https://www.haaretz.com/2007-10-25/ty-article/dr-naveh-or-how-i-learned-to-stop-worrying-and-walk-through-walls/0000017f-db53-df9c-a17f-ff5ba92c0000.

effective. Collectively, these three features shaped the adoption of another purported RMA, particularly in the form of network-centric warfare (NCW) during the 1990s and early 2000s.

The apparent success of the precision revolution, as embodied in NCW, fostered over-confidence in these new systems, which collided with the complex realities of 21st century warfare. Coalition forces in Afghanistan and Iraq, as well as Israel in Lebanon, struggled to achieve decisive military-strategic outcomes in their respective counterinsurgency (COIN) campaigns. During the 1990s and early 2000s, the Israeli Defense Forces (IDF) adopted US-style approaches such as "effects based operations" (EBO), emphasizing airpower to rapidly target and eliminate enemy forces in hopes of resolving conflicts quickly.[22] However, during the Second Israel-Lebanon War of 2006, the IDF engaged in a 34 day ground campaign that ended in a stalemate as Israeli airpower failed to eliminate Hezbollah fighters who were well-concealed. The Winograd Commission, established by the government to assess the war, concluded that "the expectation by some members of the IDF's leadership that the nation's precision stand-off capability could decide the outcome of the war without a major supporting ground action was 'wrong.'"[23]

The years 2014 and 2015 marked a collective intellectual crisis for many Western armed forces. With the resurgence of a revanchist Russia and an increasingly assertive China, defense planners turned towards the emerging concept of great power competition. Russia's annexation of Crimea and the subsequent war in eastern Ukraine, the rise of ISIS in both Iraq and Syria, and heightened Chinese military activity in the South China Sea redirected attention to larger-scale (conventional) military operations. For some within Western militaries, the refocusing was a welcome relief after years of deadlocked counterinsurgency in the Middle East—akin to the US Army's pivot after Vietnam.[24]The return to developing concepts for state-based opponents seemed more intuitive and strategically clear.

---

22 Avi Korber, 'The Israel Defense Forces in the Second Lebanon War: Why the Poor Performance?', *Strategic Studies* 31, no. 1 (2008), https://www.tandfonline.com/doi/full/10.1080/01402390701785211.

23 As cited in Benjamin S. Lambeth, *Air Operations in Israel's War Against Hezbollah: Learning from Lebanon and Getting It Right in Gaza* (Santa Monica, CA: RAND Corporation, 2011), 213; Martin van Creveld, 'Israel's Lebanese War: A Preliminary Assessment', *The RUSI Journal* 151, no. 5 (1 October 2006): 40–43, https://doi.org/10.1080/03071840608522872.

24 Micah Zenko, 'America's Military Is Nostalgic for World Wars', *Foreign Policy* (blog), 13 March 2018, https://foreignpolicy.com/2018/03/13/americas-military-is-nostalgic-for-great-power-wars/.

Both China and Russia had closely analyzed the performance of Western forces and adapted accordingly. Concepts such as China's "systems confrontation warfare" and Russia's "reconnaissance strike complexes" were developed to specifically target perceived vulnerabilities in Western forces.[25] A growing concern among Western defense planners was the development of states of Chinese and Russian anti-access/area-denial (A2AD) "bubbles." These zones integrate advanced sensors, missile systems and electronic warfare capabilities to restrict freedom of maneuver for opposing military forces across all domains. Studies quickly proliferated that foresaw China's A2AD capabilities in the South China Sea and Taiwan Strait aimed to deter US intervention, while Russia's A2AD zones in Kaliningrad and Crimea threatened NATO's operational flexibility in Eastern Europe.[26]

Russia and China's adoption of lessons from NATO operations reflects the dynamics of military-strategic evolution. The strategic military ecosystem in which these doctrines develop creates evolutionary pressures that drive adversaries to adapt and innovate.[27] This process has led to mirror-imaging—a two-way dynamic—where Russian notions like the reconnaissance-strike complex (rooted in Soviet thinking from the 1980s) parallel Western thinking on cross-domain precision fires. Similarly, China's systems destruction approach emphasizes paralysis and disintegration through the combination of kinetic strikes and non-kinetic means.

The roots of the first explicitly "multidomain" concepts appeared in 2015, with a speech by then US Deputy Secretary of Defense Robert Work at the US Army War College in the context of developing a new "offset strategy" against Russia and China. He argued that "the real essence of the third

---

25 Jeffrey Engstrom, 'Systems Confrontation and System Destruction Warfare: How the Chinese People's Liberation Army Seeks to Wage Modern Warfare' (RAND Corporation, 1 February 2018), https://www.rand.org/pubs/research_reports/RR1708.html; Oscar Jonsson, *The Russian Understanding of War: Blurring the Lines between War and Peace* (Washington, DC: Georgetown University Press, 2019).

26 David A. Shlapak and Michael Johnson, 'Reinforcing Deterrence on NATO's Eastern Flank: Wargaming the Defense of the Baltics' (RAND Corporation, 29 January 2016), https://www.rand.org/pubs/research_reports/RR1253.html; Mark F. Cancian, Matthew Cancian, and Eric Heginbotham, 'The First Battle of the Next War: Wargaming a Chinese Invasion of Taiwan', CSIS International Security Program (Center for Strategic and International Studies: Washington, D.C., January 2023), https://csis-website-prod.s3.amazonaws.com/s3fs-public/publication/230109_Cancian_FirstBattle_NextWar.pdf?VersionId=WdEUwJYWIySMPIr3ivhFolxC_gZQuSOQ; Roger Cliff, *China's Military Power: Assessing Current and Future Capabilities* (Cambridge, UK: Cambridge University Press, 2015).

27 Rafe Sagarin, *Learning From the Octopus: How Secrets from Nature Can Help Us Fight Terrorist Attacks, Natural Disasters, and Disease* (New York: Basic Books, 2012); David Kilcullen, *The Dragons and the Snakes: How the Rest Learned to Fight the West* (London: Hurst Publishers, 2020).

offset strategy is to find multiple different attacks against opponents across all domains so they can't adapt...."[28] The US Army followed this initiative with the public release of "Multi-Domain Battle" in 2018. Described as an evolution of AirLand Battle, it emphasized a "system of systems, increased operational options, integration, and speed," revisiting key elements from earlier US concepts.[29] From this initial conceptual phase, individual armed services and ministries of defense launched ostensibly collective but often disjointed efforts to develop their own service-specific or national joint warfighting concepts.[30]

Based on this historical review, six major themes emerge as critical to the successful development of warfighting concepts: the importance of straightforward language and ideas; the centrality of clear threats within a broader national (political-military and inter-service) system; realism about the maturity of technology; the necessity of ensuring a coherently argued internal logic that includes defeat mechanisms; and finally the relevance of recognizing risks.[31] From AirLand Battle to the high optimism of NCW, through the comprehensive approaches of the COIN era, and on to its most recent evolution into hybrid warfare and gray zone conflict, concept development has proliferated widely and has been no stranger to controversy.[32]

What Makes For A Good Warfighting Concept?

The six factors identified above form the analytical framework for the remainder of this chapter, inverting elements that contribute to failure to identify those that favor success. The first factor is whether a concept's

---

28 Bob Work, 'Army War College Strategy Conference', US Department of Defense, 8 April 2015, https://www.defense.gov/News/Speeches/Speech/Article/606661/army-war-college-strategy-conference/.

29 'Multidomain Battle' (Washington, D.C.: US Army Science Board, January 2018), https://asb.army.mil/Portals/105/Documents/2010s/2017%20A%20MDB%20Report.pdf?ver=bhWh5nT9fIwNANl0jW3wGQ%3D%3D.

30 Kelly McCoy, 'The Road to Multidomain Battle: An Origin Story', Modern War Institute, 27 October 2017, https://mwi.westpoint.edu/road-multidomain-battle-origin-story/.

31 Davis Ellison and Tim Sweijs, 'Empty Promises? A Year Inside the World of Multidomain Operations', War on the Rocks (blog), 22 January 2024, https://warontherocks.com/2024/01/empty-promises-a-year-inside-the-world-of-multidomain-operations/.

32 Donald Stoker and Craig Whiteside, 'Blurred Lines: Gray-Zone Conflict and Hybrid War – Two Failures of American Strategic Thinking', Naval War College Review 73, no. 1 (2020); Lukas Milevski, 'When Does Gray Zone Confrontation End? A Conceptual Analysis', Joint Force Quarterly 112 (Quarter One 2024).

language is sufficiently clear, as different states, services, and even individuals use similar terms interchangeably, describe overlapping concepts, or make minor adaptations to definitions leading to confusion akin to a "Tower of Babel". The second factor is whether a concept aligns with a state's overall defense apparatus—culturally, institutionally and procedurally—and is accessible beyond a small community of concept developers. The third examines whether the technology underpinning a concept is sufficiently mature and available in sufficient quantities within a state and its armed forces. The fourth factor is clarity about the threat. States facing immediate existential threats often develop concepts to specific scenarios, while those without such threats tend to rely on concepts with only vague descriptions. The fifth factor, central to this report's logic, is the presence of a clear argument for how a concept will achieve success—a 'theory of success'—including defeat mechanisms that support this argument. Finally, the sixth factor considers whether and how conceptual work has addressed risk. Each of these six factors will be addressed in turn below.

(Un)clarity of Language

The military concept development world is rife with what insiders refer to as "buzzword bingo," where new terms, acronyms, and entire concepts proliferate year after year.[33] Organizations such as the UK's Development, Concepts and Doctrine Centre (DCDC) and the US Army's Training and Doctrine Command (TRADOC) are professionally tasked with driving this proliferation.[34] NATO alone has 30 Allied Joint Publications (AJPs) governing military doctrine, which are themselves subordinate to overarching concepts like the NATO Warfighting Capstone Concept (NWCC) and the Concept for the Deterrence and Defence of the Euro-Atlantic Area (DDA).

This extensive conceptual development creates a fragmented mixture of language and ideas, often developed in isolation and frequently repackaging preexisting concepts. While this is not inherently problematic—assuming

---

33 Kate Bateman, 'War on (Buzz) Words', *Proceedings*, August 2008, 20–23; Elena Wicker, 'Full-Spectrum Integrated Lethality? On the Promise and Peril of Buzzwords', *War on the Rocks*, 17 May 2023, https://warontherocks.com/2023/05/full-spectrum-integrated-lethality-on-the-promise-and-peril-of-buzzwords/.
34 David Morgan-Owen and Alex Gould, 'The Politics of Future War: Civil-Military Relations and Military Doctrine in Britain', *European Journal of International Security* 7 (2022): 551–71.

that the underlying assumptions are sound—the real challenge arises from the coexistence of new thinking alongside older, established concepts. This leads to multiple terms and ideas being used interchangeably or inconsistently, without a shared understanding.[35] Over time this recycling of language further muddles the field, complicating efforts to achieve clarity and cohesion in military doctrine.

Across the cases we researched, there is significant variation in terms and meanings associated with multidomain concepts. The term multidomain is followed by various descriptors, chiefly "operations" (Denmark,[36] NATO,[37] US[38]), "integration" (UK[39]), "manoeuvre" (Israel[40]) and "deterrence" (Taiwan[41]). France uses Multimilieux / multichamps,[42] while Germany refers to Multidimensionalität.[43] The meanings of "domain" and "dimension" also differ: some countries (Denmark, France, Germany, Israel, US) explicitly limit the term to the five military domains (air, sea, land, cyber, space) while others (UK and Taiwan) adopt a broader interpretation that includes other government functions. This linguistic complexity deepens outside English-speaking contexts, where "domain" and "dimension" are sometimes interchangeably used, as seen in German and Hebrew. Exercises range

---

35 Stephen Peter Rosen, *Winning the Next War: Innovation and the Modern Military*, Cornell Studies in Security Affairs (Ithaca, NY: Cornell University Press, 1991), 34–36.

36 'Multi-National Capability Development Campaign - Multidomain Multi-National Understanding Report Annex A' (Norfolk, VA: NATO, November 2022).

37 'Alliance Approach to Multidomain Operations' (NATO Allied Command Transformation, 2022).

38 General James C. McConville, 'Army Multidomain Transformation: Ready to Win in Competition and Conflict' (Washington, D.C.: Headquarters, Department of the Army, 16 March 2021).

39 Ministry of Defence, 'Joint Concept Note 1/20, Multidomain Integration' (Ministry of Defence, November 2020), https://assets.publishing.service.gov.uk/government/uploads/system/uploads/attachment_data/file/950789/20201112-JCN_1_20_MDI.PDF.

40 Lappin, 'The IDF's Momentum Plan Aims to Create a New Type of War Machine'.

41 'Taiwan National Defense Report 2021' (Taipei: Ministry of National Defense, October 2021), https://www.ustaiwandefense.com/tdnswp/wp-content/uploads/2021/11/Taiwan-National-Defense-Report-2021.pdf; 'Taiwan 2021 Quadrennial Defense Review' (Taipei: Ministry of National Defense, 2021), https://www.mnd.gov.tw/.

42 Philippe Gros et al., 'Intégration Multimilieux / Multichamps : Enjeux, Opportunités et Risques à Horizon 2035' (Fondation pour la recherche stratégique, 2022), https://www.frstrategie.org/sites/default/files/documents/publications/recherches-et-documents/2022/102022.pdf.

43 Generalleutnant Alfons Mais, 'Mittlere Kräfte - Operative Reaktionsfähigkeit Und Motor Der Modernisierung' (Köln, 26 April 2023).

from single-service operations involving one unit to whole-of-government approaches.[44]

What emerges from this era of concept development demonstrates a continued and even increased proliferation of terms and ideas that do not have a shared meaning. Importantly, even when a new concept is formally agreed upon, its implementation often diverges from the original definitions. As seen in the cases above, terms like "multidomain" are adapted to fit different contexts.

Poor Regime Fit

In many cases, the MDO concept does not integrate well within existing government and military structures. This is particularly evident when a new concept places the military or defense in a leading or coordinating role for other ministries or departments. The British concept, which positions the military as a central coordinator for all security affairs, has been at odds with both the Foreign, Commonwealth and Development Office and Parliament.[45] The concept has been less digestible within the British political-military system as it proposes an outsized, and arguably inappropriate, role for the military.

In the US, inter-service rivalry has directly impacted efforts to institutionalize multidomain operations across the armed forces. The US

---

44 Judah Ari Gross, 'In 1st Drill, IDF's Ghost Unit Tests out New Tactics with Jets, Tanks and Robots', 23 July 2020, https://www.timesofisrael.com/in-1st-drill-idfs-ghost-unit-tests-out-new-tactics-with-jets-tanks-and-robots/; Tania Donovan, 'Lightning Edge 21: 25th Infantry Division Exercises Multidomain Task Force Capabilities', US Army, 14 May 2021, https://www.army.mil/article/246417/lightning_edge_21_25th_infantry_division_exercises_multi_domain_task_force_capabilities; Orlandon Howard, 'US Army Tests New Multidomain Ops Doctrine in Warfighter Exercise', US Army, 18 October 2022, https://www.army.mil/article/261239/us_army_tests_new_multidomain_ops_doctrine_in_warfighter_exercise; Armée Française, 'Press Kit: ORION 23' (Press notice, Exercise ORION 23, Paris, February 2023), https://www.defense.gouv.fr/sites/default/files/operations/20230228_Press_Kit_Orion.pdf; 'NATO Exercise STEADFAST JUPITER 2022 Concludes', NATO Joint Warfare Centre, 20 October 2022, https://www.jwc.nato.int/articles/steadfast-jupiter-2022-concludes; 'BALTOPS 22, the Premier Baltic Sea Maritime Exercise, Concludes in Kiel', US Navy, 17 June 2022, https://www.navy.mil/Press-Office/News-Stories/Article/3066830/baltops-22-the-premier-baltic-sea-maritime-exercise-concludes-in-kiel/ https%3A%2F%2Fwww.navy.mil%2FPress-Office%2FNews-Stories%2FArticle%2F3066830%2Fbaltops-22-the-premier-baltic-sea-maritime-exercise-concludes-in-kiel%2F.
45 Morgan-Owen and Gould, 'The Politics of Future War: Civil-Military Relations and Military Doctrine in Britain'.

Army and Air Force have led competing development efforts,[46] while the Navy and Marine Corps developed their own service-specific approaches tailored to the Western Pacific.[47] Differing service processes in training, budgeting, and procurement further hinder joint efforts by locking implementation into service-specific channels.

At the lowest level, tensions exist within individual services as well. Many countries pursuing multidomain operations—including the United States, United Kingdom, Germany, France,[48] and Israel,[49]—have envisioned the army division as the most appropriate echelon for MDO. However, this has not been consistent across cases. Both the United States and Israel have smaller multidomain units such as the US Multidomain Task Force (brigade-sized formations in Europe and the Pacific) at the theater level,[50] while Israeli Ghost unit[51] operates as an experimental special forces battalion. In Israel, debates over who "does" multidomain operations reflect long-standing rivalries within the army between the airborne and armored corps.[52]

Dissent exists across all three levels-between government ministries, between services and within individual services. Concepts stressing a "whole-of-government" type approach risk civil-military tension over command,[53] while those that emphasize precision-strike systems risk inter-service rivalries over who owns these new capabilities or who leads in command. Although these debates are largely confined to peacetime, they often persist in wartime, making them critical to address.

---

46 Sean M. Zeigler et al., 'Aligning Roles and Missions for Future Multidomain Warfare' (Washington, D.C.: RAND Corporation, 19 August 2021), https://www.rand.org/pubs/research_reports/RRA160-1.html.

47 Will Spears, 'A Sailor's Take on Multidomain Operations', *War on the Rocks*, 21 May 2019, https://warontherocks.com/2019/05/a-sailors-take-on-multidomain-operations/.

48 Anthony King, *Command: The Twenty-First Century General* (Cambridge, UK: Cambridge University Press, 2019).

49 Herzi Halevi, 'Multidomain Defense', IDF, 20 October 2001, https://www.idf.il/en/mini-sites/dado-center/vol-28-30-military-superiority-and-the-momentum-multi-year-plan/multidomain-defense-maj-gen-herzi-halevi/.

50 Donovan, 'Lightning Edge 21'.

51 Gross, 'In 1st Drill, IDF's Ghost Unit Tests out New Tactics with Jets, Tanks and Robots'.

52 Zaki Shalom, 'The "War of the Generals" after the Yom Kippur War', *Strategic Assessment* 24, no. 3 (2021).

53 Lawrence Freedman, *Command: The Politics of Military Operations from Korea to Ukraine* (Oxford, New York: Oxford University Press, 2022).

Technological Immaturity

Much of the technological emphasis in MDO concepts is on speed: speed in communication, speed in action, and speed in movement. From one perspective, this focus makes sense, assuming that war can be resolved early and quickly.[54] However, constantly pushing for speed risks ignoring tempo, potentially causing operations to spiral out of control for both commanders and political leaders.[55] Additionally, the assumed linkage between tactical and operational speed and the beneficial strategic outcomes does not hold true in all situations. Compounding this issue is the material reality that many game-changing capabilities are simply not yet available. This is particularly evident in European forces,[56] which continue to face significant shortages in capabilities such as command, control, communications, computers, intelligence, surveillance, and reconnaissance capabilities (C4ISRs) and deep precision strike.[57]

Most MDO concepts fall prey to this technological overconfidence, especially regarding communications. Assured connectivity in combat is central to nearly all MDO frameworks. Nations like the United States, United Kingdom, France, Germany, Israel, Taiwan, and NATO place next-generation C4ISR capabilities at the core of their concepts, assuming robust networks will be available near future. However, despite significant attention—particularly informed, by experiences in Ukraine—this level of assured connectivity

---

54 Andrew Metrick, 'Rolling the Iron Dice: The Increasing Chance of Conflict Protraction' (Washington, D.C.: Center for a New American Security, November 2023), https://www.cnas.org/publications/reports/rolling-the-iron-dice.

55 Nina A. Kollars, 'War at Information Speed: Multidomain Warfighting Visions', in *War Time: Temporality and the Decline of Western Military Power, Edited by Sten Rynning, Olivier Schmitt, and Amelie Theussen*, Chatham House Insights Series (London: Royal Institute of International Affairs, 2021).

56 Hugo Meijer and Stephen G. Brooks, 'Illusions of Autonomy: Why Europe Cannot Provide for Its Security If the United States Pulls Back', *International Security* 45, no. 4 (20 April 2021): 7–43, https://doi.org/10.1162/isec_a_00405.

57 Ben Barry et al., 'The Future of NATO's European Land Forces: Plans, Challenges, Prospects' (London: International Institute for Strategic Studies, 2023), https://www.iiss.org/research-paper/2023/06/the-future-of-natos-european-land-forces/.

remains far from realistic.[58] Russian,[59] Chinese,[60] and Iranian[61] forces have heavily invested in electronic warfare capabilities aimed at degrading their opponents' battlefield connectivity over the past decades,[62] exposing a critical blind spot.

The assumption of technological maturity inherent in ongoing conceptual work presents a serious flaw. There is little reason to believe that the envisioned speed and decisiveness of multidomain operations are either technologically feasible or capable of delivering the desired outcomes. High-level efforts in recent years such as the US Joint All-Domain Command and Control (JADC2) system have yet to achieve significant progress. This persistent hubris within Western thinking about warfare poses a real risk.

Vague Threats

It seems self-evident that a military concept should designate a specific adversary. If the intended result is to compel an enemy to do your will, it follows that the adversary and the threat it poses must be explicitly considered. Many concepts, however, fail to single out adversaries and instead offer only vague threat descriptions. While the United Kingdom, France, Germany, and Denmark are implicitly focused on Russia and NATO's eastern front, this does not translate into detailed threat descriptions based on enemy approaches. NATO benefits from greater specificity due to its 2022 Strategic Concept, which explicitly identifies two threats: Russia and terrorist organizations.[63] For the United States this task is more complex as its efforts must address global interests. The US Army's multidomain operations concept is implicitly designed around both a Baltic and Taiwan scenario,[64]

58 Maggie Smith and Jason Atwell, 'A Solution Desperately Seeking Problems: The Many Assumptions of JADC2', Modern War Institute, 3 May 2022, https://mwi.westpoint.edu/a-solution-desperately-seeking-problems-the-many-assumptions-of-jadc2/.

59 Duncan McCrory, 'Electronic Warfare in Ukraine' (Joint Air Power Competence Centre, 7 October 2023), https://www.japcc.org/articles/electronic-warfare-in-ukraine/.

60 Engstrom, 'Systems Confrontation and System Destruction Warfare'.

61 'Iran Stages "electronic Warfare" Drills against Mock Enemy Drones, State TV Reports', Reuters, 25 August 2023, https://www.reuters.com/world/middle-east/iran-stages-electronic-warfare-drills-against-mock-enemy-drones-state-tv-2023-08-25/.

62 James Black et al., 'Multidomain Integration in Defence: Conceptual Approaches and Lessons from Russia, China, Iran and North Korea', n.d.

63 'NATO Strategic Concept 2022' (NATO, 2022), https://www.nato.int/strategic-concept/.

64 Andrew Feickert, 'Defense Primer: Army Multidomain Operations (MDO)' (Washington, D.C.: Congressional Research Service, 21 November 2022), https://sgp.fas.org/crs/natsec/IF11409.pdf.

obliquely framing the main challenge as ensuring maneuverability in a missile-dominated environment.

For countries facing direct threats at their borders, such as Taiwan or Israel, this issue is less problematic. Their concepts identify adversaries and contain detailed threat descriptions. In contrast, major NATO states lack this specificity, which creates challenges for both strategy-making and defense-planning. Detailed threat descriptions remain a perennial difficulty for concept developers, particularly within multinational structures like NATO. The most detailed assessments are often classified, and achieving a common inter-service and intra-alliance view of threat specificity is an ongoing challenge. Nevertheless, the need persists for an adequately granular problem definition against which a theory of success can be formulated—aligning ways and means against something tangible.

(No) Theory of Success

Very few actors explicitly formulate a theory of success in their military conceptual work. Among the cases examined in this study, only France, Israel, and Taiwan attempt to make a causal argument about how their proposed multidomain operations concepts lead to defeating an opponent.

For instance, building on Peter Viggo Jakobsen's framework, such an argument might state: "IF NATO forces adopt a multidomain operations approach that incorporates long-range precision fires alongside forward defensive systems, THEN these forces can effectively defeat a Russian attack along the eastern front, BECAUSE these divisions can effectively target rear-echelon targets while blunting assaults by frontline Russian units."[65] This logic mirrors the Cold War-era AirLand Battle concept.

The example above illustrates what Eado Hecht, a prominent Israeli military analyst, describes as a "defeat mechanism"—a process that explains how specific actions cause the damage necessary to defeat an enemy.[66] Such a mechanism is testable and can be refined through simulations and training exercises. However, some approaches, such as those adopted by the UK and NATO, rely on broad comprehensive strategies that lack clear causal claims. These "whole-of-government" approaches aim for deterrence or victory but

---

65 Peter Viggo Jakobsen, 'Causal Theories of Threat and Success – Simple Analytical Tools Making It Easier to Assess, Formulate, and Validate Military Strategy', *Scandinavian Journal of Military Studies* 5, no. 1 (9 September 2022): 177–91, https://doi.org/10.31374/sjms.164.
66 Eado Hecht, 'Defeat Mechanisms: The Rationale Behind the Strategy', *Military Strategy Magazine* 4, no. 2 (2014): 24–30.

remain scientifically unfalsifiable since they do not specify how coordination across agencies and nations directly leads to defeating a specific adversary.

No theory is perfect or guaranteed to succeed. Even the NATO example above could fail. However, it can be tested through joint exercises, simulations, and analysis of contemporary conflicts to refine its validity. What matters most is that such theories are rationally argued, include clear cause-and-effect relationships against specific opponents, and align with broader policy and strategic goals.

Opaque Risks

A key aspect often overlooked in new warfighting concepts—and in many assessments of them—is the inherent risk involved in adopting a new approach. Every new concept entails implicit trade-offs that carry risks.[67] By prioritizing certain threats, selecting specific capabilities, or proposing new organizational structures, choices are made whose drawbacks are rarely made explicit. When risks are acknowledged, they are often framed solely as the dangers of not implementing and funding the concept—a calculation shaped as much by bureaucratic considerations as by threat perceptions.

There are at least four significant risks associated with an uncritical approach to developing MDO-type concepts that receive insufficient attention: the risk of commanders being overwhelmed by an overly broad span of control;[68] an over-reliance on connectivity; a mechanistic, over-engineered approach that becomes top-heavy;[69] and the assumption that the whole is inherently greater than the sum of its parts. If left unaddressed, these risks have and will continue to derail progress in concept development.

Context is particularly important in identifying where trade-offs have been made. The timing of development efforts often implies the deprioritizing of other activities. For example, AirLand Battle not only addressed specific needs but also shifted the US Army's focus away from counterinsurgency. This highlights a broader risk: militaries do not always control how their capabilities will ultimately be employed. Overpreparing for a narrowly preferred scenario risks undermining readiness for other

---

67 Mark F. Cancian, 'Military Force Structure: Trade-Offs, Trade-Offs, Trade-Offs' (Washington, D.C.: Center for Strategic and International Studies, 26 February 2018), https://www.csis.org/analysis/military-force-structure-trade-offs-trade-offs-trade-offs.
68 Steve Maguire, 'Multi Domain Operations Below the Division', *Wavell Room* (blog), 3 September 2021, https://wavellroom.com/2021/09/03/tactical-mdo/.
69 Mattis, 'USJFCOM Commander's Guidance for Effects-Based Operations'.

possible tasks. Therefore, identifying internal risk is essential both during concept development and in evaluating adopted concepts.

Conclusions

After years of engagement with multidomain operations (MDO) operators and strategists, the authors of this chapter express concern that MDO risks remaining a fashionable concept without being implemented at scale. The comprehensive approach associated with MDO is increasingly intertwined with RMA-style techno-optimism, while various sub-components, and government sectors interpret these concepts in ways that align with their institutional interests. The "why" and especially the "how" of MDO lack a clear or fully convincing argument. In its current state, MDO resembles an aspiring emperor with barely any clothes. While it is possible to refine these concepts moving forward, the current trajectory appears unpromising.

This situation is particularly troubling for small and middle powers in Europe pursuing a MDO concept. With the United States is explicitly reducing its role in European security, the US military capabilities that European states have relied upon in their multidomain thinking are eroding. In the short term, European armed forces must prioritize mastering the basics of combined arms maneuver and improving joint operations, rather than chasing the elusive ideal of multidomain operations.

That said, militaries worldwide—from the Donbas to the Caucasus and from the Red Sea to the South China Sea—are transforming themselves to capitalize on some of the promises of this new generation of conceptual thinking. Technologies such as drones, C4ISR assets, deep precision missiles, and hypersonic glide vehicles have decisively influenced the battlefields in recent wars to the advantage of both attackers and defenders. As European militaries focus on perfecting combined arms warfare and joint operations, they should incorporate lessons from contemporary conflicts. These include leveraging distributed units, reconnaissance strike complexes, and the coordinated use of unmanned systems on land, sea and in air.

Military history demonstrates that successful military transformations are typically evolutionary rather than revolutionary. They arise from adopting new technologies, developing operational concepts, and adapting organizational structures. Therefore, combined arms maneuver and military transformation should complement each other rather than work at cross purposes.

13

Multidomain by Design
The Australian Defence Force and MDO

Andrew Carr[1]

Introduction

In 1976 the Australian Defence Force (ADF) was created. This was a historic change in Australian military organization. The intention was to bring together the three separate services and create a force that was more than the sum of its parts. With limited resources and a vast continent to defend, Australia needed a defense force that could work as one across sea, air and land to protect its interests.

Over the decades since, the ADF has transformed into a force that is now proudly "joint by design."[2] This chapter offers readers an account of this evolution, as a way of highlighting the opportunities and challenges at the heart of the US Army's Multidomain Operations (MDO) concept.[3] For MDO to succeed it will need to emulate, at a vastly larger scale, many of the organizational, intellectual and cultural changes the ADF has undertaken in its pursuit of what Australians call "jointery."

The chapter also explores how Australia views MDO, both as an operational concept, and as a framework for military cooperation in the

---

1 The use of Australian English is used in this article only for proper nouns.
2 Robbin F. Laird, *Joint by Design: The Evolution of Australian Defence Strategy* (2020).
3 United States Army, *The U.S Army in Multidomain Operations 2028*, TRADOC Pamphlet 525-3-1, December 6, 2018.

US-Australia alliance. On the surface, MDO offers a compelling vision. There are many similarities between Australian and US military assumptions about the nature and demands of modern warfare.[4] There is also a widespread desire in both Canberra and Washington D.C. for the Australia, New Zealand, and United States (ANZUS) alliance to "move beyond interoperability to interchangeability" and "operate seamlessly together, at speed."[5] There are however barriers to overcome. First, MDO is not well-aligned with the particular strategic problem set facing Australia in the 2020s. MDO emphasizes an offensive counterpunch against aggressive authoritarian states, while Australia is concerned with establishing deterrence and war time denial and resilience as the southern bastion of US efforts in the Indo-Pacific. Second, the ANZUS alliance lacks the institutional and strategic mechanisms required to navigate such divergences and enable the long term and seamless cooperation inherent to MDO. There are also notable technical barriers, particularly relating to data sharing.

Australia should be an easy case for an MDO partner, given both the closeness with the US and the widespread embrace by the ADF of joint and coalition operations. Yet, because of how MDO is framed and who it is framed for, there remains a disconnect between those ambitions and achieving the political, institutional and technical structures needed to make it the right operating concept for the US and its allies and partners in the Indo-Pacific.

The Australian Defense Force — Reorganization for Jointery

For the first seven decades after Federation, Australia's armed forces operated largely as independent services.[6] They often had more in common, and greater habits of cooperation with their British counterparts, than they did their fellow countrymen. Although their strategic purpose was firmly set by Canberra, their organizational structures, equipment, training, doctrine

---

4 See for example, Australian Army, *Accelerated Warfare: Futures Statement for an Army in Motion* (Canberra: Australian Army, Australian Defence Force, 2018).
5 Richard Marles, "Address: Center for Strategic and International Studies (CSIS), Washington D.C.," ed. Department of Defence (Canberra: Commonwealth of Australia, 2022).
6 For the best accounts of these early origins, see Jeffrey Grey, *The Australian Army*, The Australian Centenary History of Defence (Melbourne: Oxford University Press, 2001).; David Stevens, *The Royal Australian Navy*, The Australian Centenary History of Defence (Melbourne: Oxford University Press, 2001).; and Alan Stephens, *The Royal Australian Air Force*, The Australian Centenary History of Defence (Melbourne: Oxford University Press, 2001).

and views of warfare heavily reflected overseas coalition partners.[7] The services enjoyed distinct governance structures as well. From 1915 onwards both the Army and Navy had their own Government Departments, each with a Minister in the Federal Cabinet to represent their interests. In 1939, an Air Force Department was added. That same year the establishment of the Chiefs of Staff Committee helped to bring some coherence to military advice to the government. However, the central Department of Defence was, for much of the 20th century, a small department which focused on coordinating the services and their industrial and logistical requirements.[8] Without a well-staffed "strategic center" or central command structures for the forces, powerful service chiefs commonly held very divergent views on defense policy and capability.

In 1972, the government of Gough Whitlam inaugurated a review of Australia's defense organizations. This was a moment of significant strategic change. The Cold War in Asia was rapidly declining in threat, and the two main poles of the Western bulwark against Communism in the region, the United Kingdom and the United States were stepping back. Australia no longer feared toppling dominos of communist aggression. Instead, it needed a force to defend its homeland, in a self-reliant fashion—that is, without expecting the combat assistance of the United States.

Under the resulting 1973 "Tange Reforms," developed by the powerful Secretary of the Department of Defence, Sir Arthur Tange, a new structure was established, designed explicitly for cooperation. The three services retained their domain-based roles and identities; however they now sat within a single national defense force. As Australia's foremost military historian David Horner has explained, "The ADF was formed so that it could develop command, operations, logistics and training structures that might allow it to conduct joint operations in the most effective manner possible."[9]

Command structures for this new force were slow to emerge.[10] Amidst recriminations over the war in Vietnam, and a declining defense budget,

---

7 David Horner, *Making the Australian Defence Force,* vol. IV, The Australian Centenary History of Defence (Melbourne: Oxford University Press, 2001), 1.

8 David Horner, *Defence Supremo: Sir Fredrick Shedden and the Making of Australian Defence Policy* (Sydney: Allen & Unwin, 2000), 6.

9 *Making the Australian Defence Force,* IV, 1.

10 David Horner, *Strategy and Command: Issues in Australia's Twentieth-Century Wars* (Cambridge: Cambridge University Press, 2022), 274-76.

many saw amalgamation as a pathway to greater civilian control.[11] Others questioned the pace and logic of how it was implemented. As one prominent former Army official has argued:

> In every other comparable country the creation of a joint chiefs of staff and/or joint-Service, strategic-level headquarters preceded the unification of the defence group of departments or ministries. Only Australia combined the separate Defence, Navy, Army, Air and Supply Departments first … and, after a long delay, created a joint-Service, strategic-level military headquarters.[12]

It would take four decades for the ADF to build that headquarters and emerge as a genuinely joint force. Along the way many compromises were needed to bring everyone along.[13]

Representative of the fitful movement towards jointery was the acquisition of capabilities that sat across domains. In 1987, there was a coup in Fiji. This challenge confronted the Government with the ADF's limited capabilities for regional operations. Yet, which part of Defense was supposed to fund them? As the historian John Blaxland writes of attitudes during this period," Navy saw amphibious ships as serving the Army's needs and therefore not a budget priority for Navy, whereas Army saw the funding of amphibious ships as Navy's responsibilities."[14] Partly this reluctance was, and remains, a function of size. Australia's armed forces are small compared to many other major industrial states—just 58,200 active personnel and a budget of around $36.4bn in $USD in 2025—which forces each service to specialize narrowly in their primary domain.[15] The turmoil also reflected problems with the translation of the "Defence of Australia" policy into a coherent national military strategy during the 1980s.

---

11 Neil James, 'Reform of the Defence Management Paradigm: A Fresh View' (Australian Defence Studies Centre: Australian Defence Force Academy, University of New South Wales, 2000), 13.
12 "Reform of the Defence Management Paradigm: A Fresh View," 13.
13 John Blaxland, *The Australian Army: From Whitlam to Howard* (Melbourne: Cambridge University Press, 2013), 28.
14 Blaxland, *The Australian Army: From Whitlam to Howard*, 66.
15 IISS, "The Military Balance," (Singapore: International Institute for Strategic Studies, 2025), 231.

Cultural and institutional change towards a joint identity was also gradual and sometimes driven as much by the desire for efficiencies as coherence. The creation of joint education structures, such as the Australian Joint Service Staff College, Joint Services Wing, and Australian Defence Force Academy helped break down service differences.[16] In 1996, the creation of Headquarters Australian Theater as the senior cross-domain headquarters, was a crucial reform. In 2004, Headquarters Joint Operations Command (HQJOC) was established at a specially built facility near the nation's capital to enable local and global joint operations. Finally, the position of Vice Chief of the Defence Force has also emerged as a vital leadership role for ensuring joint doctrine, logistics, education, and ensuring capability acquisitions meet the joint force's requirements.

The ADF and Multidomain Operations Today

The Australian Army's reforms between 2023 to 2025 provide an illustrative case study that highlights the ongoing work to align the service with the ADF's need for multidomain operations. Unlike in the US, single services in Australia are simply too small to operate across domains on their own, hence the 1970s push for a "joint" approach. But this comes with the challenge of balancing each service's traditional domain preferences and habits, with the needs of a cohesive national force.

For much of the 20th century, the Australian Army believed itself to be at the center of the nation's strategic policy. National days of honor celebrated its alliance-affirming expeditionary campaigns in places such as Gallipoli in World War I, Tobruk in World War II, and Korea and Vietnam in the Cold War. However, by the late 20th century, such operations were a fading memory, and the Defence of Australia policy emphasized the defense of the "Air-Sea Gap" north of the country. The Army feared a future where their role was relegated to mopping up forces which might get through and dealing with low-level harassment campaigns. In the 2000s, the Army's spirits and operational tempo increased as regional demands such as the 1999 United Nations mission into East Timor and the post-2001 Global War on Terror placed heavy demands upon its personnel. Yet being busy with these modern expeditionary conflicts would not suffice as an answer to a gnawing question

---

16 Blaxland, *The Australian Army: From Whitlam to Howard*, 69.

for Australia's land forces: What was the Army's "strategic role" within a joint and self-reliant national defense strategy?[17]

In 2023, the Australian Government released a landmark review of its armed forces titled "National Defence: Defence Strategic Review" (DSR). Conducted by a former Defence Minister and former Chief of the Defence Force, it made the stark finding that the ADF was no longer "fit for purpose."[18] Instead of a "balanced force" that was able to provide government with a range of military options as had long been prioritized, the review argued that "The ADF must evolve into a genuine integrated force which harnesses effects across all five domains: maritime, land, air, space and cyber and prioritize solutions to realistic scenarios agreed to by the Government."[19]

At its core, the DSR established a new problem-set for Australia's armed forces. Rather than awaiting a hypothetical defense of the continent, the ADF would pursue what has been called "Archipelagic Deterrence," a pro-active shaping and deterring endeavor in the archipelagos to the nation's north, against both conventional and gray-zone threats.[20] That common problem set, tied to a challenge of national significance has made it much clearer and easier for the Army to accept the significant changes now required of it. In many cases this meant implementing ideas put forward by its own people in previous years, but which had not had the opportunity to come to the fore without this requirement for fundamental change.[21]

Australia's Army has been tasked to become "optimized for littoral operations in our northern land and maritime spaces and provide a long-range strike capability."[22] As official Army documents have noted, "This direction is transformational for the Australian Army. It will see the largest recapitalisation of Army's equipment in generations…It will see us integrated by design across the ADF."[23] The task has been taken up with

---

17 Brendan Sargeant, "The Australian Army through the Lens of Australian Defence White Papers since 1976," in *An Army of Influence: Eighty Years of Regional Engagement*, ed. Craig Stockings and Peter Dennis (Melbourne: Cambridge University Press, 2021), 40.

18 DOD, *National Defence: Defence Strategic Review*, Department of Defence (Canberra: Commonwealth of Australia, 2023), 7.

19 DOD, *National Defence: Defence Strategic Review*, 19, 53.

20 DOD, *National Defence: Defence Strategic Review*; Andrew Carr, "Australia's Archipelagic Deterrence," *Survival* 65, no. 4 (2023).

21 Australian Army, *Army in a Joint Archipelagic Manoeuvre Concept*, Discussion Paper (Canberra: Commonwealth of Australia, 2014).

22 DOD, *National Defence: Defence Strategic Review*," 7.

23 Australian Army, *The Australian Army Contribution to the National Defence Strategy*, Australian Army (Canberra: Commonwealth of Australia 2024), 1.

vigor by the Chief of Army, Lieutenant General Simon Stuart, who observes that since 2023:

> We have rewritten the land domain concept, the land operating concept (which translates the joint or the integrated force concept into the land force component), and translated the Chief of Joint Operations plans into force structure and readiness requirements for the Army... We have re-organized the Army. We have changed its disposition, and we are getting after the reorganization of units at brigade and at battlegroup level.[24]

As a practical example of this, prior to 2023, the Australian Army sought to purchase 450 infantry fighting vehicles. Through the DSR this number was cut to 129 with the Army's leadership supporting the change against internal opposition.[25] In its place, the Army began acquiring systems designed for mobility and long-range strike in the archipelagos, including Littoral Maneuver Vessel's, HIMARS, and UH-60M Blackhawk helicopters. Reflecting the need for a force focused on operating in the archipelagos to the country's north, Australia is spending 38 percent of its defense budget on the maritime domain, against just 16 percent for land and 14 percent for air.[26]

In June 2024 the new capstone doctrine for the ADF, *Australian Military Power*, proclaimed that, "The ADF has evolved its organisation into an integrated multidomain force."[27] That's largely right, although there is still a lot of work needed for this to continue. As Robbin Laird, a long-time observer of the ADF has noted, for all the language about multidomain operations and common sense of an archipelagic operating environment, "Each service focuses primarily on a particular domain and sees its role in terms of a maritime strategy from their perspective."[28] Still, the acceptance of a common

---

24 Robbin F. Laird, "Australian Army Chief of Staff on Force Transition," Laird, Robbin, https://defense.info/interview-of-the-week/australian-army-chief-of-staff-on-force-transition/.

25 Anthony Galloway, "'Army Are on the Nose': Australia Warned over Slashing of Tanks," *The Sydney Morning Herald*, 29/04/2023 2023.

26 DOD, *2024 Integrated Investment Program*, ed. Department of Defence (Canberra: Commonwealth of Australia, 2024), 11.

27 "Australian Military Power: Adf Capstone Doctrine," ed. Department of Defence Australian Defence Force (Canberra: Commonwealth of Australia, 2024), 29.

28 Robbin F. Laird, "Multidomain Operations in Australia's Maritime Strategy: The Army, Navy, and Air Force Orient Their Efforts," https://defense.info/multidomain-dynamics/2024/04/multidomain-operations-in-australias-maritime-strategy-the-army-navy-and-air-force-orient-their-efforts/.

problem set has been essential for enabling significant adaption and reform in the name of joint and multidomain cooperation. This is however a foundation which MDO struggles to provide for US-Australia integration.

Challenge 1: MDO and Australia's Strategic Problem Set

Although MDO was designed as a response to the military threat posed by the People's Republic of China—an issue foremost in the concerns of Australian strategists—there is a significant divergence in diagnosis of what the specific strategic problem from China is, and in turn, how best to respond.[29] The primary military problem which MDO seeks to address in the Indo-Pacific is China seeking a *fait accompli* strike against a nearby state, rapidly establishing an anti-access/area denial (A2AD) bubble, and then forcing the US and allies to punch their way back in.[30] At least, that's what most analysts take the document to mean. As Huba Wass de Czege has observed that the concept paper, "Does not articulate a well-developed theory of the problem," a shortfall which then impedes both the deterrence of adversaries and reassurance of allies.[31] The most plausible real-world scenario for what MDO envisages is a Chinese attack on Taiwan. Other possible scenarios could see attacks on Japanese, South Korean or Philippines territory, although the political logic of why and where is harder to discern.

An attack on Taiwan would deeply trouble Australia; however, retaking that island is not the strategic problem set which Australian forces are designed to address. Instead, there are three other problem sets, all tied to China, which are far more significant in shaping Australian military strategy and in turn its compatibility with MDO. These are: (1) Establishing regional conventional deterrence, (2) establishing inner-arc gray-zone deterrence, and finally, (3) protecting a US-Australia southern bastion during a major war.

---

29 MDO is also tied to the threat from Russia, though this chapter will focus just on the application in the Indo-Pacific. To understand the language of problem sets in strategy see, Andrew Carr, 'Strategy as Problem Solving', *Parameters: US Army War College Quarterly*, 54 (1), Spring 123-137, 2024.

30 United States Army, *The U.S Army in Multidomain Operations 2028*, TRADOC Pamphlet 525-3-1, December 6 2018, viii-ix.

31 Huba Wass de Czege, *Commentary on the "US Army in Multidomain Operations 2028"* (Carlisle, PA: Strategic Studies Institute, US Army War College, 2020), xix.

Regional Conventional Deterrence

First, how does MDO help to prevent conflict in Northeast Asia from occurring? For deterrence to work, Beijing needs to believe it is unable to achieve its military goals, and/or would find the results of seeking them so unpleasant or unacceptable as to not be worth attempting. MDO seems to take its inspiration from the post-Cold War problem set, where the US would gear up to cross the oceans and fight their way in, rather than the Cold War problem set of deterring war, defending allies and handling escalation.[32] MDO's positioning of the US to retake Taiwan is one way to deter, but it would be far better if they never had to do so.

Historically, strategies which persuade the adversary that they will be denied their military ambitions tend to be more effective than those which threaten costs in response.[33] MDO however does not clearly promise either of these traditional paths. Instead, MDO seeks to deter through a promised counter-punch—the People's Liberation Army (PLA) may achieve its initial military objective but can't hold it. This carries all the problems of a punishment-based logic—relying on a future US President's willingness to initiate a major war—while offering few of the direct benefits of punishment by implying it would leave the Chinese Communist Party and Chinese homeland largely intact. This is also a deterrent threat which entirely falls upon the US to prosecute, since even nearby allies such as Japan, as well as distant ones like Australia, will not have the capacity to offer a counterpunch to the PLA in these scenarios. The other implicit deterrent element of MDO is the claim that it will result in a much more effective and efficient US warfighting capacity. That may be true, however improvements of degree to armed forces are a weak deterrence signal, compared with approaches which present the adversary with new operational or strategic problems which they must respond to.

It may well be that promising a more effective US counterpunch is the right move. Perhaps there is no obvious way to prevent the initial capture of Taiwan, although six years after MDO's public release, major studies still suggest it could be defended.[34] Or, perhaps intelligence reveals that the PLA's

---

32 "Commentary on the "US Army in Multidomain Operations 2028"," xvii.

33 Robert A. Pape, *Bombing to Win: Air Power and Coercion in War* (Ithaca: Cornell University Press, 1996).

34 Andrew S. Erickson, Conor M. Kennedy, and Ryan D. Martinson, eds., *Study No. 8, Chinese Amphibious Warfare: Prospects for a Crossstrait Invasion*, Cmsi Studies in China Maritime Development (Newport, Rhode Island: US Naval War College, 2024).

primary fear is that they would be unable to hold their claim on Taiwan, hence why we should loudly demonstrate our ability to push on that pressure point, disrupting the PLA's "theory of victory." If a counterpunch is indeed the right way to deter, the detailed case for this logic needs to be made to allies, along with the evidence to justify it. Deterrence is at heart an empirical problem—one that must be based on detailed assessments of the specific adversary and specific situations. Until that evidence is provided, MDO appears weakly placed to deter a war from occurring, while offering little substantive role for allies and partners in the Indo-Pacific.

Gray-Zone Deterrence

A second distinctive problem set for Canberra is that the PLA is increasingly well placed to pursue harassment campaigns against Australia in the same way it does to Taiwan, the Philippines and Japan on a daily basis. MDO's promise of further alliance integration could be useful in countering this threat. For instance, along the lines of the Center for Strategic and Budgetary Assessments' proposed "deterrence by detection" framework.[35] However, given MDO's primary focus on war fighting, along with changing attitudes in Washington D.C towards allies and burden sharing, it remains unclear whether the US would have much capacity or intention to fundamentally adjust how it supports its allies in deterring gray zone threats.

For the US Army in particular there is a risk that MDO cedes one of the main ways in which land power directly supports deterrence. There is an inherent tension between mobility and deterrence.[36] By promising to potentially be anywhere, the adversary may well judge our forces will be elsewhere. Or that nervous civilian leaders won't send them into harm's way. By contrast, static, highly visible land forces are "politically entangling" for communicating deterrence and reassurance alike.[37] Likely, some combination of visible forward presence will be required for effective deterrence, even as the MDO counterpunch is continually sharpened in coming years.

---

35 Thomas G; Mahnken et al., *Implementing Deterrence by Detection: Innovative Capabilities, Processes, and Organizations for Situational Awareness in the Indo-Pacific Region* (Washington D.C: CSBA, 2021).

36 Bryan Frederick et al., "Understanding the Deterrent Impact of US Overseas Forces," (Santa Monica, California: RAND Corporation 2020).; Stephan Fruehling and Andrew Carr, *Forward Presence for Deterrence: Implications for the Australian Army* (Canberra: Australian Army Research Centre, Australian Army, Commonwealth of Australia, 2023).

37 Colin S. Gray, *Modern Strategy* (Oxford: Oxford University Press, 1999), 213.

Protecting the Southern Bastion

Third, if deterrence does fail, Australia will operate as a "southern bastion" for US military power in Asia.[38] In this context, Australia faces a problem precisely the reverse of the one MDO focuses on—establishing its own A2AD bubble and support base for US operations out of northern Australia. While northern Australia is currently beyond the range of many of China's medium and intermediate ballistic missiles (such as the DF-26, DF-21A) and some of its cruise missiles (i.e., the HN 2 and HN 3), the country could still easily be hit by ICBM strikes as well as cyber, electronic warfare, subversion and other countermeasures by the PLA.

What does MDO offer Australia in this circumstance? Will joint integration between the US and Australia remain a priority if the ADF's role is to stay close to the Australian continent, while the US military operates far forward into North Asia? Likewise, what does MDO offer Australian politicians as a basis on which to explain to a worried public how the nation is being protected? In such a conflict, it is highly likely that iconic Australian structures, such as Sydney Harbor Bridge or Parliament House in Canberra will be hit by long range missiles. How easily can an Australian Prime Minister point to MDO and assure the public that the logic of alliance cooperation is strong? Because MDO's central idea is a counterpunch to an adversary's defensive zone, rather than protecting allies from an initial strike, it could easily look inappropriate.

At the core of many of these issues is the question of size. What the US Army can do alone as a single service, is a task that is often at the outer limits of what other whole-of-nation armed forces can achieve. And often then with an assist from the US in key areas such as Space. For the US MDO concept to become a structure which can drive US-Allied coalitions in the Indo-Pacific, the realities of the diversity of scope, capacity and problem sets across the region will need to be given greater account.

Challenge 2: Integrating Australia and US forces under MDO

None of the issues thus far raised are beyond the capacity of the US and Australia to overcome. There are always differences in how partners diagnose the strategic problems they face at the start of a conflict, and such

---

38 Carr, "Australia's Archipelagic Deterrence."

tensions can remain throughout a conflict, even on the winning side. What matters however is that there are appropriate institutions for identifying common ground as well as the practical and technical protocols for the resulting cooperation to occur. Yet such structures are surprisingly weak in the US-Australian alliance.

Australian and US leaders say all the right things about cooperation. In 2018 the two countries celebrated a "centenary of mateship" since their initial operations during WWI. In 2021, Canberra, Washington, and London captured global attention with the Australia, United Kingdom, and United States (AUKUS) military technology sharing agreement. And in 2022 Australia's Defence Minister Richard Marles announced his government's intention to "move beyond interoperability to interchangeability…we will ensure we have all the enablers in place to operate seamlessly together, at speed."[39] Yet despite this, the ANZUS alliance between the US and Australia is far from the integrated structure which MDO will require.

Two quick examples illustrate the work to be done. At the strategic level, as Stephan Frühling observed in 2018, "What makes ANZUS so remarkable today is…the near-complete absence of understandings and mechanisms to address the political–military questions of how the allies will 'coordinate their efforts for collective defence for the preservation of peace and security.'"[40] Revisiting that judgement in 2024, Frühling assessed that despite a few big shifts such as AUKUS, "changes over the years since have been significant more for their novelty than for their overall effect on the U.S force posture in the Indo-Pacific or Australia's national defense effort."[41]

At a practical level, even when officials want to cooperate on multidomain and multi-national operations, many technological impediments remain in their way. For example, the capacity for the US and Australia to share the data for joint strike operations remains limited. There are not sufficient or enduring multi-national networks to test and practice through. Data collected by one system often cannot be fed directly into another or must be laboriously

---

39 Richard Marles, 'Address: Center for Strategic and International Studies (CSIS)' July 12, 2022, https://www.minister.defence.gov.au/speeches/2022-07-12/address-center-strategic-and-international-studies-csis.

40 Stephan Frühling, "Is ANZUS Really an Alliance? Aligning the US and Australia," *Survival* 60, no. 5 (2018): 200.

41 Stephan Fruhling, "US-Australia Alliance Force Posture, Policy, and Planning: Toward a More Deliberate Incrementalism" (Washington D.C: Carnegie Institute, 2024).

translated in ways that significantly impede sharing and integration by allied commanders. [42]

Partly these are technical and resource issues, but they also reflect intentional policy barriers, erected in the name of caution and sovereignty. These impede the US and Australian from training as an allied force today how they would like to fight tomorrow.[43] As Jon R. Lindsay has compellingly shown, when the, "information practice" of a military force is poor, the battlefield result is likely to be similarly dire.[44]

Significant work is underway to address all these problems, and offsets—such as thorough pre-planning of who and how to strike based on plausible targets within scenarios—could ease the data crunch. Many remain confident that in a genuine conflict these sources of friction will fall by the wayside. However, it remains notable how little major Australian strategic planning documents, such as the 2024 National Defence Strategy refer directly to joint US-Australia operations close to Australian shores.[45] For MDO to achieve its stated ambitions, its advocates need to be aware that even alliances as close as the one between the United States and Australia will need significant intellectual and organizational change to enable the concept to succeed.

Conclusion

This chapter has explored the creation of the Australian Defence Force in 1976 and the half-century-long effort to build a genuinely "joint" force. This is useful for readers interested in MDO for two reasons. First, it represents an illustrative case study of the challenge, duration and opportunities which a move towards a multidomain operating approach requires. Though Australia's Army is small compared to many Indo-Pacific nations, Australia's investments in advanced capability (particularly in the Air Force and Navy)

---

42 This is far from an alliance problem alone, the US military struggles even within its own networks. Sydney Freedberg JR, 'With dreams of JADC2, Pentagon relaunches AI-driven command & control experiments', Breaking Defence, February 01, 2023, https://breakingdefense.com/2023/02/with-dreams-of-jadc2-pentagon-relaunches-ai-driven-command-control-experiments-gide/.

43 Alan W. Throop and Sam Fairall-Lee, 'US-Australia information sharing: a self-inflicted Achilles' heel', *The Strategist* (Canberra: Australian Strategic Policy Institute), November 10, 2022, https://www.aspistrategist.org.au/us-australia-information-sharing-a-self-inflicted-achilles-heel/.

44 Jon R. Lindsay, *Information Technology and Military Power* (Ithaca, New York: Cornell University Press, 2020).

45 DOD, *2024 National Defence Strategy*, Department of Defence (Canberra: Commonwealth of Australia, 2024).

along with the integrated and professional quality of the forces help make the country a significant regional military power. The ADF is a force which is today genuinely more than the sum of its parts, just as MDO's authors hope the US military and its allies will ultimately become as well.

The second reason for detailing this history is that it also speaks to an important risk which US allies must face with the move to multi-national "jointery". If Australia embraces the internationally-integrative logic of MDO alongside the US, there is a risk the services will return to their habits of closer cross-national collaboration against domestic integration. Unified national command and coordination structures may fray as the demands of coalition operations and allies take precedence. Given the speed of modern war, allies will face an uncomfortable "all in" or "all out" choice. Political demands for picking and choosing when and where sovereign forces or infrastructure are drawn upon will be an increasingly risky, perhaps even unacceptable military burden.

Such a shift may be a worthy price to pay if the practical benefits of the MDO concept is sufficiently strong. All concepts, including something as prized as "joint" service cooperation are only valuable to the extent they make our armed forces more capable of overcoming the strategic problems faced today. However, asking a nation to give up five decades of reform, as well as much of their national command during war is a very high price for the US to ask of its allies. Especially at a time when questions of sovereignty have returned to center-stage in Australian political debates. It is a cost which will only be seriously contemplated when allies are thoroughly convinced that the conceptual logic of future cooperation is a persuasive and effective way of addressing their nation's specific strategic challenges.

As highlighted, that is not yet the case. To meet this goal, the US Army's MDO concept will need to better clarify how it directly contributes to the strategic problem sets facing US allies in Asia. It will also need to help the US and its allies to decide where they can cooperate and where clear delineations of responsibility must be drawn. Only then, with clarity about what multidomain integration is and does, can it be said that the United States and its allies will be joint, and by design.

14

Educating Officers for Joint and Multidomain Operations
Is There a Difference?

Thomas Crosbie and Holger Lindhardtsen[1]

The lecturer, red in the face:
> *"Multidomain Operations is the future– it's everything, everywhere, all at once – it's the new essence of war!"*

An officer on a staff course, yawning:
> *"Ok,° sure… but why should I care?"*

The fictional exchange above highlights one of the fundamental preoccupations of the present volume: what is meaningfully different about the new concepts of war, and what do these differences (if any) mean, in tangible, practical terms, for the majority of officers? There is an alarming degree of cognitive dissonance between the many things Multidomain Operations (MDO), Joint All-Domain Operations (JADO), and new military concepts suggest will change in the future of war. Yet, at the same time, almost imperceptible changes in education actually filter down to most officers.[2]

---

1 Thomas Crosbie, Associate Professor, Institute for Military Operations, Royal Danish Defence College. Holger Lindhardtsen, Researcher, Institute for Military Operations, Royal Danish Defence College.

2 The new concepts of war that we refer to here can be roughly classified as that class of Western military operational concepts which evolved as an alternative to counterinsurgency, state-building, and stability operations, and which focus on the return to near-peer and peer adversaries competing in large-scale conventional operations (LSCOs), among other operational settings. For a deeper understanding of the term "Operational Concept", see Antulio J. Echevarria II, "Operational Concepts and Military Strength," in *2017 Index of US Military Strength*, ed. Dakota Wood (Washington, D.C: The Heritage Foundation, 2016).. For an insight

While a cursory reading of MDO / JADO texts would make one think that military education is in need of an almost complete overhaul, the reality is that educators struggle to identify what, if anything, they need to do differently. Accordingly, for many of our overworked officer-students, the wonders of MDO / JADO, while impressive, are nevertheless quickly consigned to that mental black hole of "things that won't appear on the exam."

This chapter takes seriously the dilemma faced by the instructors and students in Professional Military Education (PME) programs who struggle to identify what makes MDO / JADO different.[3] Difference is, of course, relative. There are clear differences between preparing officers for an MDO / JADO future and preparing them to counter insurgencies, build states, and stabilize societies, as was the focus of military education during the Global War on Terror Era.[4] Likewise, an MDO / JADO focus is clearly different than a hybrid threats focus, since the latter focuses attention lower in the scale of escalation.[5] Indeed, few military educators would fail to recognize that whatever else it is, MDO / JADO signals a shift away from the "new wars" approach of the expeditionary era, and something like a return to the large-scale conventional operations (LSCOs) that occupied the planning and preparation of Western militaries during the Cold War.[6]

---

into the various ways these concepts have been developed, see: Nicholas E. Bixby, "Joint All-Domain Operations (JADO): The Maneuver Concept for Future Conflict," *Over The Horizon Journal*, November 22, 2024, https://othjournal.com/2024/11/22/joint-all-domain-operations-jado-the-maneuver-concept-for-future-conflict/; Development, Concepts and Doctrine Centre, "Multidomain Integration" (UK Ministry of Defence, 2020); Paolo Giordano, "Multidomain Operations in NATO - Explained," NATO's ACT, October 5, 2023, https://www.act.nato.int/article/mdo-in-nato-explained/; Shaun Cannon, "The Alliance's Transition to Multidomain Operations," *Journal of the JAPCC* 37 (May 28, 2024): 16–26.

3 Like MDO / JADO, the term PME is taken to mean slightly different things for different communities. In the US, the concept is explicated most recently in Joint Staff, "Officer Professional Military Education Policy," Chairman of the Joint Chiefs of Staff Instruction (Washington, D.C: Joint Staff, April 15, 2024)., with a classic criticism articulated by Joan Johnson-Freese, *Educating America's Military* (Routledge, 2013).. The European perspective has been widely debated, as for example in the work of Libel Tamir Libel, *European Military Culture and Security Governance: Soldiers, Scholars and National Defence Universities* (London ; New York: Routledge, Taylor & Francis Group, 2016); Tamir Libel, "Professional Military Education as an Institution: A Short (Historical) Institutionalist Survey," *Scandinavian Journal of Military Studies* 4, no. 1 (April 15, 2021), https://doi.org/10.31374/sjms.79.

4 Todd Greentree and Craig Whiteside, "Teaching for Irregular Warfighting Competencies," *War Room - US Army War College* (blog), October 25, 2024, https://warroom.armywarcollege.edu/articles/competencies-4/.

5 Heather S. Gregg, "Teaching Hybrid Threats and Hybrid Warfare," *War Room - US Army War College* (blog), October 31, 2024, https://warroom.armywarcollege.edu/articles/competencies-5/.

6 James Campbell, "The Joint Approach to Intermediate PME: Ensuring Warfighting Competency," *War Room - US Army War College* (blog), October 24, 2024, https://warroom.

Accordingly, when we refer to a difference in focus, we mean different than joint operations. More specifically, do these emerging concepts imply the end of the era of joint operations? This is a timely question from the perspective of military educators, especially instructor and program owners at the intermediate / staff course level.[7] Triggered by the combined effects of the withdrawal of Western militaries from distant theaters and the sudden recognition of Russia's aggressive posture following its invasion of Ukraine, many such programs are currently undergoing a period of dramatic change in what and how they educate their student officers, and for many this raises questions about whether the "joint operations" focus retains any validity. Perhaps programs should be completely reimagined, based on the emerging theories of JADO, MDO, or other national variation? This chapter argues that, on the contrary, much of officer education should remain the same, at least for now.

The first part of the chapter focuses on untangling what exactly we mean by joint operations. The second part shifts to consider what we mean by multidomain or Joint All-Domain operations. The final part of the chapter reflects on what the difference between "joint and multi" heralds for Western militaries' ongoing PME reforms.

Joint Operations

The word "joint" provides our initial stumbling block in unraveling the difference between joint operations and MDO / JADO. The word is defined in current US military doctrine as an adjective which "connotes activities, operations, organizations, etc., in which elements of two or more Military Departments participate."[8] In NATO's agreed terminology, it is an "adjective used to describe activities, operations and organizations in which elements of at least two services participate ,[9] and it serves in NATO doctrine as the preferred choice over the closely related term "multiservice."

armywarcollege.edu/articles/competencies-3/; Katrine Lund-Hansen and Jeffrey M Reilly, "The Multidomain Operations Approach to Intermediate PME," *War Room - US Army War College* (blog), November 1, 2024, https://warroom.armywarcollege.edu/articles/competencies-6/.
7 Thomas Crosbie and Holger Lindhardtsen, "What Are the Options? A Competency Development Approach for Professional Military Education," *War Room - US Army War College* (blog), October 3, 2024, https://warroom.armywarcollege.edu/articles/competencies-1/.
8 *DOD Dictionary of Military and Associated Terms* (Washington, D.C: US Department of Defense, 2021), 113.
9 "Joint," NATO Term, 16 1999, https://nso.nato.int/natoterm/Web.mvc.

At the basic level of organizational policy, "joint" is used in American military circles to refer to activities that involve more than one military department (Army, Navy, and Air Force), which would suggest that activities that combined elements from the Navy and Marine Corps (both services within the Department of the Navy) or the Air Force and Space Force (both services within the Department of the Air Force) are not in this strict sense joint.[10] Meanwhile, in NATO circles, "joint" more loosely refers to any activity including multiple services (which might pedantically include the US Navy and US Marine Corps), but with the added ambivalence of whether this refers to services of one country or of multiple countries. However, while a NATO activity that includes elements of the US Army and Danish Army might fit the definition of joint, as would a NATO-sanctioned activity included elements of the Danish Army and Danish Air Force, in practice the word "combined" is used to refer to any operation involving more than one country.

These sorts of definitional exercises provide relatively little insight, except to raise the question of why terms are needed in the first place. The answer to that is immediately evident to those who work in military organizations: virtually every Western military is profoundly shaped by the role of services, which both shape the broader organization of military power through path dependency and repeatedly shape the formation of personnel through early career training. The overcoming of service differences (cultural, technological, and otherwise) is very often a significant challenge. [11]

Despite being challenging, joint operations are not modern phenomena, and joint warfare (the planning and conduct of joint operations) is as old as history. The reason is simple: joint collaboration is often advantageous on the battlefield. Accordingly, as long as states and empires have divided their military power into services, there have been periodic attempts to gain advantages by encouraging (or forcing) the leaders of military units from different service backgrounds to work together. Since human conflict was limited to the land and maritime domains for most of our history, the history of jointness is a history of land-maritime coordination. A classic example is the Battle of Salamis, where Greek forces embarked light infantry on

---

10 The British prefer "jointery" to "jointness", but the concept remains fundamentally the same.
11 Carl H. Builder, *The Masks of War: American Military Styles in Strategy and Analysis* (Santa Monica: RAND Corporation, 1989); Jeffrey W. Donnithorne, *Four Guardians: A Principled Agent View of American Civil-Military Relations* (Baltimore, Md.: Johns Hopkins University Press, 2018), https://www.amazon.com/Four-Guardians-Principled-Civil-Military-Relations/dp/1421425424; Sharon K. Weiner, *Managing the Military: The Joint Chiefs of Staff and Civil-Military Relations* (Columbia University Press, 2022).

their warships, and after a naval battle against the Persian fleet, conducted amphibious landings, disembarking the infantry forces to destroy the remaining Persian forces on land. The Athenian campaign to conquer Sicily during the Peloponnesian War, which also saw the use of naval and land forces integrating to conduct amphibious landings, is another paradigmatic example.[12] From the Roman amphibious invasion of Britain in 43 AD to Grant's Vicksburg Campaign in the American Civil War, battlefield commanders have periodically launched operations relying on the coordination of army and navy elements. [13] The common denominator in these examples is that they were conducted in the littoral environment and mostly using naval capacities to support land-based warfare either through naval bombardments of land-based defenses or amphibious landings (or both).

World War I heralded the arrival of a new warfighting domain, namely the air domain made accessible through the introduction of aircraft in battle.[14] The emergence of a new domain and the revolutionary potential of the new aviation technologies forced military organizations to rethink the mechanics of coordination between assets operating in different domains. Partly, this was done to deconflict activities in the land and air domains, in order to minimize the risk of friendly fire, as well as to unlock the obvious benefits to reconnaissance offered by aircraft.[15] However, it was equally obvious to early air power proponents that better air-land coordination would improve fighting efficacy: neither the Allied nor the Central Powers could afford to neglect the potential of airpower once presented to the battlefield.[16]

---

12 Jeffrey M Reilly, "Multidomain Operations: A Subtle but Significant Transition in Military Thought," *Air & Space Power Journal* 30, no. 1 (2016): 61–73.

13 Steven Iacono, "The Roman Amphibious Invasion of Britain," *Naval History* 2023 (December 1, 2023), https://www.usni.org/magazines/naval-history-magazine/2023/december/roman-amphibious-invasion-britain; B.A. Friedman, "Vicksburg: The Past and Future of Amphibious Operations," The Strategy Bridge, July 4, 2017, https://thestrategybridge.org/the-bridge/2017/7/4/vicksburg-the-past-and-future-of-amphibious-operations.\\uc0\\u8220{}The Roman Amphibious Invasion of Britain,\\uc0\\u8221{}{\\i{}Naval History} 2023 (December 1, 2023

14 Carl Spaatz, "Evolution of Air Power: Our Urgent Need for an Air Force Second to None," *Military Affairs* 11, no. 1 (1947): 2, https://doi.org/10.2307/1982685; Jeremy Black, *Air Power: A Global History* (Rowman & Littlefield, 2016).{\\i{}Air Power: A Global History} (Rowman & Littlefield, 2016

15 Andrew Whitmarsh, "British Army Manoeuvres and the Development of Military Aviation, 1910—1913," *War in History* 14, no. 3 (July 2007): 325–46, https://doi.org/10.1177/0968344507078378.

16 Edward R. Lucas and Thomas A. Crosbie, "Evolution of Joint Warfare," in *Handbook of Military Sciences*, ed. Anders McD Sookermany (Cham: Springer International Publishing, 2021), 1–11, https://doi.org/10.1007/978-3-030-02866-4_21-1.

In the remainder of our discussion of joint operations, we focus on the American case, since it has provided the template for military organization and doctrine adopted by most Western countries.[17] In the United States, the interwar period was marked by intense interservice rivalry between the Army and Navy, albeit with a few exceptions (including collaboration in cryptanalysis).[18] The Army planned for operations involving an unopposed landing on friendly shores. The Navy planned to conduct few if any amphibious operations. The potential for Army-Navy joint operations was considered slim, but sufficiently possible that very basic joint doctrine was drafted in 1927, with the modest goal of describing how such operations might occur. The doctrine, titled *Joint Action of the Navy and the Army*, "assumed that operations would *normally* be conducted through mutual cooperation" (meaning, planned in parallel, without unity of command, and aligning only at the moment of attack).[19]

As World War II dragged on in Europe, American military leaders began to recognize that coordination with allies, should the United States join the war effort, would likely require better coordination between the two services, since it would require huge numbers of Army units to be transported across a militarized Atlantic with the possibility of a contested landing at various points. For the Army, joint cooperation was suddenly a prerequisite for joining the fight at all. Meanwhile, since the start of the war, German military leaders recognized that early successes were explained in

17 Kjell Inge Bjerga and Torunn Laugen Haaland, "Development of Military Doctrine: The Particular Case of Small States," *Journal of Strategic Studies* 33, no. 4 (August 2010): 505–33, https://doi.org/10.1080/01402390.2010.489707; David P Calleo, "The American Role in NATO," *Journal of International Affairs* 43, no. 1 (1989): 19–28.the development of the Norwegian Armed Forces' Joint Doctrine of 2007 is analysed in order to illustrate some distinct characteristics of the development of small state doctrines. First, small states have limited freedom and limited institutional capacity to realise their own ideas about the use of their military forces. Furthermore, their contribution of forces to multilateral military operations signals political support for an institution (NATO

18 Samuel P. Huntington, "Interservice Competition and the Political Roles of the Armed Services," *American Political Science Review* 55, no. 1 (March 1961): 40–52, https://doi.org/10.2307/1976048; "Army and Navy Get Their Act Together," National Security Agency/Central Security Service, August 20, 2021, https://www.nsa.gov/History/Cryptologic-History/Historical-Events/Article-View/Article/2740672/army-and-navy-get-their-act-together/http%3A%2F%2Fwww.nsa.gov%2FHistory%2FCryptologic-History%2FHistorical-Events%2FArticle-View%2FArticle%2F2740672%2Farmy-and-navy-get-their-act-together%2F.

19 Charles E. Kirkpatrick, "Joint Planning for Operation Torch," *Parameters* 15, no. 1 (1991): 75; "Joint Action of the Army and the Navy" (Government Printing Office, Publication Office Records, 1927), Box 45, Folder 28, Naval War College Archives.

part by their effective coordination of air, land, and at times, naval assets at the tactical level. [20]

Following the American declaration to join the Allied war effort in December 1941, the primary need for allied military coordination became self-evident, and by January the Combined Chiefs of Staff structure was established to oversee American and British forces.[21] To facilitate the work of the Combined Chiefs of Staff, the American services were coordinated under a new Joint Chiefs of Staff structure.[22] By August 1942, General Dwight D. Eisenhower was named Commander-in-Chief of the American Expeditionary Force, and was directed by the Combined Chiefs of Staff to gain control of North Africa. His planning for this mission was conducted by a joint and combined staff, meaning teams of officers from all three services of both the United States and United Kingdom.[23]

The success of the resulting Operation Torch demonstrated the extraordinary benefits (but also the great challenges) that come with close collaboration between multiple services and countries.[24] Later examples from the Pacific Theater showcased the alignment of air, land, and maritime forces to gain control of key islands through amphibious assaults, described by Williamson Murray as the "highpoint" of World War II jointness.[25]

Accordingly, World War II provided the template for contemporary joint operations, although the post-war era saw a long period of equivocation and reluctance to embrace the apparent benefits to the services of working more closely together.[26] Indeed, the United States saw a decline in organizational and operational jointness in both the Korean and Vietnam Wars.[27] The decline culminated in the 1980s, where the failures of Operations Eagle Claw and

---

20 Lucas and Crosbie, "Evolution of Joint Warfare," 3.
21 David Rigby, *Allied Master Strategists: The Combined Chiefs of Staff in World War II* (Naval Institute Press, 2012).
22 Kirkpatrick, "Joint Planning for Operation Torch," 78.
23 Kirkpatrick, 78.
24 John Gordon IV, "Joint Power Projection: Operation Torch," *Joint Force Quarterly*, 1994, 60–69; Carrie A. Lee, "Operation Torch at 75: FDR and the Domestic Politics of the North African Invasion," *War on the Rocks*, November 8, 2017, https://warontherocks.com/2017/11/16075/; Carrie A. Lee, "Military Politics on the Battlefield: Strategy and Effectiveness at War," in *Military Politics: New Perspectives*, ed. Thomas Crosbie (New York: Berghahn Books, 2023), 165–85, https://www.berghahnbooks.com/title/CrosbieMilitary.
25 Williamson Murray, "The Evolution of Joint Operations," *Joint Force Quarterly*, 2002, 34.
26 Alastair Finlan, Anna Danielsson, and Stefan Lundqvist, "Critically Engaging the Concept of Joint Operations: Origins, Reflexivity and the Case of Sweden," *Defence Studies* 21, no. 3 (July 3, 2021): 356–74, https://doi.org/10.1080/14702436.2021.1932476.
27 Murray, "The Evolution of Joint Operations," 35–36.

Urgent Fury showcased the cost of the continued lack of cooperation between services.[28]

These incidents were part of the decision to implement the 1986 Goldwater-Nichols Act, which is a decisive moment the development of joint forces and joint operations in Western militaries.[29] Goldwater-Nichols greatly increased the power of the Chairman of the Joint Chiefs of Staff, and installed various forcing mechanisms to ensure that the US Department of Defense become a truly joint organization. Of particular note, joint education was embedded throughout the PME structure, and officers were required to work joint postings before being eligible to be promoted to general or flag officer level. Since the Act became law, all American officers have been (repeatedly) taught about the benefits of joint operations throughout their careers, and (in principle) all general and flag officers have worked in joint posts, at least to some degree. In this sense, we must look upon the post-1986 American Way of War as fundamentally defined by the joint (inter-service, multiple-domain) perspective of its officer corps. The American military still sent out units supplied by the services primarily into the domains of which those services specialize (for example, the Army still fought on land), but it did so within a framework of understanding that situates those domain activities in relation to potential effects within other domains (for example, Army officers having some awareness of how the maritime and air domains potentially affect the land domain).

How, then, do we understand joint operations today? As already noted, the Department of Defense has a strong joint leadership structure, thanks to the Goldwater-Nichols Act. All five uniformed services are headed by general and flag officers who (for the most part) have had direct personal experience of joint staff positions. The entire American officer corps is educated in joint warfighting (to some degree) from Second Lieutenant to Brigadier General (and their naval equivalents) and above.[30] The seven regional commands, four functional commands, and the joint staff are all joint organizations, and their activities are defined through the voluminous literature of Joint Doctrine, including the Joint Publication (JP) 3-0 *Joint Campaigns and Operations* series,

---

28 Finlan, Danielsson, and Lundqvist, "Critically Engaging the Concept of Joint Operations," 360. It is worth noting that these operations, which were so important for generating political will to press the services to embrace the principles of jointness, were light, special operations, and not the large conventional warfighting affairs we associate joint operations with today. The significance of this peculiarity merits further inquiry.
29 Weiner, *Managing the Military*; Lucas and Crosbie, "Evolution of Joint Warfare."
30 Joint Staff, "Officer PME Policy."

and its subordinate doctrine.[31] This doctrine provides the most authoritative source of information about what joint actually means in practice in the American military context.

The current version of JP 3-0 describes its purpose as both providing doctrine to govern forces in their planning, preparation and execution of joint campaigns and operations and providing guidance to Joint Force Commanders (JFCs) on how to exercise their authority. It stresses that while its doctrine is intended to govern forces' activities, much leeway remains to the JFC in organizing the force and executing the mission "in a manner the JFC deems most appropriate."[32] In this same publication, "Joint Operations" are understood to be "military actions conducted by joint forces and those Service forces employed in specified command relationships with each other, which, of themselves, do not establish joint forces."[33] Likewise, the Joint Force is "composed of significant elements, assigned or attached, of two or more Military Departments under a single JFC."[34]

Evidently, the key component of a joint operation is the Joint Force, and the key actor in the Joint Force is the JFC, an individual that the doctrine is eager to avoid restricting or limiting in any way. A Joint Force is composed of two or more services under a single unifying commander, but naturally that commander is an officer from a particular service background, and his or her military judgment will be largely shaped by those experiences and service outlook. The doctrine aims to reinforce the JFC's professional judgement (rooted in expertise in one domain), while supporting it with the expertise of a subordinate level of commanders, namely the component commanders, each representing a different domain or function. The planning directly overseen by the JFC's command team is abstracted to the operational level, but it builds on subordinate unit planning from the domain components at the tactical level. Typically, the JFC's staff sets a battle rhythm within which the components fit their own battle rhythms, creating a nesting doll of organizational processes overseen by the JFC.

A key tenet of joint operations is the tight coordination and unification of military services and their assets, to exert military power through joint actions. The adjective "joint" is not bureaucratic pabulum, but a critically important distinction referring again to the authority of the JFC. Tactical

---

31 "Joint Campaigns and Operations" (Joint Chiefs of Staff, 2022).
32 Ibid.
33 Ibid.
34 "Joint Campaigns and Operations."

actions are conducted by units under the subordinate (domain-specific) commands, but these actions are conceptualized through the planning process as supporting operational actions, which are joint in nature. What this means is that the JFC is expected to think and act through the conceptual framework of the so-called "joint functions." Joint functions are the fundamental categories of operational activities through which commanders integrate and synchronize services into a Joint Force.[35] In practice, they constitute a heuristic tool that JFCs are expected to continuously reflect upon in order to achieve a balance across the enormous range of activities happening within an operation.[36] In other words, joint operations are unlocked in the mind of the JFC: it is his or her responsibility to imaginatively align the many disparate activities of the components (coordinated through the planning process) into sequences of actions that are complementary and mutually reinforcing. The JFC does so in principle by conceiving of the battlefield through the *joint* functions, rather than through component-provided products. In practice, of course, these activities are not done solely or even primarily in the mind of the JFC but constitute the voluminous staff work of the joint headquarters' staff.

When we think of joint operations through a service-centric view that emphasizes the alignment of functions and command, we highlight a latent historical trajectory underlying the doctrinal developments. As new technologies have emerged and become integrated into military practice, they tend to trigger a response in military doctrine and organization intended to increase alignment between military services, in order to optimize fighting power. When this alignment is resisted, the price is paid in blood, as seen in Operation Eagle Claw, which ultimately led to the loss of eight US servicemembers.[37] Therefore, one important way to understand modern day joint operations is through service coordination: joint operations are, in effect, the use of a Joint Force through the joint functions to coordinate the efforts of units provided by multiple services into focused collective action under the leadership of a single commander. Like a reverse prism, the Joint

---

35 Thomas Crosbie, "Getting the Joint Functions Right," *Joint Force Quarterly* 94, no. 3 (2019): 96–100; Matthew J Tackett, "The Joint Functions: Theory, Doctrine, and Practice," *Joint Force Quarterly* 115, no. 4 (2024): 117–24.

36 The Joint Functions appear in both US and NATO joint doctrine, with an important difference being that NATO doctrine adds Civil-Military Cooperation (CIMIC) to the list of functions. The exact language of each function has tended to vary over time, but the current US list includes: command and control, information, intelligence, fires, movement and maneuver, protection, and sustainment.

37 "Operation Eagle Claw," US Airborne & Special Operations Museum, March 7, 2021, https://www.asomf.org/operation-eagle-claw/.

Force takes bands of colors (signifying the components and their services) and blends them into white light (signifying the unitary warfighting entity directed by the JFC). At least, that is the ambition.

Multidomain Operations / Joint All-Domain Operations (MDO / JADO)

As noted earlier, one of the first thinkers to explore what have become the central concepts of MDO / JADO was Jeffrey M. Reilly [38] in his article "Multidomain Operations: A Subtle but Significant Transition in Military Thought." Reilly's article explores multidomain operations as such. that is, military operations in history that involve multiple physical domains. This should immediately bring to mind our earlier discussion of joint operations, and indeed in documenting the history of joint operations, we have already presaged much of the history of multidomain operations, at least in Reilly's initial framing.

The difference lies in the analytical focus. When we think about joint operations, we focus on multiple organizations acting together (much like a joint in a human body, which at its simplest is a connection linking two bones). Since in most human societies for most of human history, military organizations are structured around specializing in either land or maritime warfare, then joint military operations have traditionally been multidomain (in this literal sense).[39] By contrast, when we think about the challenges of conducting or countering multidomain operations (as Reilly encouraged his readers to do), our primary focus is not on the all-too-common bureaucratic obstacles of coordinating interservice activities, but rather on the warfighting challenges associated with projecting military power across those domains. This shift in focus is essential to unraveling the distinctive features of MDO / JADO thinking, and why there are meaningful differences.

Notably, Reilly was not the only one questioning if the joint operations doctrine that was current during the drawdown of American forces in Iraq and Afghanistan was fit for purpose. One of the triggers motivating Reilly's reflections was a question by General Martin Dempsey, then-Chairman of the

---

38 Reilly, "Multidomain Operations: A Subtle but Significant Transition in Military Thought."
39 Examples of non-joint multidomain operations and non-multidomain joint operations are not hard to imagine, but have not figured prominently in human history, nor do they merit much consideration today. Viking raids might be considered an example of the former, since Viking war parties had no domain-specific military organization. Operations involving army and marine units on land are an example of the latter.

Joint Chiefs of Staff, posed to the Military Education Coordination Council: "what's after joint?".[40] Dempsey was not daydreaming about rolling back Goldwater-Nichols and unravelling 50 years of military developments. Rather, his concern was with encouraging the Department of Defense (DOD) to reflect on a new suite of challenges to American military supremacy: "after joint" really meant the next step *beyond* joint, rather than the next trend after joint operations had run its course. Indeed, a basic understanding of the MDO / JADO development requires that we unequivocally start with the modern—very joint—foundation of the American and NATO professional military: a system of domain-specific armed services delivering units to larger organizational structures that are designed to counteract the siloing tendency of services while also capitalizing on the domain-expertise of services.

In 2015, Deputy Secretary of Defense Bob Work challenged the Army to revise its operations doctrine in response to what Dempsey and others perceived to be the new set of challenges.[41] The US Army's Training and Doctrine Command (TRADOC), inspired perhaps by Reilly's language, sought to answer Dempsey's question and Work's challenge through reorienting operations doctrine around the idea of "Multidomain Battle."[42] TRADOC commander Gen. David Perkins published three articles in *Military Review* to outline what the new concept meant for the Army's doctrine, organization, and future warfighting.[43] Two important TRADOC publications followed in rapid succession: the 87-page *Multidomain Battle: Evolution of Combined Arms for the 21st Century* in 2017, quickly superseded by the 100-page pamphlet 525-3-1, *The US Army in Multidomain Operations 2028*, in 2018, [44] the latter signed off by Perkins's replacement Gen. Stephen J. Townsend.

---

40 Reilly, "Multidomain Operations: A Subtle but Significant Transition in Military Thought," 61.
41 Kelly McCoy, "The Road to Multidomain Battle: An Origin Story," Modern War Institute, October 27, 2017, https://mwi.westpoint.edu/road-multidomain-battle-origin-story/, https://mwi.westpoint.edu/road-multidomain-battle-origin-story/.
42 McCoy.
43 David G. Perkins, "Multidomain Battle: Driving the Change to Win the Future," *Military Review*, 2017, 6–12; David G. Perkins, "Multidomain Battle The Advent of Twenty-First Century War," *Military Review*, 2017, 8–13; David G. Perkins, "Preparing for the Fight Tonight: Multidomain Battle and Field Manual 3-0," *Military Review*, 2017, 6–13.
44 David G. Perkins, "Multidomain Battle: Evolution of Combined Arms for the 21st Century" (US Army Training and Doctrine Command, December 2017); Stephen J. Townsend, "The US Army in Multidomain Operations 2028" (US Army Training and Doctrine Command, December 8, 2018).

Reilly's article and the TRADOC documents produced under Perkins and Townsend share many commonalities. Most notably, these documents are all focused on preparing for a rapidly changing operating environment, one expected to continue to change even after the doctrine has been published (in other words, none of the documents are oriented toward describing a future, stable state, but instead all focus on managing persisting change). The nature of that change is driven by near-peer or peer adversaries, generally named as Russia and China (but sometimes including North Korea and Iran).

The reason for the sudden and persisting volatility in the operating environment is largely due to the convergence of two factors. First is the nature of the technologies now being fielded by those adversaries: autonomous and unmanned platforms, artificial intelligence (A.I.)-assisted targeting and control systems, hypersonics, and the weaponization of space and the electromagnetic spectrum (EMS) all contribute to an interactive conflict environment in which human decision-making bound by International Humanitarian Law and other concerns for preventing harm to civilians is simply too slow to provide a convincing deterrent when matched against systems that are not equally limited. Joint doctrine directs the JFC to consult with teams of experts (including the components' domain experts) but also limits the JFC to a decision space within both legal norms and aligned with the political level's (often murky) strategic intent. More tangibly, as RAND Corporation discovered in its series of wargames in 2014-2015, NATO was incapable of resisting a rapid Russian invasion that seized the Baltic States.[45]

Second is the changing perception of adversaries' risk-acceptance with respect to escalating the standard competition between states to levels that may be perceived as aggressive. Russia's decision to invade Crimea dispelled the comforting notion that Europe had finally left its war-ridden past behind. A collection of new concepts quickly emerged in the academic and practitioner environments to describe the new logics of brinkmanship between states, with concepts such as "hybrid warfare" and "gray zone conflict" emerging as the favored formulations (although not without their critics).[46]

---

45 David A. Shlapak and Michael Johnson, "Reinforcing Deterrence on NATO's Eastern Flank: Wargaming the Defense of the Baltics" (RAND Corporation, January 29, 2016), https://www.rand.org/pubs/research_reports/RR1253.html; Guillaume Lasconjarias, "NATO's Response to Russian A2AD in the Baltic States: Going Beyond Conventional?," *Scandinavian Journal of Military Studies* 2, no. 1 (August 21, 2019), https://doi.org/10.31374/sjms.18.
46 Donald Stoker and Craig Whiteside, "Blurred Lines: Gray-Zone Conflict and Hybrid War— Two Failures of American Strategic Thinking," *Naval War College Review* 73, no. 1 (January 24, 2020): 14–48.

MDB / MDO were advanced as to the solution to a changing operating environment in which America's adversaries are acquiring capabilities advanced enough to challenge American military dominance across all operating environments. In other words, US superiority is no longer a guarantee.[47] From a joint operations perspective, the JFC could no longer feel assured that the Joint Force could shape the battlefield through extensive air and naval strikes before deploying land troops. At the same time, the Joint Force would deploy "on an expanded battlefield that is increasingly lethal and hyperactive" [48]: the comforting assumption of a protected rear area, and the strong preference for fixed command nodes, were suddenly questioned. The blurring distinctions between "below armed conflict" and "armed conflict" further undermine military readiness in the sense that the legal justification for the use of force, the trigger for most planning processes to truly begin, may be absent or challenged in ways the military finds difficult to address. In short, the early thinking about MDB / MDO was focused on the many dilemmas that emerge when thinking about countering an adversary (or multiple adversaries) bent on using any means, including those other than what is traditionally considered war, to achieve strategic objectives and weaken the United States. This includes the use of unconventional warfare, diplomatic and economic actions, influence campaigns and information warfare, either directly or by proxy.[49]

As previously stated, one driving force behind the development of joint operations as we know it today was the invention of new technology, heavier-than-air flight, which created a shock to the system by adding a new domain, the air domain, through which military actions could now be conducted. MDO was also developed due to a delayed "shock" in the system, the recognition that already-existing technologies would allow suddenly inclined adversaries to exploit the space and electromagnetic domains (understood as physical environments) as part of multifaceted campaigns intended to stress democratic states where they are most vulnerable.

Army thinking has continued to evolve, but in surprising ways. Let us compare the definitions used by TRADOC in 2018 and 2022. In 2018, MDOs are defined as:

---

47 Reilly, "Multidomain Operations: A Subtle but Significant Transition in Military Thought," 61; Townsend, "The US Army in Multidomain Operations 2028," 6.
48 Townsend, "The US Army in Multidomain Operations 2028," 6.
49 Townsend, 9.

> Operations conducted across multiple domains and contested to overcome an adversary's (or enemy's) strengths by presenting them with several operational and/or tactical dilemmas through the combined application of calibrated force posture; employment of multidomain formations; and convergence of capabilities across domains, environments, and functions in time and spaces to achieve operational and tactical objectives.[50]

The 2022 definition, taken from Field Manual (FM) 3-0 *Operations*, is, surprisingly, less focused on domains: "Multidomain operations are the combined arms employment of joint and Army capabilities to create and exploit relative advantages that achieve objectives, defeat enemy forces, and consolidate gains on behalf of joint force commanders."[51] Reading into the doctrine of FM 3-0, this however, seems to be an intentional choice, as it states unequivocally that "all operations are multidomain operations."[52] This odd statement is explained in terms of the fundamental interconnectedness of armed services:

> Army forces employ organic capabilities in multiple domains… Lower echelons may not always notice the opportunities created by higher echelons or other forces that operate primarily in other domains; however, leaders must understand how the absence of those opportunities affects their concepts of operations, decision making, and risk assessment.[53]

"All operations are multidomain operations" only really makes sense when the traditional understanding of military operations is entirely upended. The logic is that no service will ever again enjoy the luxury of ignoring what the other services are doing, and no JFC can look on his or her domain components as discrete universes fused in the mind of the JFC alone.

So far, our focus has been on US doctrine, with a special emphasis on the US Army. The challenges presented to the operational environment are

---

50 Townsend, "The US Army in Multidomain Operations 2028."
51 "Operations" (Washington, D.C.: Headquarters, Department of the Army, 2022).
52 Ibid.
53 Ibid.

not exclusive to the Army alone, however, and as such, other institutions and nations have adopted similar approaches, but with a wide range of variations in terminology and emphasis. While the Army has acted as a first-mover in titling MDO as the solution to the observed changes in military operations, the other services have come up with their own concepts. The Navy developed a concept called Distributed Maritime Operations,[54] the Marine Corps developed Expeditionary Advanced Base Operations (EABO),[55] a joint Navy-Marine concept was named Littoral Operations in Contested Environments,[56] and on top of this another joint maritime concept was named Integrated All-Domain Naval Power. [57] The two services in the Department of the Air Force describe their role in what they refer to as Joint All-Domain Operations (JADO),[58] a term also used by the Department of Defense.[59] JADO is supposed to be the overarching concept to combine joint force capabilities, in short, MDO on a joint level. JADO however, does not appear to be exclusive from the Joint Warfighting Concept, which aims to synchronize the fires coming from all domains and services.[60]

These are just the American concepts. If we go further into the NATO alliance, national concepts have been or are being developed from multiple countries, including France's M2MC (Multi-Milieux and Multi-Champs), and the United Kingdom's MDI (Multidomain Integration). More countries either have published or are developing their own concepts as well.[61] Finally, the Alliance itself has been especially preoccupied with conceptualizing MDO. Published in 2023, NATO's definition is the one that most NATO states and partner states will have to adjust to: "the orchestration of military activities, across all domains and environments, synchronized with

---

54 Ronald O'Rourke, "Defense Primer: Navy Distributed Maritime Operations (DMO) Concept" (Congressional Research Service, November 20, 2024).

55 "Tentative Manual for Expeditionary Advanced Base Operations" (Department of the Navy, Headquarters, United States Marine Corps, May 2023).

56 "Littoral Operations in a Contested Environment" (Department of the Navy; United States Marine Corps, 2017).

57 Kenneth J. Braithwaite, "Advantage at Sea: Prevailing with Integrated All-Domain Naval Power" (Department of the Navy, December 2020).

58 William G. Holt and Bratton, "The Department of the Air Force Role in Joint All-Domain Operations" (US Air Force, US Space Force, November 19, 2021).

59 Benjamin Selzer, "Taking Cues From Complexity: How Complex Adaptive Systems Prepare for All-Domain Operation," *Joint Force Quarterly*, no. 113 (2024): 4–13.

60 Selzer, 5.

61 Murielle Delaporte, "Supporting Multidomain Operations: Thinking and Living Through MDO," EUROSATORY, March 7, 2024, https://www.eurosatory.com/en/supporting-multidomain-operations-thinking-and-living-through-mdo/.

non-military activities, to enable the Alliance to deliver converging effects at the speed of relevance."[62]

The NATO definition does share some similarities with the original definition of MDO from TRADOC's pamphlet from 2018, such as the convergence of effects, and the emphasis on multiple domains. One key difference, however, is the inclusion of "non-military actors", which is not present in any of the US definitions and gives the NATO concept a notably different emphasis. What this also shows, however, is that the many different definitions, concepts, doctrines and discussions can serve to confuse rather than clarify our understanding of how multidomain operations will address the pressing issues of a changing operating environment. Across the concepts, doctrines, and definitions of MDO, are some shared key tenets, that would help explain how MDO will address the future operating environment, and thereby also how it distinguishes itself from traditional joint operations.

First is the addition of domains, the most distinctive feature of this entire class of concepts. While the names of the domains can differ from concept to concept, most of them include space and cyber as new added domains on top of air, land, and sea / maritime. This means that the sorts of future military operations that Western militaries are preparing to fight are those conducted across five domains instead of three. This is entirely intuitive: military operations will certainly involve the space domain through the presence of satellites and likewise will involve the cyber domain or EMS. Notably, this could simply mean that those areas are denied: the denial of space or cyber assets would, after all, be a major defining factor. Furthermore, the recognition of additional domains contributes to a displacement of the joint operations logic of components mapping onto services mapping onto domains. MDO / JADO thinking aims to shift that series of logical connections to a new way of thinking. In an MDO approach, combat effects are shifted from being service-centric to domain-centric.

The second important commonality is convergence or converging effects, terms which describe a process in which military effects have to align across domains at the so-called speed of relevance to ensure maximum efficiency. The speed of relevance is, as the name suggests, relative.[63] If an

---

62 Cannon, "The Alliance's Transition to Multidomain Operations."
63 Speed of relevance does not necessarily equal as fast as possible. In some cases, the goal may be to speed up to overtake the adversary's tempo. In other cases, the goal may be to slow down the adversary.

adversary uses or is suspected of using automated systems, then the speed may be faster than our standard, human-staffed systems can manage.

This leads to the third commonality, which is the transformation of the command and control systems as we know them. To shift from a service-focus to a domain-focus requires, according to NATO, a more agile system, since the current model (designed for joint operations) lacks cross-domain integration and sufficient agility to achieve said effects. [64] Presently, the only obvious way to achieve this is through technological advancements, particularly greater use of artificial intelligence and machine learning, which are expected to play a key role in the adaptation of how fast military systems can process information and effectively shorten decision cycles ensuring rapid decision making and rapid delivery of effects.[65]

To shortly summarize, the first key tenets shared in common across most MDO concepts concerns the addition of domains. This is not speculative, nor is it a normative commitment to do things differently. Rather, it is a descriptive acknowledgement of a fact of life that has been the case since at least the 1990s. The reason the space and cyber domain or EMS are now considered so important has to do with the second and third commonalities: the adversaries of democratic states are able and (it seems) willing to create and sustain pressure across five domains simultaneously, creating converging effects that we must be able to counter; and thus, we must quickly change our command and control systems to account for these changes (or else, lose the capacity to convincingly deter through our conventional forces).

Our Boring, Mundane, Everyday, Multidomain World

The first section of this chapter stressed how the introduction of air power led politicians to slowly recognize that interservice rivalry created too many vulnerabilities on the battlefield and required that things change. Therefore, the services developed ways of working together. Western states largely copied the American solution, which was to continue to structure the fighting force around services that specialize in single domains, creating alignment through the empowering of a JFC. The JFC's singular vision would

---

64 "Next Generation Command and Control," NATO Allied Command Transformation, December 6, 2023, https://www.act.nato.int/article/next-generation-c2/.
65 Dan G. Cox, "Artificial Intelligence and Multidomain Operations," *Military Review*, no. May-June (2021): 76–91.

be informed by constant expert advice from domain experts running domain components, in addition to a variety of other sources of valued input.

The second section told a different story. As Western militaries withdrew from distant theaters where they practiced various forms of irregular warfare, they began to recognize the need to counter a new generation of threats from Russia, China, and other non-democratic states. Recognizing that these adversaries were emboldened to launch long degrading campaigns against Western powers, and that they now had sufficiently powerful conventional forces to seriously stress Western military capabilities across all warfighting domains, military leaders began to focus on the vulnerabilities created by the joint operations model. A flurry of conceptual developments evolved around the closely linked concepts of MDO and JADO. These thinkers built on a joint operations foundation, taking for granted the persistence of services, the continued benefit of having components conduct tactical planning, and a JFC fusing military power on the battlefield. However, a consensus emerged that some things needed to change.

The changes posited by MDO / JADO relative to joint operations can be divided into three categories. First is mindset. As the US Army now stresses, all operations are multidomain operations, and even if at the lower levels only one domain is evident, work is being done to explore options of multiple domains even at the tactical level, as is evident with the innovation of activities on air-ground littoral.[66] This is because all modern operations will be defined in part by the ability or inability to exploit domains (since space and cyber are so deeply integrated into military operations in the land, air and maritime domains). The recognition of interconnectedness pulls MDO from the stratosphere of corps-level convergence mechanisms down to the individual officer's mindset.[67] A notable difference between preparing officers for joint operations versus MDO / JADO deployments is that our new doctrine and theories of war indicate officers should have a much deeper level of insight into how their tactical actions affect and are affected by assets in other domains.

The second change concerns what in NATO language is termed the orchestration of military activities, in other words, how the Joint Force

---

66 Amos Fox, "Army Aviation and Decisiveness in the Air-Ground Littoral," AUSA, August 23, 2024, https://www.ausa.org/publications/army-aviation-and-decisiveness-air-ground-littoral.
67 "Operations"; Rebecca Segal, "Who 'Does' MDO? What Multidomain Operations Will Mean For—and Require Of—the Army's Tactical Units - Modern War Institute," Modern War Institute, March 10, 2023, https://mwi.westpoint.edu/who-does-mdo-what-multidomain-operations-will-mean-for-and-require-of-the-armys-tactical-units/.

works. So far, the new doctrine relies on the same structure of components feeding tactical-level planning and execution into a JFC's operational-level vision. What differs is our preparation for a rapid change in how this actually works on the battlefield, following the anticipated introduction of AI-assisted decision-making. This must not be a hot potato we pass forward until one unlucky JFC is forced to figure it out. Logically, preparing for this new development starts now, with preparing officer students for a deeper level of understanding of how domains interact and how technologies affect operations in order to prepare them to support JFCs of the future.

Finally, NATO thinking highlights an area of importance which American military thinking has hesitated to explore. This concerns the JFC's currently very limited ability to coordinate with non-military actors. Long-standing taboos in the American military profession around the symbolic pollution of politics (taboos which are not present in many other NATO states) makes this aspect of military education particularly challenging. Civil-Military Operations (CMO) or Civil-Military Cooperation (CIMIC), viewed by many as an unrewarding aspect of the expeditionary warfare era which has been left behind, remains a powerful tool for JFCs, and the most logical tool to structure more effective ties with non-military actors, including, for example, satellite providers. [68] In NATO doctrine, CIMIC is one of the joint functions, and thus remains a much more visible element of the battlefield in NATO than in an American perspective. Preparing all officers to support JFCs in future MDOs should logically include a much greater emphasis on developed skills in synchronizing military and non-military actions, likely through the CIMIC function.

This chapter has tried to take seriously the dilemma faced by the instructors and students in PME programs who struggle to identify what makes MDO / JADO different from joint operations. We have taken an (at times painfully close) look at what exactly constitutes joint operations and what the global professional community expects to be the defining features of MDO / JADO. Doing so provides us with an answer, at last, to the callow officer in our vignette, who casually acknowledged the PME instructor's exhortation of the merits MDO / JADO, but then asked, "why should I care?" The answer that future PME instructors—as well as policymakers, researchers and analysts, military leaders, and all others closely engaged with military affairs—are hereby offered is this: you should care about MDO / JADO

---

68 Christoph Harig, "The Future of Civil-Military Operations," In-Depth Briefing (Centre for Historical Analysis and Conflict Research, November 2024).

because you are now expected to develop a mindset that places you outside your service's traditional way of thinking, while developing skills to support a commander operating in a much faster, much more challenging operating environment, in part through working closely with civilian actors who you may have no formal mechanisms to control. In other words, you should care, because it is the next, very difficult step in your career.

15

More Cowbell
Is Multidomain Operations Just Another Concept?

William F. Owen

No reader should mistake what follows as an exercise in philosophical wordplay where the mere act of debating and discussing the concepts is the desired output. If multidomain operations do not materially prepare an army for war better than other competing ideas, it wastes time and resources. Ideas can only produce benefits in action. The judgment at stake here is that there are good and bad ideas, and we can judge which is which in terms of material results. This work assumes an agreed-upon definition of words and a version of military history. If we do not have that, then military science and military thought are merely recreational philosophical excursions that provide forums for reputational utterances. In other words, we await the next big military concept, lessons learned, or the best way to win wars from the loudest clown driving the craziest in the broken circus. Be aware that the US Army is a concepts and doctrine-based Army, which is merely the formal recognition of ideas and teaching. There are both good and bad ideas. The same goes for teaching.

For our discussion, Multidomain Operations (MDO) is assumed to be the concepts the Congressional Research Service described as of October 1st, 2024.[1] It is assumed to be both accurate and authoritative. If it were not, then there are legislative and possible legal implications. As in FM-3 *Operations*, the actual written doctrine is also highly relevant.

---

1 https://sgp.fas.org/crs/natsec/IF11409.pdf.

What will be discussed here only lightly touches upon the MDO. Instead, it looks more like the story of failed or stillborn concepts that have flowed from the US defense community to NATO and allies, mostly to deliver an overall negative effect. The overall risk is that, at its heart, MDO might be a set of obvious and banal statements masquerading as a new idea. The flaws lie in the failure of previous ideas and ignorance of what works. This covers various reasonably poorly defined concepts such as maneuver warfare, the operational level of war, and Effects-Based Operations (EBO). Notably, and in almost every case, the arguments surrounding these ideas eventually undermine the idea to the point where its usefulness or applicability is substantially reduced. Some endure through organizational need, but that is not to say their usefulness is either apparent or proven. They exist as words uttered without definition or, in the words of one British General, "word salad."

A concept is nothing more than an idea, but once used within a bureaucracy, it becomes an idea with advocates, a plan, and a budget, essentially a sales force and advertising designed to bring the idea into being as something both done and taught. This is why concepts and doctrine, despite being different things, are often seen to go hand in hand. Someone has an idea, and that idea then needs to be taught to people so that it will become effective practice when resourced with the right training, equipment, and organization. Yet, these fundamental requirements often upend concepts into confusing and often nonsensical word salads, whose complexity disguises banal and facile ideas as world-changing philosophies that should guide all concerned. MDO may be the opposite, but its stable has bred few winners.

Maneuver Warfare

The inherent belief behind maneuver warfare, in all its guises, is that there is a better, more efficient way to conduct warfare at a lower cost. This way is rooted in supreme skill and excellence, where shock, surprise, and precision render the enemy psychologically or physically helpless, unlike the unthinking application of blunt and bloody force, supposedly best demonstrated by the Western Front of WWI. The historical reality is that no such distinction exists unless negligent conduct is considered an alternative to skilled conduct. Given the challenge to produce an example of "non-maneuver warfare," the "maneuverist" will always cite unsuccessful or badly conducted operations.

Maneuver warfare may have had honest intentions, but over time, it transformed into a hugely problematic set of imprecise ideas, the most useful of which were merely statements of the obvious. Perhaps more damaging was the associated concept of mission command, which became lock-stepped with maneuver warfare as being the method of command required to apply maneuver warfare in practice. If history was the evidence used to create maneuver warfare, then it was badly done, because good historical research provides the most damming body of evidence against it as a distinct idea. Despite the term lingering in the collective consciousness of many NATO Armies and Allies, the current debates about maneuver warfare and mission command are still fixed on a flawed understanding of what most armies historically taught as best practice. In 1907, Sir Douglas Haig wrote "Cavalry Studies: Strategical and Tactical," which, based on how British cavalry was trained and employed, showed that maneuver and mission command were the norm. In 1907, there was only "command." Full divisional orders, most likely written in the saddle and as reproduced in Haig's book, were short, concise, and one sheet of paper. They said what needed to be done. Not how it should be done. If a command can either be "restrictive" or "directive, " surely doctrine should just suggest when and under what circumstances each style should be employed.

The fact that debates about the veracity or usefulness of such ideas have lingered and been squabbled over for more than 40 years should signify a failure of thought and clarity of expression. Maneuver, in its correct military context, merely means to seek a position of advantage relative to an adversary. The definitions of an attack on an enemy's will and cohesion avoid recognizing that killing and wounding many men reduces will and cohesion better than anything else.

The greatest flaw in the original expression of the idea was that maneuver warfare and attritional warfare are distinct from one another, with the strongly implied, yet sometimes explicit guidance of "maneuver good— attrition bad." As many have explained, including this author, the opposite of maneuver is not attrition. The converse of maneuver is immobility or stasis— semantics matter. It would be unfair to wholly associate maneuver warfare with either the USMC or the US Army. The whole root of the concept mostly emanates from Basil Liddell Hart, and his near obsessive belief in defeating the enemy via the complete avoidance of battle, authors such as Brian Holden Reid and Alex Danchev had identified. Liddell Hart was also not beyond fraud and plagiarism to promote himself. Still, when men like Colonel John Boyd read and repeated his work, it gained prominence and false provenance for being original and insightful. Yet it was anything but. The origins of maneuver warfare as an idea, instead of something most good armies have taught, are well covered in the current academic literature. Still, it is mostly academic, as in something removed from the work that practitioners worry about. What works matters more than adherence to style and fashion.

For all the argumentation and nugatory debate about maneuver warfare, one question is seldom asked. If maneuver warfare was just warfare done well, and attrition was warfare done badly, where was the insight or value to what was advertised as a theory or doctrine? Warfare aims to break the collective enemy's will to persist in combat. As a point of doctrine, that should be done for the least possible cost in blood and treasure. Mutual attrition is bad; no sane military mind has ever spoken in its defense, bar extremely rare conditions of extremes such as Verdun. However, visiting attrition upon the enemy without suffering loss is the acme of tactics. Even when mutual attrition has been intended, it is a condition of acceptance forced by circumstances, not a preferred method of operation.

As Robert Leonhard showed so well in his 1991 book, "The Art of Maneuver," in the eyes of the US Army, maneuver warfare morphed into AirLand Battle, where it became a tool of attrition.[2] Removed from a *maneuver versus attrition* debate, AirLand Battle was a completely concrete concept that was proven in application in the First Gulf War, albeit removed from its original context of a Soviet invasion of Western Europe. The perception of strong similarities between AirLand Battle and MDO was strong enough to produce an element of debate.[3] Is MDO just AirLand Battle "on steroids?" If true, does that make it good or bad? Does it validate it as an approach?

Suppose the supposed need for MDO was that "enemies" understood the AirLand Battle concept and had inoculated themselves against it. In that case, this is essentially groundless when the historical record is examined with any rigor. However, to really examine MDO in that context, we first need to examine AirLand Battle in detail. AirLand Battle produced usable guidance, plus the attendant forces structures and guidance that provided a concrete basis for success. AirLand Battle differed considerably from maneuver warfare because it was a concrete solution to a very real problem, as in the defense of West Germany against a Soviet attack. It was not an abstract doctrinal concept based on poor history and a false dichotomy.

---

2 Leonhard, Robert, *The Art of Maneuver*, Ballantine Books 1991.
3 Shmuel Shmuel, *War on the Rocks*, June 2017. David E Johnson Shared Problems, Rand, August 2018.

However, the problem created by poor historical understanding permeated AirLand Battle significantly, not in terms of its practical conduct, but rather the later concepts it accidentally spawned.

The broad narrative of AirLand Battle flows from the perceived risk that the 1976 "Active Defense" lacked a solution to the idea that the US and NATO might win the first series of defensive battles. The popular narrative said active fefense did not make or was ignorant of the need to maneuver, and thus, did not use maneuver warfare as its bedrock. However, true or not, this was it, as will become clear, it was an irrelevant criticism.

In reality, the key enabling activity of AirLand Battle was *follow-on forces attack* or FOFA. The supposed solutions were articulated in the 1982 and 1986 editions of FM 100-5 *Operations*.

What most didn't know, but would have been evident given any degree of historical and operational research, was that active defense was not the problem it was advertised as by the civilian critics. If FOFA was the solution, it should not have been advertised as a novel approach. The use of aircraft to attack the enemy reserve dated to at least the last years of WW1.[4]

Perhaps more problematic for active defense, as outlined in the 1976 version of FM 100-5, was that it was often advertised as a direct outcome of what were seen as "lessons learned" from the 1973 October War between Israel and Egypt, Syria, Iraq, Jordan, and Morocco. The detail and reality made such an assertion less certain. How the US came to believe what it did about that war and the outcomes and decisions that were created remains inadequately studied and could form a book in and of itself. Still, the fact remains that FM 100-5 (1976) was based on first-hand observation of that War.[5] The later criticism of FM 100-5 (1976) studiously avoided challenging any lessons and insights that TRADOC and its research team gleaned and instead focused on more abstract issues.

Despite narratives to the contrary, the main failing of FM 100-5 was identified by one of its creators, General Don Starry.[6] What was wrong with the 1976 edition was not identified by civilian critics who lacked the training and experience to understand where the idea fell short.[7] More simply put, where active defense failed had little to do with the criticisms outlined by the

---

4 Jones. HA. The War in the Air 1937.
5 Herbert Paul. Deciding What Has to be Done. Leavenworth Paper Number 16 July 1988.
6 YouTube. Accessed October 27, 2024. https://www.youtube.com/watch?v=QgwqT4fqU2E.
7 Bronfeld, Saul Did TRADOC Outmanoeuvre the Manoeuvrists? A Comment WAR &SOCIETY, Volume 27, Number 2 (October 2008).

likes of Bill Lind, later author of the Maneuver Warfare Handbook.[8] Lind's solution to what he saw as the problems with active defense was to suggest adopting the idea of Fuller and Liddell Hart. The problem was, and still is, that Lind and others did not know how to translate Liddell Hart's abstract fantasies into operational plans and training. Lind and other civilian critics also did not understand the real problem. Advocating the wrong solution is usually a clear indicator of not understanding the problem.

When TRADOC set out to write the 1976 edition of FM 100-5, being largely informed by the lessons of the 1973 Yom Kippur War, the man charged with validating the lessons into practical guidance was the Major General commanding the US Army's Armor Center at Fort Knox. This was Don Starry. In 1976, Don Starry was promoted to Lieutenant General. He took over the US V Corps, which had the potential mission of defeating a Soviet Army in the area of the Fulda Gap in Germany. Based on a detailed examination of the terrain, the mission, rehearsals, and over 100 wargames and simulations, Don Starry was forced to conclude that what he had advocated for in FM 100-5 did not work in the context of the Corps he commanded and the threat it faced. Yet, the doctrine he developed had been intended to do exactly that. Discussing the practical problems of applying FM 100-5 to the specific problem of defending the Fulda Gap would require substantially more space than can be addressed here. Still, a major part of the problem was assuming that what worked on the Golan Heights in 1973 somehow traveled to a completely different problem set, the Fulda Gap. In simple but not simplistic terms, while the US could win the first battle, and maybe even the second or third, sooner or later, it would lose because the Soviet reserve would keep coming while the US had run out of ground to fall back on. The explicit and fundamental problem with what FM 100-5 (1976) taught was that you would be overwhelmed if faced with overwhelming odds. In mid-1977, the only people who knew the problem's "why" and "how" were Starry and his staff. Everyone else was guessing. More importantly, Starry was the only person with the credentials or qualifications to solve the problem. In June 1977, Don Starry replaced General William DuPuy as the 2nd commander of TRADOC.

The primary solution by Starry was to target the oncoming Soviet divisions before they could be fed into the fight to reinforce wherever the Soviets believed they were successful. This was neither new nor novel and was well-founded in the historical record. In fact, its routes could easily be

---

8 Lind, Bill Maneuver Warfare Handbook, Westview Special Studies in Military Affairs, 1985.

traced to the British doctrine from the 1860s, where the "strategic intent" was to degrade the enemy before he came to battle.[9] Espoused by Edward Hamely and many later writers, including Henderson, it was an admixture of Jomini and the British success in the Crimean campaign of destroying the Russians logistically, forcing their withdrawal from the theater.[10] The idea was applied in various forms throughout the 20th century, and using air power and long-range fires to attack, wear down, and attrit the enemy before contact was far from novel. The Israelis had more than proven this in 1967, but in 1973, the Egyptian Air Defense umbrella had blunted their ability to target the Egyptians crossing the canal. Before the advent of air power, the British intended to use divisional-level cavalry maneuvers, something they applied to great effect in Palestine in 1918. Notably, the doctrinal frameworks articulated by the Soviets in the inter-war years were only slight variations of what the British had done in Palestine in 1918.[11]

While ultimately, the success of the US Doctrine in the Gulf War of 1991 can be squarely laid at the feet of both DuPuy and Don Starry, the fact remains that the AirLand Battle was never tested in the context in which it was intended. The more interesting fact is that what the Israelis learned from 1973 was visibly different from what the US chose to learn. Most obviously, the Israelis went on to cover the Golan Heights with minefields, fortified infantry positions, and anti-tank obstacles. Even the meagre minefields and anti-tank obstacles in place in October 1973 had complicated and constrained the Syrians despite their breakthrough in the south, which an armored reserve had blocked.

One of the main instruments used to attack the Soviet follow-on forces was the A-10, which was developed largely from the lessons of Korea and Vietnam. Based on the lessons of 1967 and 1973, the Israelis rejected the A-10 as too slow and vulnerable. They developed heavy APCs in preference to lightly-armored IFVs such as the Bradley, and most importantly, they doubled the size of their army. They also put a vast amount of thought and resources into destroying the Syrian air defense networks, which they achieved within hours in 1982. Again, it is fair to ask why the Israelis saw things so differently from the US if the US was drawing lessons from the

---

9 Hamley, Edward Operations of War. 1866.

10 Merrill James M. British-French Amphibious Operations in the Sea of Azov, 1855 *Military Affairs*, Vol. 20, No. 1 (Spring, 1956), 16-27.

11 Povlock Paul A. Deeo Battle in World War One: The British 1918 Offensive in Palestine, Faculty of the Naval War College 1997.

Israeli experience. There was, however, strong agreement on the need to destroy the enemy air defense network to bring the full effect of air power to bear. In 1976, the IDF procured US Lance missiles with cluster munition warheads to target air defense systems. The problem posed by enemy air defense was agnostic of terrain, and the problem in Europe was likely the same in the Middle East. This all highlights that looking at one conflict and attempting to extrapolate those insights to a conflict that has not occurred but would be under significantly different conditions adds more than a slight level of complexity. The British Army put a lot of resources into studying the Russo-Japanese War of 1904-05. It is the source of considerable academic debate on how well those observations informed the conduct of WWI or the formulation of the 1909 Field Service Regulations.

The most obvious insight into FM 100-5's evolution from 1976 to 1982 and even 1986 is that a doctrine founded on solid fundamentals and reinforced by research is unlikely to be wrong. AirLand Battle did not burn down active defense. It evolved it. If there is any doubt as to this, Chapter 8 of FM 100-5 1976 is titled AirLand Battle. It contains a detailed discussion of suppressing enemy air defense to enable close air support, but notably, not interdiction. What the 1982 edition did was expand that to battlefield interdiction, which needed to be conducted as part of what was called *Deep Attack*. To provide a C3I framework for this idea, the 1982 edition created a Deep, Close, Rear framework, which led to the widespread use of the term *Deep Battle*. Combined with other choices, this would become a treacle-like conceptual speed break within the US and even the UK doctrine, which exists even today. While arguably extremely useful as a conceptual tool, the use of identical wording in the Soviet doctrine emanating from the 1930s caused, and still causes, much confusion to those unfamiliar with the idea's US origins.

By far, the most far-reaching and overwhelmingly decisive difference the 1982 edition applied was the creation or adoption of an operational level of war. Never explicitly mentioned in 1976, by 1982, it was defined as:

> The operational level of war uses available military
> resources to attain strategic goals within a theater of war.
> Most simply, it is the theory of larger unit operations.
> It also involves planning and conducting campaigns.
> Campaigns are sustained operations designed to defeat
> an enemy force in a specified space and time with

simultaneous and sequential battles. The disposition
of forces, selection of objectives, and combined arms
complement and reinforce each other, greatly magnifying
their individual effects. In AirLand Battle doctrine,
synchronization applies both to our conventional forces
and, when authorized, to nuclear and chemical weapons.
It also characterizes our operations with other services and
allies.[12]

There are two critical points here. Firstly, the operational level of war is intended to "attain strategic goals within a theatre of war." Implicitly, this tied the operational level of war to theater command.

Secondly, "It also involves planning and conducting campaigns. Campaigns are sustained operations designed to defeat an enemy force in a specified space and time with simultaneous and sequential battles." Again, implicitly, this tied campaigns to theaters and the enemy's defeat within that theater. The 1976 edition had dealt with none of this. The 1976 edition only made four mentions of the word campaign, and none in the context related to the 1982 edition. The 1968 edition made it very clear that the whole purpose of FM 100-5 was:

The doctrine contained herein applies to all levels of
command in a theatre of operations and particularly to
levels above division. Military operations are actions, or
the carrying out of strategic, tactical, service, training,
or administrative military missions.

In 1968 and 1976, there was no operational level of war. It was as non-existent as the laptop computer. There was only strategy and tactics. Further careful reading of the 1968 edition shows that the US Army's definition of strategy was far removed from the simple Clausewitzian definition, which would be much cited in later years, and in the minds of many, this author included, still forms the most solid and coherent definition in existence.[13]

The important point is that what occurred in US doctrine between 1976 and 1982 had little to do with the evolution of active defense into AirLand

---

12 FM100-5 1982 Edition Section 2-3.
13 FM 100-5 1968 Edition Section 1-4, 1-5, 1-6.

Battle. The seismic shift created a new framework for how the US sought to conduct warfighting at the theater level.

The reason for this was entirely logical, and arguably, it did not and should not have required the invention of an operational level of war. In 1917 and 1918, General Allenby had twice defeated the Turks at the theater level, only using strategy and tactics, with no intervening level and simultaneously employing air, sea, land, and electronic Warfare. Arguably, so had every Western theater Commander in WWII, including Field Marshal Archibald Wavell in Burma.

The concept of deep attack or follow-on forces attack did not require the operational level of war because identical concepts had achieved the same effect before the operational level was invented. To put this in an absolute context, Britain fought and won the Falklands War in 1982 without any reference to the operational level of war anywhere in its doctrine or teaching at staff college. Striking an enemy reserve or echelon from the air dated back to WWI, so it was not a new idea. The operational level of war and its subsequent conflation with the same term used in Soviet doctrine were both unnecessary and mistaken.

Effects-Based Operations?

The origin of EBO is derived directly from the perceived success of AirLand Battle, in an attempt to expand on and formalize the effects of air power seen in Desert Storm in 1991. Laid out in the 2001 "Effects Based Operations: Change in the nature of warfare" by Brigadier Devid Deptula, the document attempted to set forth the vision that technologies such as stealth and precision weapons could leverage such high levels of effectiveness and efficiency that an enemy could be so severely degraded that land campaigns would be conducted with far greater speed and decision than was previously possible. It was proposed to have a radical enhancement of the air element of the AirLand Battle, which was expanded to a theater-wide concept. The problem was faux novelty, overstatement, hyperbole, and poor grammar. The acid test of poor grammar was to take any sentence explaining or defending EBO and remove the words "effects-based." If the sentence remained broadly true and correct, what was the purpose of defining something as effects-based? This point was not as trivial as it might appear. Effects are the results of actions. Why not action-based operations? What EBO was constantly altered and morphed over time. The UK coined the term effects-based approach.

The whole debate about EBO could well consume a PhD thesis, and indeed has, but the real point about EBO was that it was an early attempt at a multidomain concept. Combined with network-enabled capability and the Revolution in Military Affairs, it ticked all the boxes that all the services and Congress wanted to see in a concept. The document categorically stated that "Jointness is the use of the most effective force for a given situation."[14]

The celebration period for EBO ended abruptly in August 2008, when General James Mattis, commanding US Joint Forces Command (JFCOM), declared EBO as "fundamentally flawed." This logically leads to the question of whether Mattis and JFCOM understood the idea, and what had suddenly made it flawed, which was not obvious some years before? Had JFCOM broken a workable idea? A debate has raged, ebbed, and flowed on military forums and journals ever since. Still, there has been little material improvement regarding what the idea brings regarding training, equipment, and organization. Its practical expression, versus other choices, has never been clear or well-explained. Deptula blamed "land-centric JFCOM" for not understanding the concept.

EBO now MDO

Much of the military thought and science surrounding these debates can be seen as academic, philosophical, and abstract compared to the hands-on need to develop plans, write orders, and execute operations that define military success. If the ambition of any US concept is to somehow harness the ideals of maneuver warfare with the perfection of precision targeting enabled by networks, all would seem to be well, assuming that the carpenter knows how to use all the tools he has correctly.

Given what we know about warfare in the 20th century, including many institutional failures to simply apply what worked well in 1918 to be relearned in 1944-45, which were observed again in 1973, if we accept this as true (and some may not), it seems fair to ask what changed to make MDO the answer, more than any other concept advocated. For example, MDO's operational framework is tied to battlespace with divisional and corps frontages with evolved or novel ideas about the proven *close, deep, and rear*, which can be just as easily temporal and conditional and not just geographic. From most perspectives, MDO looks like a collection of every US concept brought

---

14 Deptula, David *Effects Based Operations: Change in the Nature of Warfare.* 2001 Page 23.

Term	FM 100-5 (1982)	FM 100-5 (1993)	FM 3-0 (2022)	+/−
Maneuver	278	200	221	+21
Interdiction	28	37	13	−24
Air interdiction	7	5	4	−1
Mission Command	0	0	29	+29
Attrition	6	6	1	−5
Operational Level	11	26	24	−1
Operational Art	0	20	58	+38

Table 15.1 Doctrine Buzzwords by Year and Publication
Source: Author.

together from the last 45 years, with connection and coherence supplied by additional text, discussion, and explanation. The 1986 edition of FM 100-5 was 186 pages, excluding the index. The FM 3-0 (2022), which is planned to introduce MDO, is 239 pages. FM 100-5 (1993) is 151 pages, excluding the index and glossary.

While possibly simplistic, Table 15-1 does have a fraction of insight.

Maneuver, mission command, and operational art are the winners in terms of MDO. Are they the cowbells the US Army wanted more of? Interdiction and air interdiction have crashed in terms of relevance. The problem is that all those ideas have not attracted the best explanation or insights regarding the US doctrine to date.

In Conclusion, So What?

The greatest risk with MDO, as with all organizational military concepts, is that someone comes up with an idea and then tries to work out how to make it work. The better way would have been to identify what can be done simply and effectively, and then resource the training, equipment, and organization that provides that solution. Additionally, *multidomain* implies complexity and a technocentric approach. It means a greater need for control, and thus will probably expand planning times and complicate sustainment. If the claim is that it will do the opposite, then this is the acme of skill, but it needs to be shown and measured to justify that claim.

It is easy and cheap to throw rocks and be the man not in the arena, but that man has had to rely on many not-very-good ideas for quite a long time.

Did anything about all the debates since 1976 materially change how the US Army fought successfully? If it did, then it was those things that should endure. It should not be abstract philosophical, navel-gazing, and cloud-shoveling so beloved by those debating doctrine and concepts.

What if a US partner or ally rejects the MDO assumptions as too complicated and impractical? For example, during recent Operations in Gaza, an IDF officer opined that with multiple divisions and brigades in contact, it is far easier to deconflict than synchronize.[15]

The greatest risk of MDO is that it over-promises to policymakers something that military concepts have never been able to deliver, and that is to achieve the political end state that the war or conflict seeks. Armed forces exist to destroy and defeat other armed forces to achieve military objectives. Military objectives are entirely separate from political objectives. You can win all the battles and still lose the war because no military concept can compensate for governments committing an armed force to military action, which cannot attain the desired outcome. The more complicated and obscure the military concept, the harder it will be to see its shortcomings because its defenders and advocates will always have additional caveats and excuses to wield in its support. Any concept not anchored in concrete advice on how to equip, train, and organize, regardless of where, when, why, or against whom you might fight, may render you less prepared for the war you get rather than the one you imagined.

---

15 Personal communication in March 2024.

CONCLUSION

Amos C. Fox and Franz-Stefan Gady

Multidomain Operations: The Pursuit of Battlefield Dominance in the 21st Century is a call to action. As concurrently constructed, MDO is the military equivalent of a Potemkin Village—it is all façade and no filler. To be sure, throughout this book the reader has encountered the reality of MDO's development, implementation, theoretically and applied shortcomings. What's more, the book has addressed how the US's allies and partners think about and try to implement MDO. A set of themes regarding these shortcomings have emerged throughout the book.

Flawed Concept and Doctrine Development Process

First, as many of the book's contributors, to include Jesse Skates, Bryan Quinn, Davis Ellison, Tim Sweigs, and Wilf Owens rightly note, the concept and doctrine development process used by the US Army to develop MDO was flawed. Ellison and Sweigs assert that sound concepts and doctrine should be anchored on a realistic theory of success and defeat mechanism(s). What's more, Ellison and Sweigs posit that sound concepts and doctrine must possess articulate language, cultural alignment, technological reality, concrete threats, a theory of success, and a clear framing of the concept regarding risk. MDO, both within the US military, and for the US's allies and partners, lacks nearly everyone of those defining features.

Instead, Heather Venable reminds the reader that MDO is full of chock full of bureaucratic Pentagon-ese that over-estimates what technology can deliver against a scatter-shot array of threats, while not appropriately accounting for risks. Risk here is greater than the usual military definition focused on risks to mission and risks to force. MDO, and the entire stand-off warfare and strike-as-strategy ethos, fails to account for strategic risk.

For instance, does the concept appropriate account for each contributor's industrial base and its capacity to keep pace with wars that aren't won quickly?

During Operation Inherent Resolve's battle of Mosul (October 2016 – July 2017), for instance, the US ran out of precision guided munitions on multiple occasions during the nine-month campaign.[1] Likewise, the US was fighting significant MDO-esque stand-off warfare campaigns in Syria and The Philippines and continuing its counterinsurgency efforts in Afghanistan. The strain of strike-driven strategy across the Middle East and Indo-Pacific put such strain on the US industrial base, that the US faced strategic munition shortages across the board.[2] The pattern of munitions shortfalls also plagues both Ukraine and Russian in their ongoing war for territorial and political control of Ukraine.[3] What's more, in less than a week Israel, in their ongoing conflict with Iran, Israel burned through its interceptor missiles and had to dip deep into its stocks to keep their missile defense shield functioning.[4]

As military analyst Tyler Hacker reflects, this is not just a thought experiment or theoretical debate, but a looming strategic crisis for both the US and its allies and partners.[5] Moreover, the US House Select Committee on China has made similar findings. Citing production capacity, existing stockpiles, and forecasted expenditure rates associated with MDO, testimony

---

1 Jeff Daniels, "ISIS Fight Shows US Military Can Use Lower-Cost Weapons with Lethal Results," *CNBC*, 19 July 2017, available at: https://www.cnbc.com/2017/07/19/isis-fight-shows-us-can-use-lower-cost-weapons-with-lethal-results.html.

2 "Army May Not Have Enough Munitions," *Association of the United States Army*, March 17, 2017, available at: https://www.ausa.org/news/army-may-not-have-enough-munitions; Jen Judson, "Army Concerned Over Shrinking Munitions Stockpile," *DefenseNews*, March 8, 2017, available at: https://www.defensenews.com/digital-show-dailies/global-force-symposium/2017/03/08/army-concerned-over-shrinking-munitions-stockpile.

3 David Vergun, "Air Force General Says Ukraine Needs Ammunition," *US Department of Defense*, February 28, 2024, available at: https://www.defense.gov/News/News-Stories/Article/Article/3689717/air-force-general-says-ukraine-needs-ammunition/; Andrew Kramer, "Dwindling Ammunition Stocks Pose Grave Threat to Ukraine," *New York Times*, April 5, 2024, available at: https://www.nytimes.com/2024/04/05/world/europe/ukraine-ammunition-shells-russia.html.

4 Shelby Holliday, "Israel Is Running Low of Defensive Interceptors, Official Says," *Wall Street Journal*," June 18, 2025; "Israel Running Low on Arrow Interceptors, US Burning Through Its Systems Too," *Times of Israel*, June 18, 2025, available at: https://www.timesofisrael.com/israel-running-low-on-arrow-interceptors-us-burning-through-its-systems-too-wsj/.

5 Tyler Hacker, "Money Isn't Enough: Getting Serious About Precision Munitions," *War on the Rocks*, April 24, 2023. Available at: https://warontherocks.com/2023/04/money-isnt-enough-getting-serious-about-precision-munitions/; Tyler Hacker, "How the Arsenal of Democracy Can Get Back on Its Feet With Some Help from the Past," *War on the Rocks*, June 19, 2025. Available at: https://warontherocks.com/2025/06/how-the-arsenal-of-democracy-can-get-back-on-its-feet-with-some-help-from-the-past/.

---

during a December 2024 the committee found that in a potential conflict with China the US would run out of munitions within days—not weeks, months, or years.[6] Quite alarming. Yet, MDO doctrine and MDO-related force structure changes, to include gutting land forces, and removing the redundancy and resilience of the remaining forces in the name of efficiency, continues its voracious march forward.

Moreover, these cuts to land forces, deemed by US Secretary of Defense Pete Hegseth as a means to make the US military "leaner" and "more lethal," by investing those dividends into long-range precision fires, unmanned systems, air and missile defense, cyber, electronic warfare, and counter-space capabilities is a strategic risk to the US's ability to fight and win wars.[7] Hegseth refers to formations such as armor and manned aviation as "outdated" and impediments to deterrence and the US's ability to fight and win conflicts in the 21st century.[8] This is a bold and ahistorical assertion. Time will tell if assuming significant degrees of strategic risk for the sake of promoting a poorly constructed doctrine will pay off for the US and its allies and partners.

Thus, policymakers, strategists, and concept developers, doctrine writers, and force designers must appreciate the factor that cheap mass and threats not losing quickly can rapidly undue MDO. Failing to appreciate these realities will leave the US and its allies and partners left listing in the lurch. MDO advocates, grounded in the belief that stand-off warfare and high-technology applications will dominate in 21st century wars, mustn't forget the continuities of war. As scholars like Cathal Nolan and Williamson Murray find, wars are won through industrial output, blood, steel, and attrition.[9] Likewise, scholar Ben Connable's research, which examines 400 cases of warfare and finds that, "But even the most prolific use of drones in the history of warfare (Ukraine) did not stop or…consistently or radically alter the machine-on-machine, soldier-to-soldier fighting taking place under

---

6 "Rebuilding the Arsenal of Democracy: The Imperative to Strengthen America's Defense Industrial Base and Workforce," *The Select Committee on the CCP*, December 5, 2024. Available at: https://selectcommitteeontheccp.house.gov/committee-activity/hearings/rebuilding-arsenal-democracy-imperative-strengthen-americas-defense.

7 Pete Hegseth, "Memorandum For Pentagon Leadership; Subject: Army Transformation and Acquisition Reform," *US Department of Defense*, April 30, 2025. Available at: https://media.defense.gov/2025/May/01/2003702281/-1/-1/1/ARMY-TRANSFORMATION-AND-ACQUISITION-REFORM.PDF.

8 Hegseth, "Memorandum For Pentagon Leadership."

9 Cathal Nolan, *The Allure of Battle: A History of How Wars Have Been Won and Lost* (Oxford: Oxford University Press, 2017), 576-580; Williamson Murray, *The Dark Path: The Structure of War and the Rise of the West* (New Haven, CT: Yale University Press, 2024), 374.

their gaze."[10] Similarly, scholar Michael O'Hanlon finds that, "Yet in all likelihood, technological innovation will not fundamentally change ground warfare in the decades ahead, especially in the close fight, and in complex environments."[11]

As a result, to dominate in 21st century warfare, the US and its allies and partners must not divest themselves from seemingly outdated, large, and redundant ground combat capabilities. As empirical evidence illustrates, the possession of those systems are what helps weather the rigors of war's strife. Likewise, the absence of these systems is generally what cause victory to be frittered away as small boutique forces give way under relentless combat. To parry quick destruction and to obtain battlefield (and battlespace) dominance, policymakers must invest in developing more responsive and more robust land combat systems to couple with multidomain warfighting systems.

For Want of a Theory of Victory

Neither the US Army, nor NATO's MDO doctrine provides a cogent theory of victory.[12] As Ellison, Sweigs, Quinn, and Jeff Meiser state, a theory of victory should be (a) a clear, logical, and linear articulation of how a combatant accounts for a host of competing, hostile, and counteracting variables and (b) attempts to neutralize and overcome those variables to (c) achieve or obtain their desired outcome in war, or (d) the contributing conditions that fuel strategic victory. Rather, the MDO relies on a rather flimsy set of tenets—agility, convergence, integration, and synchronization—as its north star. The idea of cross-domain synergy to create complimentary effects in other domains are tightly linked to the tenets of MDO.[13]

In the absence of a theory of victory, MDO anchors on convergence—MDO's *primus inter pares*—for how it delivers victory. Convergence seeks to achieve "rapid and continuous integration of all domains across time,

---

10 Ben Connable, *Ground Combat: Puncturing the Myths of Modern War* (Washington, DC: Georgetown University Press, 2025), 263.
11 Michael O'Hanlon, *The Future of Land Warfare* (Washington, DC: Brookings Institution Press, 2015), 163.
12 See FM 3-0, *Operations* (Washington, DC: Government Printing Office, 2022); TRADOC Pamphlet 525-3-1, *The US Army in Multi-Domain Operations 2028* (Fort Eustis, VA: US Army Training and Doctrine Command, 2018); *NATO 2022 Strategic Concept* (Brussels, Belgium: NATO, 2022); AJP-01, *Allied Joint Doctrine* (Brussels, Belgium: NATO Standardization Office, 2022).
13 FM 3-0, *Operations*, 48-49.

space, and capabilities to overmatch the enemy."[14] Moreover, convergence is illogically conceived as both an outcome and a process. The US Army, for instance, posits that, convergence is, "An outcome created by the concerted employment of capabilities against combinations of decisive points in any domain to create effects against a system, formation, or decision maker, or in a specific geographic area."[15] This is a self-referential circular logic and certainly not a theory of victory.

The other glaring issue with MDO and convergence is the ranking of needs. A good concept and doctrine should prioritize a threat and then everything else flows from there. MDO and convergence, however, place orienting on the application of technology and operating in all domains ahead of being threat focused. That is, the goal of MDO and convergence is to use technology in all domains and then worry about how that relates to a threat; instead of identifying the threat and demonstrating a clear path to defeat that threat. That isn't to imply that defeating a threat is a linear process. To be sure, war exists in an adversarial environment in which all participants work toward their own self-interest, whether that means winning or just not losing. However, that lens; that is, the non-linearity of competitive competition, is more apropos to applied strategy and plans, and not concept and doctrine development.

Therefore, it is prudent for policymakers, strategists, concept developers, doctrine writers, and force designers to rethink MDO and its absence of a theory of victory. As it stands, MDO possesses a theory of technology application, but no theory of victory. Operating with new technology in new domains in new ways is absolutely appropriate. However, these ideas must be bound in a logical, linear pathway geared toward defeating a clearly defined threat. This pathway mustn't be self-referential, nor jargon-laden with the institutional babble found in most doctrine today. The ideas must be clearly articulated and show direct and secondary causality.

Make It Matter

The third major flaw with MDO is that it is challenging to see its tactical application. And if a doctrine doesn't possess tactical relevance, the doctrine isn't relevant. This is because though strategy drives tactics, tactics deliver

---

14 TRADOC Pamphlet 525-3-1, *The US Army in Multi-Domain Operations*, iii.
15 FM 3-0, *Operations*, 247.

the military wins or losses that inform applied strategy and help shape policy decisions regarding a war.

MDO lacks tactical clarity because it is essentially a strike and technology-integration concept and doctrine for the operational level of war, and not a warfighting concept or doctrine. As FM 3-0 states, corps are the primary action elements of MDO, while divisions also enable corps.[16] Yet, the brigade combat team, as William Murray and Robert Rose noted, is all but left out of the discussion. To be sure, FM 3-0 punts the BCT in stating that, "Divisions set conditions for subordinate brigade combat teams (BCTs) to accomplish their mission in close combat."[17] More to the point, FM 3-0 states that BCTs, and their subordinate elements, apply battle drills and create effects from massed artillery and mortars. MDO and convergence are all but irrelevant in the BCT's space.

Yet, BCTs—and comparable formations in ally and partner military forces—are actually the forces that fight and win wars. The lack of depth in BCTs, and redundancy of BCTs across a force are how wars are lost. Although Ukraine is generating upwards of 50,000 drones a month, their inability to muster more manpower and generate additional BCTs has hampered its ability to retake its annexed terrain from Russia[18]. As of June 2025, analysts Yurri Clavilier and Michael Gjerstad offer than Ukraine needs between 50,000 and 100,000 soldiers to make a significant impact on Russia's hold of eastern and southern Ukraine.[19] Likewise, Russia's inability to generate their equivalent to BCTs forced the Kremlin to authorize the Wagner Group open Russia's prison gates to cull criminals to outfit units needed to stave off defeat and win decisive battles in places like Bakhmut.[20] What's more, as former Pentagon Deputy Press Secretary Sabrina Singh stated, Russia's manpower challenges forced it to rely on upwards of 11,000 North Korean soldiers to

---

16 FM 3-0, *Operations*, 47.

17 FM 3-0, *Operations*, 47.

18 David Axe, "As the Ukrainians Fling 50,000 Drones a Month at the Russians, The Russians Can't Get Their Drone Jammers to Work," *Forbes*, February 2, 2024. Available at: https://www.forbes.com/sites/davidaxe/2024/02/16/as-the-ukrainians-fling-50000-drones-a-month-at-the-russians-the-russians-cant-get-their-drone-jammers-to-work/.

19 Yurri Clavilier and Michael Gjerstad, "Combat Losses and Manpower Challenges Underscore the Importance of 'Mass' in Ukraine," *International Institute for Strategic Studies*, February 10, 2025. Available at: https://www.iiss.org/online-analysis/military-balance/2025/02/combat-losses-and-manpower-challenges-underscore-the-importance-of-mass-in-ukraine/.

20 "Russian Federation: UN Experts Alarmed by Recruitment of Prisoners by Wagner Group," *Office of the High Commissioner for Human Rights*, March 10, 2023. Available at: https://www.ohchr.org/en/press-releases/2023/03/russian-federation-un-experts-alarmed-recruitment-prisoners-wagner-group.

help it retake Kursk through 2024.[21] Israel, in its war against Hamas in Gaza, has been able to systematically defeat Hamas not through the application of stand-off warfare, or the use of land forces in ground combat, but through the adroit synergy of stand-off, multidomain operations with blistering ground combat.[22]

The reorganization of force design around corps-level (and higher) enabling, at the expense of BCTs is also noteworthy. As Murray notes, BCTs have been gutted to make cap room for the Multidomain Task Force (MDTF). So what's occurring is the US military, are removing needed capabilities from the forces that historically win or lose wars, to provide the financial and human capital to build forces that are in practice shaping and enabling forces. These moves will thus increase capacity to shape the environment for close combat and enable close combat forces, but those forces—like BCTs—will be so depleted and malnourished that they will culminate almost at the point in which they make contact with threat forces.

The policy decisions are quite easy to see. Instead of embracing a whimsical belief that fighting from afar across an array of domains will mystify an adversary and overwhelm their ability to make decisions because of the number of dilemmas they're presented with, focus on a concept and doctrine with a clear theory of victory. Yet, that theory of victory must not be just an operational strike concept and doctrine. It must have utility and resonate at the tactical level guided by field grade officers; that is, the realm of BCTs and battalions. Field grade officers—majors, lieutenant colonels, and colonels—exist to command and staff tactical formations. They bring knowledge, experience, and education to tactical formations that isn't found with junior officers. To be relevant, MDO, or joint all-domain operations (JADO), as Thomas Crosbie and Holger Lindhardsen note, must be something that field grades attending staff and war colleges can see their place in, quickly grasp, and upon graduation apply in the formations they lead.

Force design is of equal importance. For MDO to be anything more than a corps-level strike doctrine, force structure at the tactical level must be modernized in a meaningful way. This doesn't mean gutting close combat

---

21 "Deputy Pentagon Press Secretary Sabrina Singh Holds an Off-Camera, On-the-Record Press Briefing," *US Department of Defense*, November 18, 2024. Available at: https://www.defense.gov/News/Transcripts/Transcript/Article/3968387/deputy-pentagon-press-secretary-sabrina-singh-holds-an-off-camera-on-the-record/.
22 Andie Parry, "Israel is Defeating Hamas, But Destroying Hamas Will Require a Post-War Vision," *Institute for the Study of War*, September 19, 2024. Available at: https://www.understandingwar.org/backgrounder/israel-defeating-hamas-destroying-hamas-will-require-post-war-vision.

and close combat-enabling formations, nor trimming what is seemingly excess fat from the bone. These formations must continue to grow, while being augmented with new technology, lessons from ongoing conflicts, and deep thinking and analysis about the future of technology and war.

Forms of Warfare

The third theme of this book addresses forms of warfare. Aside from being a corps-level strike concept and doctrine, MDO vigorously holds onto the insistence that maneuver as the way in which the US and its allies and partners fight. Yet, the reality of war holds that this is untrue. All 21st century wars demonstrate that maneuver, and more importantly, the maneuver phase of an operation or campaign, is fleeting and only occupies a small segment of a much grander range of warfighting requirements.

Michael Kofman and Franz-Stefan Gady noted that Western military thinking tends to only examine warfare through the maneuver phase of an operation or campaign and pay short shrift to the phase in which competing forces have closed on one another. In this phase, which tends to be where most of combat exists, attritional and positional warfare dominates. Likewise, as Fox notes, domination is not something that a combatant can just show up and do. Domination requires significant resource commitment up front, and then a steady flow of additional resources to maintain dominance. The degree to which gaining and maintaining dominance drains resources is almost entirely dependent on a hostile actor's level of resistance, force of arms, ingenuity, and evasion from detection. Of no less equal importance is how environmental factors such as the terrain, to include urban and subterranean environments further complicate the matter.

In addition, it is important to note that while MDO concept and doctrines speak to maneuver, MDO is actually a combination of attritional and positional stratagems. The application of firepower, predominately through long-range precision strike, and the associated investments in fires units like the MDTF and cheap strike drones, while decreasing the size and capability of BCTs and BCT-enabling units is a surefire sign of a attritionist warfare. To be sure, if eliminating threats with fires, to then have mobile land forces occupy those areas and conduct clean up duty is the guiding principle—as it is with MDO—then you've got an attritionist's concept and doctrine. This concept is fully embraced in FM 3-0, which states, "It is essential that Army forces defeat enemy stand-off approaches while preserving enough combat power to follow through and complete the mission."[23] In this case, 'defeat enemy stand-off approaches' means using long-range strike and friendly stand-off capabilities to fight and win at the primary objective, so that land forces can follow behind and take care of an unfinished business.

On the other hand, if mobile drives toward an objective, in combination with shaping and enabling strikes is the guiding principle—which is *not* the case with MDO—then you've got a mobilist's concept and doctrine. The use of mobilist here is intention. Mobile land warfare is not inherently maneuver warfare, because within mobile warfighting, any number of forms of warfare can, and do, flourish. The relative utility of any type of warfare depends on the phase in which a force is operating, the terrain within that phase, what their objective is, and how that objective is operating in relation to those battlefield variables.[24] Positional warfare is prominent here. Positional warfare, not articulated anywhere in MDO, is the Actor A's (*A*) synchronized application of movement and firepower to purposefully lure an adversary (Actor B, or *B*) into a detrimental location, or to force *B* into maintaining ground in an injurious location.[25] The purpose of *A's* positional operations and tactics, whatever that may be, is to (1) facilitate their objectives, (2) deny *B* their objectives, or (3) the more probable combination of both (1+2) those outcomes.[26] The pursuit of positions of relative advantage is the underlying principle of positional warfare.[27] A cursory reading of MDO concepts and doctrines find this idea—the pursuit of positions of relative advantage from which to improve the odds of mission accomplishment—pervasive.

Thus, after peeling away the institutional bias for maneuver warfare, which is baked into MDO, as Gady, Kofman, Fox, and Owen have all done within this book, it is quite easy to see that MDO is actual both a concept and doctrine fundamentally attritionist and positionist, and not maneuverist. That is not inherently a bad thing. However, it is therefore prudent for policymakers, strategists, and senior military leaders to understand that attritional and positional stratagems tend to contribute to long, destructive, and costly wars. Attritionist stratagems of stand-off warfare, like MDO,

---

23 FM 3-0, *Operations*, 47.
24 Amos Fox, "Maneuver is Dead? Understanding the Conditions and Components of Warfighting," *RUSI Journal* Vol. 166, no. 6-7 (2021): 10-18. DOI: 10.1080/03071847.2022.2058601.
25 Amos Fox, "The Israel-Hamas Conflict: 'You Might Not Be Interested in Attrition, But Attrition is Interested In You,'" *Small Wars and Insurgencies* Vol. 35, no 6 (2024): 5-7. DOI: 10.1080/09592318.2024.2346128.
26 Fox, "The Israel-Hamas Conflict," 5-7.
27 Fox, "The Israel-Hamas Conflict," 7.

do not lead to quick decisive wars. Strategists and senior military leaders must clearly communicate these ideas to policymakers so that appropriate preparations can be made throughout the elements of national power.

Likewise, force design should follow. While the MDTF is paving the way for MDO's attritionist methodology, land forces need to be properly configured for the positionist approach which pervades MDO. This requires land forces that (a) are large and can wither the blistering attrition of an adversary's stand-off warfighting, (b) possess more operational and tactical mobility capacity, (c) possess more mobility-enhancement / creation capability, and (d) possess more multidomain reconnaissance capability.

For Western militaries, MDO concepts and doctrines must be revised to speak correctly and clearly about both how they are fighting, not how they prefer to fight, and how adversaries will fight. Neither of these revisions should carry forward institution preferences and narratives, because this leaves military forces uneducated about the things they will experience on the battlefield and the tough attritional and positional challenges they will have to wrestle with on the ground and across the multidomain battlespace. Rather, attritionist and positionist stratagems must be clearly defined, as well as the operations and tactics that support those stratagems. Moreover, clearly articulating the phased battlefield (battlespace) is required. As Gady and Kofman note, military thought spends far too much time thinking about the maneuver phase, when in reality, most combat occurs in other phases, such as the attritionist and positionist phases. Thus, these additions to future versions of MDO are imperative.

Allies and Partners

Lastly, policymakers, strategists, senior military leaders, concept developers, doctrine writers, and force designers must all ask themselves *regarding allied and combined MDO, who does what*? The problem with allied doctrine is it supposes an idea but those distinguish how each participant fits within the conceptual and doctrinal framework. Likewise, the concept and doctrine must be appropriately modulated to fit the capabilities and capacity of the contributing members. If the doctrine exceeds the capabilities and capacity of the contributing allies and partners, the doctrine does not pass the *feasible, acceptable, suitable, and complete* (FAS-C) test.

As currently written, NATO's MDO doctrine resonates as a solid conceptual framework for operating in multiple domains, but might not be feasible, nor complete. Some of the larger questions with the doctrine include, who is responsible for what elements of execution of MDO? Are all members supposed to possess the same MDO capabilities, just at strengths and sizes relative to their state's unique capacity? Or, do various states serve as modular forces that provide discrete 'plug-in' capabilities for the alliance?

Likewise, as international relations scholarship reminds us, we cannot forget to consider the impact of international anarchy and the need for states to account for self-help in a highly volatile international system. Thus, we need to consider how contributions to the alliance fit (i.e., compliment or conflict) within each state's own self-interest and providing first and foremost for their own national defense. As Barry Posen reminds us, if a state's self-interest is greater than that of those of the alliance, then their force structure will likely be geared to a doctrine and force structure oriented on their own grand strategy.[28]

Therefore, policymakers, strategists, senior military leaders, concept developers, and doctrine writers should take a hard look at whether MDO is a feasible, acceptable, suitable, and complete doctrine for the totality of NATO's 32 member states, as well as non-NATO allies and partners. The importance of doctrine is not to have something that looks good on PowerPoint slides and sounds slick in a briefing room. Rather, the importance of doctrine is to possess a package of ideas that supports what each contribute can truly bring to the fight, and how that package of capabilities and capacity can support realistic political-military objectives.

To conclude, current MDO concepts and doctrine present an aspirational way of thinking about how to dominate the multidomain battlefield in the 21st century. However, the concept and doctrine possesses many glaring gaps, logic fallacies, and linear pathways to support achieving and maintaining battlefield dominance. These issues with multidomain operations need to be taken serious because if Western militaries find themselves on multidomain battlefields in eastern Europe or in the Indo-Pacific, engaged in large-scale combat operations against threats like Russia or China, current MDO concepts and doctrine leave much to be desired for how to dominate those environments and defeat those adversaries.

---

28 Barry Posen, *The Sources of Military Doctrine: France, Britain, and Germany Between the World Wars* (Ithaca, NY: Cornell University Press, 1984), 24-25.

BIBLIOGRAPHY

"About USFJ." 2025. *US Forces Japan*. January 14. https://www.usfj.mil/About-USFJ.

"Air Force Doctrinal Note 1-21: Agile Combat Employment." 2022. *Department of the Air Force*. August 23. https://www.doctrine.af.mil/Portals/61/documents/AFDN_1-21/AFDN%201-21%20ACE.pdf.

Air-Sea Battle: Service Collaboration to Address Anti-Access & Area Denial Challenges. 2013. Current Strategy Forums 1. https://digital-commons.usnwc.edu/cgi/viewcontent.cgi?article=1000&context=csf.

"Air Superiority 2030 Flight Plan Enterprise Capability Collaboration Team." 2016. *US Air Force*, May. https://www.af.mil/Portals/1/documents/airpower/Air%20Superiority%202030%20Flight%20Plan.pdf.

"Army Reserve Important for Total Force." 2017. *Association of the United States Army*, October 11. https://www.ausa.org/news/army-reserve-important-total-force.

"Air Force Future Operating Concept." 2015. *Department of the Air Force*. https://www.af.mil/Portals/1/images/airpower/AFFOC.pdf.

"Air-Sea Battle." 2015. *Air-Sea Battle Office*. January 13. https://dod.defense.gov/Portals/1/Documents/pubs/ASB-ConceptImplementation-Summary-May-2013.pdf.

"The Department of the Air Force Role in Joint All-Domain Operations." 2021. *Air Force Doctrine Publication 3-99 and Space Doctrine Publication 3-99*. November 19. https://www.doctrine.af.mil/Portals/61/documents/AFDP_3-99/AFDP%203-99%20DAF%20role%20in%20JADO.pdf.

Army Doctrine Publication 1-01: Doctrine Primer. 2019. Department of the Army. https://armypubs.army.mil/epubs/DR_pubs/DR_a/pdf/web/ARN18138_ADP%201-01%20FINAL%20WEB.pdf.

Army Doctrinal Publication 3-0: Unified Land Operations. 2011. Department of the Army. https://irp.fas.org/doddir/army/adp3_0.pdf.

Field Manual (FM) 3-0: Operations. 2022. Department of the Army. https://armypubs.army.mil/epubs/DR_pubs/DR_a/ARN43326-FM_3-0-000-WEB-1.pdf.

Field Manual 3-90: Tactics. 2023. Department of the Army. https://armypubs.army.mil/epubs/DR_pubs/DR_a/ARN38160-FM_3-90-000-WEB-1.pdf.

Field Manual 4-0: Sustainment. 2024. Department of the Army. https://armypubs.army.mil/epubs/DR_pubs/DR_a/ARN41683-FM_4-0-000-WEB-2.pdf.

Field Manual 100-5: Operations. 1982. Department of the Army. https://cgsc.contentdm.oclc.org/digital/collection/p4013coll9/id/976/.

Army Techniques Publication 3-35: Army Deployment and Redeployment. 2023. Department of the Army. https://armypubs.army.mil/epubs/DR_pubs/DR_a/ARN37655-ATP_3-35-000-WEB-1.pdf.

"China's Anti-Access Area Denial." 2024. Missile Defense Advocacy Alliance. October 27. https://missiledefenseadvocacy.org/missile-threat-and-proliferation/todays-missile-threat/china/china-anti-access-area-denial.

"Command and Control Research Portal." 2023. Command and Control Research Portal. Accessed July 5. https://internationalc2institute.org.

"Defending Japan's Peace from the Skies, Evolution in JASDF." 2024. Air Self-Defense Force, Japan. December 22. https://www.mod.go.jp/asdf/en/roles/role04/page04/.

"North Korea Military Power: A Growing Regional and Global Threat." 2021. Defense Intelligence Agency. October 15. https://www.dia.mil/News-Features/Articles/Article-View/Article/2812198/defense-intelligence-agency-releases-report-north-korea-military-power/.

"Army Multidomain Transformation Ready to Win in Competition and Conflict." 2021. Chief of Staff Paper #1. March 16. https://api.army.mil/e2/c/downloads/2021/03/23/eeac3d01/20210319-csa-paper-1-signed-print-version.pdf.

"How Project Convergence and JADC2 aim to accelerate the decision loop." 2021. Defense News. undated. https://www.c4isrnet.com/native/Raytheon-intelligence-space/2021/09/27/how-project-convergence-and-jadc2-aim-to-accelerate-the-decision-loop/.

"Ukraine: a Wake-Up Call for the UK and NATO." 2024. International Relations and Defence Committee. September 26. https://committees.parliament.uk/committee/360/international-relations-and-defence-committee/news/202994/ukraine-a-wakeup-call-for-the-uk-and-nato/.

"2024, Defense of Japan." 2024. Japan Ministry of Defense. https://www.mod.go.jp/en/publ/w_paper/wp2024/DOJ2024_EN_Full.pdf.

"Trends in China Coast Guard and Other Vessels in the Water Surrounding the Senkaku Islands, and Japan's response." 2024. Ministry of Foreign Affairs of Japan. December 1. https://www.mofa.go.jp/region/page23e_000021.html.

"National Defense Program Guidelines for FY 2011 and beyond." 2010. Japan Security Council. December 17. https://japan.kantei.go.jp/kakugikettei/2010/ndpg_e.pdf.

"On Joint Operational Access Concept (JOAC)." JMSDF Command and Staff College. December 21. https://www.mod.go.jp/msdf/navcol/index.html?c=topics&id=008.

"Joint Concept for Competing." 2023. Joint Chiefs of Staff. US Department of Defense.

"Kursk Offensive: A Timeline of Ukraine's Attack and Russia's Fightback." 2015. Reuters. March 12. https://www.reuters.com/world/europe/ukraines-attack-russias-fightback-kursk-2025-03-12/.

Ministère de la Guerre. 1921. Instruction Provisoire sûr l'Emploi Tactique des Grande Unités. October.

"Fact Sheets, DCDC – Background." 2008. The National Archives. February 5. https://webarchive.nationalarchives.gov.uk/ukgwa/20080205182025/http://www.mod.uk/DefenceInternet/FactSheets/DcdcBackground.htm.

"Bōeishō AI katsuyō suishin kihon hōshin [Basic Policy for the Promotion of AI Application by the Ministry of Defense]." 2024. Japan Ministry of Defense. July 2. https://www.mod.go.jp/j/press/news/2024/07/02a_03.pdf.

"China's Activities in East China Sea, Pacific Ocean, and Sea of Japan." 2024. Japan Ministry of Defense. October. https://www.mod.go.jp/en/d_act/sec_env/pdf/ch_d-act_a.pdf.

"Defense Buildup Plan." 2022. Japan Ministry of Defense. December 16. https://www.mod.go.jp/j/policy/agenda/guideline/plan/pdf/program_en.pdf.

"Defense of Japan 2014." 2014. Japan Ministry of Defense. https://warp.da.ndl.go.jp/info:ndljp/pid/11591426/www.mod.go.jp/e/publ/w_paper/pdf/2014/DOJ2014_2-5-1_web_1031.pdf.

"Defense of Japan 2020." 2020. Japan Ministry of Defense. https://www.mod.go.jp/en/publ/w_paper/wp2020/DOJ2020_EN_Full.pdf

"Defense of Japan." 2024. Japan Ministry of Defense. https://www.mod.go.jp/en/publ/w_paper/wp2024/DOJ2024_EN_Full.pdf.

"Medium Term Defense Program (FY 2014–FY 2018)." 2013. Japan Ministry of Defense. December 17. https://warp.da.ndl.go.jp/info:ndljp/pid/11591426/www.mod.go.jp/j/approach/agenda/guideline/2014/pdf/Defense_Program.pdf.

"National Defense Program Guidelines for FY 2014 and Beyond." 2013. Japan Ministry of Defense. https://warp.da.ndl.go.jp/info:ndljp/pid/11591426/www.mod.go.jp/j/approach/agenda/guideline/2014/pdf/20131217_e2.pdf

"Progress and Budget in Fundamental Reinforcement of Defense Capabilities, Overview of FY 2025 Budget Request." 2024. Japan Ministry of Defense. https://www.mod.go.jp/en/d_act/d_budget/pdf/20241126a.pdf.

"Reiwa 4 nendo seisaku hyōkasho [FY 2022 Policy Evaluation Report]." 2022. Japan Ministry of Defense. https://www.soumu.go.jp/main_content/000854078.pdf.

"Miscalculations, Divisions Marked Offensive Planning by US, Ukraine." 2023. Washington Post. December 4. https://www.washingtonpost.com/world/2023/12/04/ukraine-counteroffensive-us-planning-russia-war/.

"Multidomain Operations in NATO—Explained." 2023. NATO Allied Command Transformation. October 5. https://www.act.nato.int/article/mdo-in-nato-explained/.

"Multidomain Operations." 2025. NATO Allied Command Transformation. January 10. https://www.act.nato.int/activities/multidomain-operations.

"National Defense Program Guidelines for FY 2019 and Beyond." 2018. Japan Ministry of Defense. Dec. 18. https://warp.da.ndl.go.jp/info:ndljp/pid/11591426/www.mod.go.jp/j/approach/agenda/guideline/2019/pdf/20181218_e.pdf.

"National Defense Strategy." 2022. Japan Ministry of Defense, Japan. Dec. 16. https://www.mod.go.jp/j/policy/agenda/guideline/strategy/pdf/strategy_en.pdf

"Reika Butai Shōkai [Introduction of the Subordinate Units]." Ground Component Command, Japan. https://sec.mod.go.jp/gsdf/gcc/hq/220-unites.html.

Reiwa 6 nendo ho-ku chū SAM butai sōgō kunren [FY 2024 Comprehensive Training of Hawk and Medium SAM Units]. 2024. 1st Antiaircraft Artillery Brigade, Ground Self-Defense Force, Japan. https://www.mod.go.jp/gsdf/nae/katudou/1aab/katudou_syoukai/butai_kunren/R6_1AAB-sougou-boukuu.html.

"Saiba-Anzen Hoshō Bun'ya de no Taiō Nōryoku no Kōjō ni Muketa Yūshikisha Kaigi [Expert Group on Enhancing Response Capabilities in Cyber Security]." 2024. Cabinet Secretariat, Japan, June 7, 2024. Last modified November 29. https://www.cas.go.jp/jp/seisaku/cyber_anzen_hosyo/index.html.

"Sustaining U.S. Global Leadership: Priorities for 21st Century Defense." 2012. Department of Defense, January 4. https://apps.dtic.mil/sti/citations/ADA554328.

"Taiwan Contingency Also One for Japan, Japan-U.S. Alliance: Ex-Japan PM Abe." 2021. Kyodo News, December 1. https://english.kyodonews.net/news/2021/12/b38433927c1e-taiwan-contingency-also-one-for-japan-japan-us-alliance-abe.html.

"Rethinking Fire and Manoeuvre across physical and non-physical aspects of domains." 2023. The Hague Center for Strategic Studies Symposium and NATO HQ SACT. September 28. https://hcss.nl/news/symposium-rethinking-fire-and-manoeuvre-across-physical-and-non-physical-aspects-of-domains/.

"France Uncovers a Vast Russian Disinformation Campaign in Europe." 2024. Economist. February 12. https://www.economist.com/europe/2024/02/12/france-uncovers-a-vast-russian-disinformation-campaign-in-europe.

"The Winograd Report: The Main Findings of the Winograd Partial Report on the Second Lebanon War." 2007. Wall Street Journal. April 30. https://www.wsj.com/public/resources/documents/winogradreport-04302007.pdf.

"AirLand battle emerges: Field Manual 100—5 Operations, 1982 and 1986 editions: TRADOC 50th anniversary series." 2023. US Army. May 22. https://www.army.mil/article/266846/airland_battle_emerges_field_manual_100_5_operations_1982_and_1986_editions_tradoc_50th_anniversary_series.

"Trial of the Major War Criminals Before the International Military Tribunal, Vol. 15." 1948. International Military Tribunal. https://www.loc.gov/rr/frd/Military_Law/pdf/NT_Vol-XV.pdf.

"Multidomain Battle: Combined Arms for the 21st Century, 2025-2040." 2017. Training and Doctrine Command. https://www.govinfo.gov/content/pkg/GOVPUB-D101-PURL-gpo129084/pdf/GOVPUB-D101-PURL-gpo129084.pdf.

"Army Modernization Strategy: Investing in the Future." 2021. US Army. https://stratml.us/pdfs/AMS.pdf.

TRADOC Pamphlet 525-3-6: Army Functional Concept for Movement and Maneuver. 2017. Training and Doctrine Command. https://adminpubs.tradoc.army.mil/pamphlets/TP525-3-3.pdf.

TRADOC Pamphlet 525-92: The Operational Environment 2024-2034: Large-Scale Combat Operations. 2024. Training and Doctrine Command. https://adminpubs.tradoc.army.mil/pamphlets/TP525-92.pdf.

TRADOC Pamphlet 525-3-1: The U.S. Army in Multi-Domain Operations 2028. 2018. Training and Doctrine Command. https://adminpubs.tradoc.army.mil/pamphlets/TP525-3-1.pdf.

"US Army Conducts First Anti-Ship Ballistic Missile SINKEX Using PrSM." 2024. Naval News. June 23. https://www.navalnews.com/naval-news/2024/06/u-s-army-conducts-first-anti-ship-ballistic-missile-sinkex-using-prsm/.

"Gen. Charles Flynn Opening Remarks at LANPAC24." 2024. US Army Pacific. May 14. https://www.usarpac.army.mil/Our-Story/Our-News/Article-Display/Article/3775247/gen-charles-flynn-opening-remarks-at-lanpac24/.

US Constitution, Section 8. https://constitution.congress.gov/browse/article-1/section-8/.

Memorandum Directing the Development of the 2025 National Defense Strategy. 2025. Memorandum from the Office of the Secretary of Defense. Washington: May 1, 2025. https://media.defense.gov/2025/May/02/2003703230/-1/-1/1/MEMORANDUM-DIRECTING-THE-DEVELOPMENT-OF-THE-2025-NATIONAL-DEFENSE-STRATEGY.PDF.

Military and Security Developments Involving the People's Republic of China. 2023. Department of Defense, Annual Report to Congress. https://media.defense.gov/2023/Oct/19/2003323409/-1/-1/1/2023-MILITARY-AND-SECURITY-DEVELOPMENTS-INVOLVING-THE-PEOPLES-REPUBLIC-OF-CHINA.PDF.

National Defense Strategy of the United States of America 2022. 2022. Department of Defense.

"Defense Acquisitions: Future Combat System Risks Underscore the Importance of Oversight." 2007. Government Accountability Office, GAO-07-672T. https://www.gao.gov/products/gao-07-672.

Joint Publication 3-0: Joint Operations. 2018. Department of Defense. https://irp.fas.org/doddir/dod/jp3_0.pdf.

"US Security Cooperation with Korea." 2025. Department of State. January 20. https://www.state.gov/u-s-security-cooperation-with-korea/.

"2018 National Defense Strategy of The United States of America." 2018. Office of the Secretary of Defense. 2018. https://www.defense.gov/News/Feature-Stories/story/Article/1656414/what-is-the-national-defense-strategy/.

Alman, David. 2020. "Bending the Principle of Mass: Why That Approach No Longer Works for Airpower." *War on the Rocks*. December 15. https://warontherocks.com/2020/09/bending-the-principle-of-mass-why-that-approach-no-longer-works-for-airpower/.

Al-Sibai, Noor. 2025. "Amazon's Drone Delivery Program Is an Utter Failure," *Futurism*. January 8. https://futurism.com/the-byte/amazon-drone-delivery-fail.

Allenby R. Braden, et al. 2011. *The Techno-Human Condition*. MIT Press.

Andresky, Nikolai and Adam Taliaferro. 2019. "Future Study Plan 2019: Operationalizing Artificial Intelligence for Multi-Domain Operations." Army Futures Command. https://apps.dtic.mil/sti/trecms/pdf/AD1084346.pdf.

Andrysiak, Peter et al. 2024. "Empowering the Combatant Commands is Critical for the Future Fight." *War on the Rocks*. December 10. https://warontherocks.com/2024/12/empowering-the-combatant-commands-is-critical-for-the-future-fight/.

Axe, David. 2024. "Russia's 1st Guards Tank Army Has Won Its First Battle in Two Years – By Advancing a Mile and Capturing a Half-Dozen Buildings." *Forbes*. January 30. https://www.forbes.com/sites/davidaxe/2024/01/30/russias-1st-guards-tank-army-has-won-its-first-battle-in-two-years-by-advancing-a-mile-and-capturing-a-half-dozen-buildings/.

Bacevich, Andrew. 1986. *The Pentomic Era: The Army Between Korea and Vietnam*. National Defense University Press.

Baker, Sinéad. 2024. "Retired NATO commander says Ukraine's Kursk Invasion Proves It Can Succeed Without Western Advice." *Business Insider*. September 3. https://www.businessinsider.com/ukraine-kursk-shows-ability-without-much-west-advice-former-nato-2024-9.

Bartos, Haleigh. 2023. "What Went Wrong? Three Hypotheses on Israel's Massive Intelligence Failure." *Modern War Institute*, October 31. https://mwi.westpoint.edu/what-went-wrong-three-hypotheses-on-israels-massive-intelligence-failure/.

Bellamy, Christopher. 2015. *The Evolution of Modern Land Warfare: Theory and Practice*. Routledge.

Benson, Bill. 2012. "Unified Land Operations: The Evolution of Army Doctrine for Success in the 21st Century." *Military Review* 92 (2): 2-12. https://www.armyupress.army.mil/Portals/7/military-review/Archives/English/MilitaryReview_20120630MC_art010.pdf.

Berg, Paul and Kenneth Tilley. 2018. "Task Force Normandy: The Deep Operation that Started Operation Desert Storm." In *Deep Maneuver: Historical Case Studies of Maneuver In Large Scale Combat Operations*, edited by Jack Kem. Army University Press.

Biddle, Stephen and Ivan Oelrich. 2016. "Future Warfare in the Western Pacific: Chinese Anti-Access/Area Denial, US Air Sea Battle, and Command of the Commons in East Asia." *International Security* 41 (1): 7-48. https://doi.org/10.1162/ISEC_a_00249.

Blackwill, Robert, et al. 2024. *Lost Decade: the US Pivot to Asia and the Rise of Chinese Power*. Oxford University Press.

Bloch, Marc. 1999. *Strange Defeat: A Statement of Evidence Written in 1940*. Norton.

Borne, Kyle. 2019. "Targeting in Multidomain Operations." *Military Review* 99 (3): 60-67. https://www.armyupress.army.mil/Journals/Military-Review/English-Edition-Archives/May-June-2019/Borne-Targeting-Multidomain/.

Bourget, P.A. 1956. *Le General Estienne, Penseur, Ingenieur et Soldat*. Berger-Levrault.

Boyd, John. 2018. *A Discourse on Winning and Losing*. Air University Press.

Brodie, Bernard, and Fawn Brodie. 1973. *From Crossbow to H-Bomb: The Evolution of the Weapons and Tactics of Warfare*. Indiana University Press.

Brown, Kerry, et al. 2019. *The Trouble with Taiwan: History, the United States and a Rising China*. Bloomsbury Publishing.

Brown, Robert. 2017. "The Indo-Asia Pacific and the Multidomain Battle Concept." *The Military Review* 97 (5): 14-20. https://www.armyupress.army.mil/Portals/7/military-review/Archives/English/BROWN_PRINT_The_Indo_Asia_Pacific.pdf.

Buley, Ben. 2007. *The New American Way of War: Military Culture and the Political Utility of Force*. Routledge.

Cáceres, Sigfrido. 2013. *China's Strategic Interests in the South China Sea: Power and Resources*. Routledge.

Campbell, Caitlin. 2024. "Taiwan: Defense and Military Issues." Congressional Research Service. IF 12481. https://crsreports.congress.gov/product/pdf/IF/IF12481.

Campbell, David and Jesse McIntyre. 2018. "A Policy of Defeat." *Small Wars Journal*. July 4. https://smallwarsjournal.com/jrnl/art/policy-defeat.

Cancian, Mark, et al. 2023. *The First Battle of the Next War: Wargaming a Chinese Invasion of Taiwan*. Center for Strategic and International Studies. https://www.csis.org/analysis/first-battle-next-war-wargaming-chinese-invasion-taiwan.

Cavas, Christopher. 2016. "CNO Bans 'A2AD' as Jargon." *Defense News*. October 3. https://www.defensenews.com/naval/2016/10/04/cno-bans-a2ad-as-jargon/.

Carvalho, Andre, et al. 2024. "Tanques na Península Coreana: uma avaliação das plataformas K2 e M2020." *Panorâmico* 3 (9). https://ompv.eceme.eb.mil.br/images/publicacoes/panoramico/panoramico-vol3-n09-set-dez2024/Tanques_na_peninsula_coreana-_uma_avaliaco_das_plataformas_K2_e_M2020_Andre_Luiz_Viana_Cruz_de_Carvalho_e_Joo_Paulo_Ribeiro_Nogueira.pdf.

Chauvineau, Louis. 1939. *Une invasion est-elle encore possible?* Berger-Levrault.

"China's Anti-Access Area Denial." 2024. *Missile Defense Advocacy Alliance*. October 27. https://missiledefenseadvocacy.org/missile-threat-and-proliferation/todays-missile-threat/china/china-anti-access-area-denial.

Clinton, Hillary. 2011. "America's Pacific Century." *Foreign Policy*. October 11. https://foreignpolicy.com/2011/10/11/americas-pacific-century.

Coles, Isabel, et al. 2024. "Behind Ukraine's Russia Invasion: Secrecy, Speed and Electronic Jamming." *Wall Street Journal*. Aug. 17. https://www.wsj.com/world/behind-ukraines-russia-invasion-secrecy-speed-and-electronic-jamming-188fcc22.

Collins, Liam, et al. 2023. "The Battle of Hostomel Airport: A Key Moment in Russia's Defeat in Kyiv." *War on the Rocks*. August 10. https://warontherocks.com/2023/08/the-battle-of-hostomel-airport-a-key-moment-in-russias-defeat-in-kyiv.

Cooper, Zack and Hal Brands. 2024. "Dilemmas of Deterrence: The United States' Smart New Strategy Has Six Daunting Trade-offs." *Center for Strategic and International Studies*. March 12. https://www.csis.org/analysis/dilemmas-deterrence-united-states-smart-new-strategy-has-six-daunting-trade-offs.

Corbett, Art. 2018. "Expeditionary Advanced Base Operations (EABO) Handbook: Considerations for Force Deployment and Employment." *U.S. Marine Corps Warfighting Lab, Concepts and Plans Division.* June 1. https://www.mca-marines.org/wp-content/uploads/Expeditionary-Advanced-Base-Operations-EABO-handbook-1.1.pdf.

Corum, James. 1997. *A Clash of Military Cultures: German & French Approaches to Technology Between the World Wars.* Joint Doctrine Division Support Group. https://apps.dtic.mil/sti/tr/pdf/ADA323798.pdf.

Corum, James. 1992. *The Roots of Blitzkrieg: Hans Von Seeckt and German Military Reform.* Kansas University Press.

Coville, Michael. 1991. "Tactical Doctrine and FM 100-5." M.M.A.S. thesis., U.S. Army School of Advanced Military Studies. https://apps.dtic.mil/sti/tr/pdf/ADA259135.pdf.

Cox, Matthew. 2021. "Army Chief Defends Long-Range Missile Effort After Air Force General's Public Attack." *Military.com.* April 13. https://www.military.com/daily-news/2021/04/13/army-chief-defends-long-range-missile-effort-after-air-force-generals-public-attack.html.

Creveld, Martin. 2006. "Israel's Lebanese War: A Preliminary Assessment." *RUSI Journal* 151 (5): 40-43. https://doi.org/10.1080/03071840608522872.

Crozier, Michael. 1964. *The Bureaucratic Phenomenon.* University of Chicago Press.

Cushman, John. 1986. *Command and Control of Theater Forces: Adequacy.* National Defense University Press.

Daniels, Jeff. 2017. "ISIS Fight Shows US Military Can Use Lower-Cost Weapons With Lethal Results." *CNBC.* July 19. https://www.cnbc.com/2017/07/19/isis-fight-shows-us-can-use-lower-cost-weapons-with-lethal-results.html.

Davis, Brett. 2025. "UAS of All Sizes Aim for Army Programs at Fall AUSA Conference." *Inside Unmanned Systems.* January 10. https://insideunmannedsystems.com/uas-of-all-sizes-aim-for-army-programs-at-fall-ausa-conference.

de Gaulle, Charles. 1941. The Army of the Future. J. B. Lippincott Company.

Dettmer, Jamie. 2023. "Ukraine's Forces Say NATO Trained Them for Wrong Fight." *Politico,* September 22. https://www.politico.eu/article/ukraine-war-army-nato-trained-them-wrong-fight/.

D'Este, Carlo. 1994. *Decisions in Normandy.* Konecky and Konecky.

Doughty, Robert. 1990. *The Breaking Point: Sedan and the Fall of France, 1940.* Stackpole Books.

Doughty, Robert. 1985. *The Seeds of Disaster: The Development of French Army Doctrine, 1919-1939*. Stackpole Books.

Easton, Ian. 2017. "The Chinese invasion threat: Taiwan's Defense and American Strategy in Asia." *Project 2049 Institute*.

Easton, Ian. 2017. "Transformation of Taiwan's Reserve Force." RAND Corporation.

Echevarria, Antulio. 2016. "Rediscovering US Military Strategy: A Role for Doctrine." *Journal of Strategic Studies* 39 (2): 231-245. DOI: 10.1080/01402390.2015.1115035.

Edwards, Dominick. 2018. "Logistics Support to Semi-Independent Operations." *Army Sustainment* 50 (1): 40-42. https://alu.army.mil/alog/2018/JANFEB18/PDF/193373.pdf.

Ellison, Davis and Tim Sweigs. 2023. "Breaking Patterns: Multidomain Operations and Contemporary Warfare." *The Hague Centre for Strategic Studies*. https://hcss.nl/wp-content/uploads/2023/09/Breaking-Patterns-HCSS-2023.pdf /.

Ellison, Davis and Tim Sweigs. 2024. "Empty Promises? A Year Inside the World of Multidomain Operations." *War on the Rocks*. January 22. https://warontherocks.com/2024/01/empty-promises-a-year-inside-the-world-of-multi-domain-operations/.

Engstrom, Jeffrey. 2018. *Systems Confrontation and System Destruction Warfare: How the Chinese People's Liberation Army Seeks to Wage Modern Warfare*. RAND Corporation. https://www.rand.org/pubs/research_reports/RR1708.html.

Erlanger, Steven. 2025. "Trump Wants Europe to Defend Itself. Here's What It Would Take." *New York Times*. March 7. https://www.nytimes.com/2025/03/07/world/europe/europe-self-defense-trump.html.

Evans-Cranny, Sam. 2023. "Russia's Artillery War in Ukraine: Challenges and Innovations." *RUSI*. August 9. https://www.rusi.org/explore-our-research/publications/commentary/russias-artillery-war-ukraine-challenges-and-innovations.

"Exercise Foal Eagle." 2025. *Seventh Air Force*. February 22. https://www.7af.pacaf.af.mil/About-Us/Fact-Sheets/Display/Article/408383/exercise-foal-eagle/.

Eyer, Kevin.2019. "Operationalizing Distributed Maritime Operations." *Center for International and Maritime Security*. March 5. https://cimsec.org/operationalizing-distributed-maritime-operations.

Feickert, Andrew. 2025. "The 2024 Army Force Structure Transformation Initiative." *Congressional Research Service*. February 5. https://crsreports. congress.gov/product/pdf/R/R47985.

Feickert, Andrew. 2024. "Defense Primer: Army Multidomain Operations (MDO)." *Congressional Research Service*. October 1. https://www.congress. gov/crs-product/IF11409.

Feickert, Andrew. 2022. "US Ground Forces in the Indo-Pacific: Background and Issues for Congress." *Congressional Research Service*. August 30. https://www.congress.gov/crs-product/R47096.

Feickert, Andrew. 2021. "US Army Long-Range Precision Fires: Background and Issues for Congress." Congressional Research Service R46721. March 16. https://www.congress.gov/crs-product/R46721.

Feldman, Yotam. 2007. "Dr. Naveh, Or, How I Learned To Stop Worrying and Walk Through Walls," *Haaretz*. October 27. https://www. haaretz.com/2007-10-25/ty-article/dr-naveh-or-how-i-learned-to-stop-worrying-and-walk-through-walls/0000017f-db53-df9c-a17f-ff5ba92c0000.

Figes, Orlando. 2010. *The Crimean War, A History*. Metropolitan Books.

Filipoff, Dmitry. 2023. "Fighting DMO, Part 1: Defining Distributed Maritime Operations the Future of Naval Warfare." *Center for International and Maritime Security*. February 23. https://cimsec.org/fighting-dmo-pt-1-defining-distributed-maritime-operations-and-the-future-of-naval-warfare.

Fogg, Rodney. 2025. "From the Big Five to Cross-Functional Teams: Integrating Sustainment into Modernization." *U.S. Army*. January 10. https://www.army.mil/article/227832/from_the_big_five_to_cross_functional_teams_integrating_sustainment_into_modernization.

Fox, Amos. 2020. "Getting Multidomain Operations Right: Two Critical Flaws in the US Army's Multidomain Operations Concept." Association of the United States Army, *Land Warfare Paper 133*: 1-11. https://www.ausa. org/sites/default/files/publications/LWP-133-Getting-Multi-Domain-Operations-Right-Two-Critical-Flaws-in-the-US-Armys-Multi-Domain-Operations-Concept.pdf.

Fox, Amos. 2024. *Obstructive Warfare: Applications and Risks for AI in Future Military Operations*. Centre for International Governance Innovation, CIGI Paper No. 307. October. https://www.cigionline.org/static/documents/Fox-Sept2024.pdf.

Fox, Amos. "Precision Paradox and Myths of Precision Strike in Modern Armed Conflict." *RUSI Journal* 169 (1-2): 62-74. DOI: 10.1080/03071847.2024.2343717.

Fox, Amos. 2021. "Russian Hybrid Warfare: A Framework." *Journal of Military Studies* 10 (1): 1-11. https://sciendo.com/article/10.2478/jms-2021-0004.

Fox, Amos. "The Mosul Study Group and the Lessons of the Battle of Mosul." *Association of the United States Army. Land Warfare Paper 130*: 1-13. https://www.ausa.org/sites/default/files/publications/LWP-130-The-Mosul-Study-Group-and-the-Lessons-of-the-Battle-of-Mosul.pdf.

Fox, Amos. 2023. "Urban Warfare, Sieges, and Israel's Looming Invasion of Gaza." *War on the Rocks,* October 27. https://warontherocks.com/2023/10/urban-warfare-sieges-and-israels-looming-invasion-of-gaza/.

Fox, Collin. 2021. "The Porcupine in No Man's Sea: Arming Taiwan for Sea Denial." *Center for International Maritime Security (CIMSEC).* August 4. https://cimsec.org/the-porcupine-in-no-mans-sea-arming-taiwan-for-sea-denial.

Francis, Ellen, et al. 2025. "For Europeans, Signal Chat Gives Unfiltered view of Trump Team's Disdain." *Washington Post.* March 25. https://www.washingtonpost.com/world/2025/03/25/trump-administration-europe-signal-chat-leak/.

Fravel, Taylor. 2020. *Active Defense: China's Military Strategy Since 1949.* Princeton University Press.

Freedberg, Sydney. 2019. "All Services Sign On To Data Sharing – But Not To Multidomain." *Breaking Defense.* February 8. https://breakingdefense.com/2019/02/all-services-sign-on-to-data-sharing-but-not-to-multi-domain/.

Freedberg, Sydney. 2020. "Army Says Long Range Missiles Will Help Air Force, Not Compete." *Breaking Defense.* July 16. https://breakingdefense.com/2020/07/army-says-long-range-missiles-will-help-air-force-not-compete/.

Frieser, Karl-Heinz. 2005. *The Blitzkrieg Legend: The 1940 Campaign in the West.* Naval Institute Press.

Fuller, J.F.C. 1926. *The Foundations of the Science of War.* Hutchinson and Company Publishing.

Graham, Allison et al. 1972. "Bureaucratic Politics: A Paradigm and Some Policy Implications." *World Politics* 24: 40-79. https://doi.org/10.2307/2010559.

"UFS 24 Successfully Concludes." 2024. *United States Forces Korea*. August 29. https://www.usfk.mil/Media/Press-Products/Press-Releases/Article/3889608/ufs-24-successfully-concludes/.

Garcia, Zanel. 2019. *China's Military Modernization, Japan's Normalization and the South China Sea Territorial Disputes*. PalgraveMacMillan.

Gawrych, George. 1995. *The 1973 Arab-Israeli War: The Albatross of Decisive Victory*. Leavenworth Papers 21. Combat Studies Institute. https://apps.dtic.mil/sti/pdfs/ADA323718.pdf.

Ghosh, P.K. "Artificial Islands in the South China Sea." 2014. *The Diplomat*. September 23. https://thediplomat.com/2014/09/artificial-islands-in-the-south-china-sea.

Gill, Jasprett. 2023. "Return of CJADC2: DoD Officially Moves Ahead With 'Combined' JADC2 in a Rebrand Focusing on Partners." *Breaking Defense*. May 16. https://breakingdefense.com/2023/05/return-of-cjadc2-dod-officially-moves-ahead-with-combined-jadc2-in-a-rebrand-focusing-on-partners/.

Gray, Colin. 2002. *Strategy for Chaos: Revolutions in Military Affairs and the Evidence of History*. Frank Cass Publishers.

Greenhalgh, Elizabeth. 2017. "Marshal Ferdinand Foch versus Georges Clemenceau in 1919." *War in History War* 24 (3): 458-497. https://www.jstor.org/stable/26393387.

Halem, Harry. 2024. "Positional Warfare: Prospects for Ukraine in 2024-27." *Marine Corps Gazette*. September 1. https://www.mca-marines.org/gazette/positional-warfare/.

Hale, Robert et al. 2024. "Defense Resourcing for the Future." *Commission on Planning, Programming, Budgeting, and Execution Reform*. March 6. Commission-on-PPBE-Reform_Full-Report_6-March-2024_FINAL.pdf.

Hayman, Tamir. 2016. "Learning in the General Staff," *Dado Center Journal*, 8. September 4. https://www.idf.il/en/mini-sites/dado-center/vol-8-the-general-staff-part-a/learning-in-the-general-staff/.

Hecker, James. 2024. "Air Superiority: A Renewed Vision," *ÆTHER: A Journal of Strategic Airpower and Spacepower* 3(2): 5-10. https://www.airuniversity.af.edu/Portals/10/AEtherJournal/Journals/Volume-3_Number-2/Hecker.pdf.

Herbert, H. Paul. *Deciding What Has to Be Done: General William E. DePuy and the 1976 Edition of FM 100-5, Operations*. US Army Command and General Staff College, 1988.

Hinata-Yamaguchi, Ryo. 2024. *Developments and Transformations in Japan's Defense Planning and Readiness.*" Amsterdam University Press.

Hitchens, Theresa. 2020. "Long-Range All-Domain Purpose Prompts Roles & Mission Debate." *BreakingDefense.* July. 9 https://breakingdefense.com/2020/07/long-range-all-domain-prompts-roles-missions-debate/.

Hlad, Jennifer. 2024. "Army Tests Next-Gen Long-Range Fires Capability in Pacific." *Defense One.* June. 27. https://www.defenseone.com/technology/2024/06/army-tests-next-gen-long-range-fires-capability-pacific/397724/.

Hornung, Jeffrey, et al. 2021. "Preparing Japan's Multidomain Defense Force for the Future Battlespace Using Emerging Technologies." RAND Corporation.

Huba Wass de Czege. 2020. Commentary on "The U.S. Army in Multi-Domain Operations 2028." *Strategic Studies Institute.* April 1. https://press.armywarcollege.edu/cgi/viewcontent.cgi?article=1908&context=monographs.

Hughes, Christopher. 2022. *Japan as a Global Military Power: New Capabilities, Alliance Integration, Bilateralism-Plus.* Cambridge University Press.

Hunter, Cameron and Bleddyn Bowen. "We'll Never Have a Model of an AI Major-General: Artificial Intelligence, Command Decisions, and Kitsch Visions of War." *Journal of Strategic Studies* 47(1): 116-146. DOI: 10.1080/01402390.2023.2241648.

Inoguchi, Takashi, et al. 2011. *The US-Japan Security Alliance: Regional Multilateralism.* Springer.

Jamison, Tommy. 2024. "Taiwan's Theory of the Fight." *War on the Rocks.* February 21. https://warontherocks.com/2024/02/taiwans-theory-of-the-fight/.

Jennings, Nathan. 2024. "Penetrate, Disintegrate, and Exploit: The Israeli Counteroffensive at the Suez Canal, 1973," *Modern War Institute.* October 31. https://mwi.westpoint.edu/penetrate-disintegrate-and-exploit-the-israeli-counteroffensive-at-the-suez-canal-1973.

Jennings, Nathan. 2022. "The 1973 Arab-Israeli War: Insights for Multidomain Operations," *Association of the United States Army* Land Warfare Paper 152. https://www.ausa.org/publications/1973-arab-israeli-war-insights-multi-domain-operations.

Jensen, Benjamin. 2017. *Forging the Sword: Doctrinal Change in the US Army.* National Defense University Press.

Judson, Jen. "Army Futures Command drafting next operating concept," *Defense News*. July 31, 2023. https://www.defensenews.com/land/2023/07/31/army-futures-command-drafting-next-operating-concept/.

Judson, Jen. 2019. "Does the US Army Have Enough Weapons to Defend Europe? Exercise Defender 20 Will Reveal All." *Defense News*. December 27. https://www.defensenews.com/land/2019/12/27/does-the-army-have-its-european-weapons-stocks-right-defender-europe-2020-will-tell/.

Johnson, David. 2022. "The Army Risks Reasoning Backwards in Analyzing Ukraine." *War on the Rocks*. June 14. https://warontherocks.com/2022/06/the-army-risks-reasoning-backwards-in-analyzing-ukraine/.

Jones, Peter, et. al. 2016. *Unclassified Summary of the U.S. Army Training and Doctrine Command Russian New Generation Warfare Study*. Training and Doctrine Command.

Jonsson, Oscar. 2019. *The Russian Understanding of War: Blurring the Lines between War and Peace*. Georgetown University Press.

Kagan, Frederick. 2006. *Finding the Target: The Transformation of American Military Policy* Encounter Books.

Kamara, Hassan. 2020. "Countering A2AD in the Indo-Pacific: A Potential Change for the Army and the Joint Force." *Joint Force Quarterly* 97: 97-102. https://ndupress.ndu.edu/Portals/68/Documents/jfq/jfq-97/jfq-97_97-102_Kamara.pdf?ver=2020-03-31-215816-687.

Kapur, Ashok. 2019. *Geopolitics and the Indo-Pacific Region*. Routledge.

Kaplan, Fred. 2016. *Dark Territory: The Secret History of Cyber War*. Simon and Schuster.

Kaplan, Robert. 2015. *Asia's Cauldron: The South China Sea and the End of a Stable Pacific*. Random House Trade Paperbacks.

Kawaguchi, Shun. 2022. "Chijō Misairu, 3 Dankai Haibi 'Hangeki Nōryoku' Shatei 1000–3000 Kiro Seifu Kentō [Ground-Launched Missiles: Government Considering 3-Stage Deployment of Ground-Based Missiles with 'Counterstrike Capability' Range of 1,000 to 3,000 km]." *Mainichi Shimbun*, November 11. https://mainichi.jp/articles/20221125/ddn/001/010/004000c.

Keefer, Edward. 2017. *Harold Brown: Offsetting the Soviet Military Challenge, 1977–1981*. Secretaries of Defense Historical Series.

Kepe, Marta. 2024. "From Forward Presence to Forward Defense: NATO's Defense of the Baltics," *RAND Corporation*. February 14. https://www.

rand.org/pubs/commentary/2024/02/from-forward-presence-to-forward-defense-natos-defense.html.

Kier, Elizabeth. 2017. *Imagining War: French and British Military Doctrine Between the Wars.* Princeton University Press.

Kiesling C. Eugenia. 1996. "If It Ain't Broke, Don't Fix It: French Military Doctrine Between the World Wars," *War in History* 3 (2): 208-223. https://www.jstor.org/stable/26004549.

Kimmons, Sean. 2016. "With Multidomain Concept, Army Aims for 'Windows of Superiority.'" *US Army.* November 14. https://www.army.mil/article/178137/with_multi_domain_concept_army_aims_for_windows_of_superiority.

Kimmos, Sean. 2018. "Army Updates Future Operating Concept." *US Army.* December 6. https://www.army.mil/article/214632/army_updates_future_operating_concept.

Kirkland R. Faris. 1992. "Governmental Policy and Combat Effectiveness: France 1920-1940." *Armed Forces & Society* 18 (2): 175-191. https://www.jstor.org/stable/45305303.

Knox, MacGregor, et. al. 2001. *The Dynamics of Military Revolution, 1300-2050.* Cambridge University Press.

Kofman, Michael. 2019. "It's Time to Talk about A2/AD: Rethinking the Russian Military Challenge," *War on the Rocks.* September 5. https://warontherocks.com/2019/09/its-time-to-talk-about-a2-ad-rethinking-the-russian-military-challenge/.

Kofman, Michael, et al. 2017. *Lessons from Russia's Operations in Crimea and Eastern Ukraine.* RAND Corporation. https://www.rand.org/pubs/research_reports/RR1498.html.

Korber, Avi. 2008. "The Israel Defense Forces in the Second Lebanon War: Why the Poor Performance?" *Journal of Strategic Studies* 31 (1): 3-40. 10.1080/01402390701785211.

Koshino, Yuka. 2022. "Meeting China's Emerging Capabilities: Countering Advances in Cyber, Space, and Autonomous Systems." *National Bureau of Asian Research.* Special Report 103. https://www.nbr.org/wp-content/uploads/pdfs/publications/sr103_meetingchinasemergingcapabilities_dec2022.pdf.

Krepinevich F. Andrew. 2010. "Why AirSea Battle." *Center For Strategic and Budgetary Assessments.* February 19. https://csbaonline.org/uploads/documents/2010.02.19-Why-AirSea-Battle.pdf.

"Kūji ni 'uchū sakusendan' senmon butai ōhaba kaihen de shinsetsu e ['Space Operations Command' to Be Established in the JASDF as Part of a Major Reorganization of Specialized Units]." *Mainichi Shimbun*, October 3, 2024. https://mainichi.jp/articles/20241003/ddm/012/010/045000c.

Kull J. Daniel. 2017. "The Myopic Muddle of the Army's Operations Doctrine." *Military Review*. May 24. https://www.armyupress.army.mil/Portals/7/Army-Press-Online-Journal/documents/Kull-v2.pdf.

Lagarde, Benoît. 2011. "Grand Quartier Général, 1914-1918," *Service Historique de la Défense*. https://www.servicehistorique.sga.defense.gouv.fr/sites/default/files/notices_files/SHDGR_REP_16NN.pdf.

LaGrone, Sam. 2015. "Pentagon Drops Air Sea Battle Name, Concept Lives On." *UNSI News*. January 20. https://news.usni.org/2015/01/20/pentagon-drops-air-sea-battle-name-concept-lives.

Lambeth, Benjamin. 2011. *Air Operations in Israel's War Against Hezbollah: Learning from Lebanon and Getting It Right in Gaza.* RAND Corporation.

Lamothe, Dan, et al. 2017. "Battle of Mosul: How Iraqi Security Forces Defeated the Islamic State." *Washington Post.* July 10. https://www.washingtonpost.com/graphics/2017/world/battle-for-mosul/.

Landler, Mark. 2022. "In Standoff with Putin, Biden Makes Sure European Allies Are With Him," *New York Times.* January 28. https://www.nytimes.com/2022/01/28/world/europe/biden-putin-ukraine-europe.html/.

Lanza, Stephen. 2021. "Fires for Effect: 10 Questions about Army Long-Range Precision Fires in the Joint Fight." *Association of the United States Army* Spotlight 21-1. August 30. https://www.ausa.org/publications/fires-effect-10-questions-about-army-long-range-precision-fires-joint-fight.

Lappin, Yaakov. 2020. "The IDF's Momentum Plan Aims to Create a New Type of War Machine." *Begin-Sadat Center for Strategic Studies.* March 22. https://besacenter.org/idf-momentum-plan/.

Lee, Connie. 2019. "News from EWC: Marine Corps Defining New Operating Concept." *National Defense.* Oct 22. *https://www.nationaldefensemagazine. org/articles/2019/10/22/marine-corps-works-to-define-new-operating-concept.*

Leonard, Steve. 2017. "Broken and Unreadable: Our Unbearable Aversion to Doctrine." *Modern War Institute,* May 18. https://mwi.westpoint.edu/broken-unreadable-unbearable-aversion-doctrine/.

Libiseller, Chiara. 2023. "'Hybrid Warfare' as an Academic Fashion." *Journal of Strategic Studies* 46 (4): 858-880. 10.1080/01402390.2023.2177987.

Li, Xiaobing. 2017. *The Cold War in East Asia.* Routledge.

Li, Xiaobing. 2019. *The History of Taiwan.* Bloomsbury Publishing.

Lopez, Daniel. 2024. "Multidomain Transformation in the Indo-Pacific." *US Army*. October 16. https://www.army.mil/article/280588/multi_domain_transformation_in_the_indo_pacific.

Lopez, C. Todd. 2024. "U.S. Intends to Reconstitute U.S. Forces Japan as Joint Forces Headquarters." *DoD News*. July 28. https://www.defense.gov/News/News-Stories/Article/Article/3852213/us-intends-to-reconstitute-us-forces-japan-as-joint-forces-headquarters/.

Luhnow, David, et al. 2025. "What Does MAGA Have Against Europe?" *Wall Street Journal*. March 28. https://www.wsj.com/politics/what-does-maga-have-against-europe-96416042.

Luttwak, Edward and Eitan Shamir. 2023. *The Art of Military Innovation: Lessons from the Israel Defense Forces*. Harvard University Press.

Lynch, Michael and Brennan Deveraux. 2024. "Landpower, Homeland Defense, and Defending Forward in US Indo-Pacific Command." July 1. https://ssi.armywarcollege.edu/SSI-Media/Recent-Publications/Article/3823848/landpower-homeland-defense-and-defending-forward-in-us-indo-pacific-command/.

Lyons, Marco and Dave Johnson. 2022. "People Who Know, Know MDO: Understanding Army Multidomain Operations as a Way to Make It Better," *Association of the United States Army*, Land Warfare Paper 151. November 14. https://www.ausa.org/publications/people-who-know-know-mdo-understanding-army-multi-domain-operations-way-make-it-better.

Mansoor, Peter, and Williamson Murray. 2019. *The Culture of Military Organizations*. Cambridge University Press.

Martin, Alexander S. 1992. *The Republic in Danger: General Maurice Gamelin and the Politics of French Defence, 1933-1940*. Cambridge University Press.

Matsuoka, Misato. 2018. *Hegemony and the US–Japan Alliance*. Routledge.

Matthews, Matt. 2008. "We Were Caught Unprepared: The 2006 Hezbollah-Israeli War." *The Long War Series Occasional Paper 26*. Combat Studies Institute Press. https://www.armyupress.army.mil/Portals/7/combat-studies-institute/csi-books/we-were-caught-unprepared.pdf

Mattis, James. 2008. "USJFCOM Commander's Guidance for Effects-Based Operations." *Parameters* 38 (3): 18–25. DOI: 10.55540/0031-1723.2437.

McCoy, Kelly. 2017. "The Road to Multi-Domain Battle: An Origin Story," *The Modern War Institute*. October 27. https://mwi.westpoint.edu/road-multi-domain-battle-origin-story/.

Meiser, Jeffrey. 2024. "Bringing Method to the Strategy Madness." *War on the Rocks*. May 2. https://warontherocks.com/2024/05/bringing-a-method-to-the-strategy-madness/.

Merglen, Alban. 1996. "La Responsabilité Du Maréchal Pétain Dans Le Désastre Militaire de Mai-Juin 1940," *Guerres Mondiales et Conflits Contemporains*. http://www.jstor.org/stable/25732385.

Michishita, Narushige. 2004. "Summary of Sessions." *National Institute for Defense Studies*. March. https://ndlsearch.ndl.go.jp/books/R100000039-I1283024.

Moita, Teixeira. "The Permanence of Land Power at the Center of Military Clashes to Conquer Territories" *Coleção Meira Mattos 16*. July, 2022. http://www.ebrevistas.eb.mil.br/RMM/article/view/11383/9122.

Montefiore, Simon. 2017. *The Romanovs, 1613–1918*. Vintage Books.

Morison, Elting. 2016. *Men, Machines, and Modern Times, 50th Anniversary Edition*. Cambridge: MIT Press.

Morison, George. 1903. *The New Epoch: As Developed by the Manufacture of Power*. The Riverside Press.

Murdock A. Clark et al. "Beyond Goldwater-Nichols (BG-N): Defense Reform for a New Strategic Era - Phase 1 Report." *Center for Strategic and International Studies*. March 1, 2004. https://www.csis.org/analysis/beyond-goldwater-nichols-phase-i-report.

Murray, Williamson. 1998. "Armored Warfare: The British, French, and German Experiences." In *Military Innovation in the Interwar Period*, edited by Williamson Murray and Allan Millet. Cambridge University Press.

Murray, Williamson. 2010. "British Military Effectiveness in World War II." In *Military Effectiveness: Volume 3, The Second World War*, edited by Allan Millett and Williamson Murray. Cambridge University Press.

Murray, Williamson. 1981. "The German Response to Victory in Poland: A Case Study in Professionalism," *Armed Forces & Society* 7 (2): 285-298. http://www.jstor.org/stable/45346229.

Nagl, John. 2023. "Learning from Russia's War on Ukraine," *Military Times*. September 1. https://www.militarytimes.com/opinion/2023/09/01/learning-from-russias-war-on-ukraine/.

Nemoto, Ryo, et al." 2022. "Chinese Missiles Land in Japan's EEZ: Defense Chief." *Nikkei Asia*, August 4. https://asia.nikkei.com/Politics/International-relations/Taiwan-tensions/5-Chinese-missiles-land-in-Japan-s-EEZ-defense-chief

Nemoto, Ryo. 2022. "Japan to Expand Fuel and Ammo Storage on Islands Near Taiwan: Defense Chief." *Nikkei Asia*, September 6. https://asia.nikkei.com/Editor-s-Picks/Interview/Japan-to-expand-fuel-and-ammo-storage-on-islands-near-Taiwan-defense-chief.

O'Brien, Luke. 2016. "The Doctrine of Military Change: How the US Army Evolves." *War on the Rocks*. July 25. https://warontherocks.com/2016/07/the-doctrine-of-military-change-how-the-us-army-evolves/.

O'Brien, Phillips. 2019. *How the War Was Won: Air-Sea power and Allied Victory in World War II*. Cambridge University Press.

Operational Test and Evaluation Director. 2021. "Precision Strike Missile (PrSM)', in FY2020 Annual Report." *US Department of Defense*. https://www.dote.osd.mil/Portals/97/pub/reports/FY2020/army/2020prsm.pdf

Ozaki, Anna. 2023. "Jikūsen Jikan no 'Hayasa' o Kachime to Shita Kidōsen no Aratana Tatakaikata [Temporal Warfare – A New Way of Fighting Maneuver Warfare Using the 'Quickness' of Time as a Winning Factor]." TERCOM, Ground Self-Defense Force, Japan, April 21, 2023. Last modified April 21. https://www.mod.go.jp/gsdf/tercom/img/file2184.pdf.

Page, Stephen. 2024. "3d MDTF Demonstrates Ability to Operate in the Indo-Pacific." *US Army*. Jun. 21. https://www.army.mil/article/277487/3d_mdtf_demonstrates_ability_to_operate_in_the_indo_pacific.

Pamuk, Humeyra, and Michelle Nichols. 2022. "US Accelerates Ukraine Diplomacy as Europe Slides into Winter," *Reuters*. December 13. https://www.reuters.com/world/europe/us-accelerates-ukraine-diplomacy-europe-slides-into-winter-2022-12-13/.

Panda, Ankit. 2020. *Kim Jong Un and the Bomb: Survival and Deterrence in North Korea*. Oxford University Press.

Perkins G. David. 2017. "Multi-Domain Battle: Driving Change to Win in the Future," *Military Review* 97 (4): 6-12. https://www.armyupress.army.mil/Journals/Military-Review/English-Edition-Archives/July-August-2017/Perkins-Multi-Domain-Battle/.

Pernin G. Christopher. 2012. "Lessons from the Army's Future Combat Systems Program." *RAND Corporation*.

Phillips, Dwight. 2023. *Multidomain Operations: Passing the Torch. The Hague Centre for Strategic Studies*. https://hcss.nl/wp-content/uploads/2023/11/Multidomain-Operations-Passing-The-Torch-HCSS-2023.

Porch, Douglas. 2000. "Military 'Culture' and the Fall of France in 1940: A Review Essay," *International Security* 24 (4): 157-180. http://www.jstor.org/stable/2539318.

Porter, Michael. 1996. "What Is Strategy?" *Harvard Business Review*. November–December. https://hbr.org/1996/11/what-is-strategy.

Porter, Tom. 2023. "Ukrainian Troops Are Abandoning US Tactics in Their Counteroffensive Because They Haven't Worked." *Business Insider*. August 3. https://www.businessinsider.com/western-trained-ukrainian-troops-are-abandoning-us-tactics-report-2023-8.

Quinn, Bryan. 2023. "Sustaining Multidomain Operations: The Logistical Challenge Facing the Army's Operating Concept." *Military Review* 103 (2): 128-138. https://www.armyupress.army.mil/Portals/7/military-review/Archives/English/MA-23/Multidomain-Operations/Sustaining-Multidomain-Operations.pdf.

Rainey, James. 2024. "Continuous Transformation: Transformation in Contact." *Military Review*. August. https://www.armyupress.army.mil/Portals/7/military-review/Archives/English/Online-Exclusive/2024/Transformation-in-Contact/Transformation-in-Contact-UA.pdf.

Ragin, R. Ronald and Christopher Ingram. 2024. "Theater Sustainment Transformation: Lessons from the Russia-Ukraine War." *US Army*. April 23. https://www.army.mil/article/274914/theater_sustainment_transformation_lessons_from_the_russia_ukraine_war.

Rayment, Stephen. 2008. "Capacity of Wireless Mesh Networks." In *Emerging Technologies in Wireless LANs: Theory, Design, and Deployment*, edited by Benny Bing. Cambridge University Press.

Reilly, Jeffrey. 2016. "Multidomain Operations a Subtle but Significant Transition in Military Thought," *Air & Space Power Journal* 30 (1): 61-73. https://apps.dtic.mil/sti/tr/pdf/AD1003670.pdf.

Richardson, John. 2016. "Chief of Naval Operations Adm. John Richardson: Deconstructing A2AD," *The National Interest*. October 3. https://nationalinterest.org/feature/chief-naval-operations-adm-john-richardson-deconstructing-17918.

Richmond, Jake. 2016. "Iraqi Forces Begin the Battle of Mosul." *DOD News*. October 17. https://www.defense.gov/News/News-Stories/Article/Article/975239/iraqi-forces-begin-battle-for-mosul/.

"Rikujō bakuryōchō no gaikoku shutchō nitsuite [Foreign Business Trip of the Chief of Staff of the Ground Staff]." 2024. *Japan Ground Self-Defense*

Force. March 1. https://www.mod.go.jp/gsdf/news/press/2024/pdf/20240301.pdf

"Rikujō Jieitai no torikumi [JGSDF's Initiatives]." 2024. *Japan Ground Self-Defense Force.* https://www.mod.go.jp/gsdf/about/structure/2023.pdf#page=2.

Roehrig, Terence. 2022. *North Korea: Gray Zone Actions in the Yellow Sea.* Routledge. 2022.

Romjue L. John. 1984. *From Active Defense to AirLand Battle: The Development of Army Doctrine, 1973–1982.* Training and Doctrine and Command. https://www.tradoc.army.mil/wp-content/uploads/2020/10/From-Active-Defense-to-AirLand-Battle.pdf

Rose, Robert. 2024. "Ending the Churn: To Solve the Recruiting Crisis, the Army Should be Asking Very Different Questions," *The Modern War Institute.* February 9. https://mwi.westpoint.edu/ending-the-churn-to-solve-the-recruiting-crisis-the-army-should-be-asking-very-different-questions/.

Rose G. Robert. 2023. "Returning Context to Our Doctrine," *Military Review.* October: 1-11. https://www.armyupress.army.mil/journals/military-review/online-exclusive/2023-ole/returning-context-to-our-doctrine/.

Ross L. Andrew. 2010. "On Military Innovation: Toward an Analytical Framework." *SITC 2010 Policy Brief.* September 1. https://escholarship.org/uc/item/3d0795p8.

Roulo, Claudette. 2015. "Offset Strategy Puts Advantage in Hands of US, Allies." *DOD News.* January 28. https://www.defense.gov/News/News-Stories/Article/Article/603997/offset-strategy-puts-advantage-in-hands-of-us-allies.

S.A. Mackenzie. 1994. "Strategic Air Power Doctrine for Small Forces." *Air Power Studies Centre.*

Sanderson, Bill. 2014. "Leaked Transcripts Reveal Putin's Secret Ukraine Attack." *NY Post.* September 21. https://nypost.com/2014/09/21/leaked-transcripts-reveal-putins-secret-attack-in-ukraine/.

Sanger, David and Michael Shear. 2025. "Trump Floats Using Force to Take Greenland and the Panama Canal." *New York Times.* January 7. https://www.nytimes.com/2025/01/07/us/politics/trump-panama-canal-greenland.html.

Santora, Mark. 2025. "How Ukraine's Counteroffensive in Russia's Kursk Region Unraveled." *New York Times.* March 16. https://www.nytimes.com/2025/03/16/world/europe/ukraine-kursk-retreat-russia.html.

Schuker A. Stephen. 1986. "France and the Remilitarization of the Rhineland, 1936," *French Historical Studies* 14 (3): 299-338. https://www.jstor.org/stable/286380.

Seidman, Harold. 1970. *Politics, Position, and Power: The Dynamics of Federal Organization*. Oxford University Press.

Sickler, Robert. 2021. 'The Technology Triad: Reimagining the Relationship Between Technology and Military Innovation." PhD diss., Arizona State University. ProQuest. https://www.proquest.com/openview/0c6e2ffe e93314f3b55127b2f6e43975/1?pq-origsite=gscholar&cbl=18750&diss=y.

Simpkin, Richard. 1985. *Race to the Swift: Thoughts on Twenty-First Century Warfare*. Sterling: Brassey's.

Shamir, Eitan. 2011. *Transforming Command: The Pursuit of Mission Command in the US, British, and Israeli Armies*. Stanford University Press.

Shinkman, Paul. 2017. "ISIS War Drains U.S. Bomb Supply." *U.S. News and World Report*. February 17. https://www.usnews.com/news/world/articles/2017-02-17/us-raiding-foreign-weapons-stockpiles-to-support-war-against-the-islamic-state-group.

Shlapak, David and Michael W. Johnson. 2016. *Reinforcing Deterrence on NATO's Eastern Flank: Wargaming the Defense of the Baltics*. RAND Corporation. https://www.rand.org/pubs/research_reports/RR1253.html.

Siboni, Gabi et. al. 2018. "The Development of Security-Military Thinking in the IDF." *Strategic Assessment* 21 (1): 7-19. https://www.inss.org.il/wp-content/uploads/2018/05/The-Development-of-Security-Military-Thinking-in-the-IDF.pdf.

Skinner, Douglas. 1988. "AirLand Battle Doctrine." Center for Naval Analyses. https://apps.dtic.mil/sti/pdfs/ADA202888.pdf.

Smith, Todd. 2019. "This 3-star Army General Explains What Multidomain Operations Mean for You," *Army Times*, August 11. https://www.armytimes.com/news/your-army/2019/08/11/this-3-star-army-general-explains-what-multidomain-operations-mean-for-you/.

Starry A. Donn. 1981. "Extending the Battlefield." *Military Review* 61 (3): 31-50. https://www.armyupress.army.mil/Portals/7/online-publications/documents/1981-mr-donn-starry-extending-the-battlefield.pdf.

Sugimoto, Yasushi. 2023. '""Kekkyoku nan nin shinu no ka" shimyure-shon ni Abe shi kikikan ["How Many People Will Die in the End?" Abe Concerned by Simulation Result].' *Sankei Shimbun*, September 29. https://www.sankei.com/article/20230929-T3JBQU3ZLVJYXKCUU44VQJFPLM/.

Sugimoto, Yasushi. 2023. "Kono mama ja Chūgoku ni katenai" kuniku no usaden ["We Can't Beat China If We Don't Do Something About It." Usaden as a Desperate Measure].' *Sankei Shimbun*, September 27. https://www.sankei.com/article/20230927-HASC5DBRVFIWZLSQYMQGOR5Q2Q/?959896.

Tainter, Joseph, 1988. *The Collapse of Complex Societies*. Cambridge University Press.

Takabatake, Futoshi. 2024. "NATO's Approach to Multidomain Operations: From the Perspective of the Economics of Alliances," *Defence and Peace Economics* 35 (3): 281–294. DOI: 10.1080/10242694.2023.2235502.

Takada, Katsuki. 2023. "Jōsetsu tōgō shireibu nitsuite [On Permanent Joint Operation Command]." *Anzen hoshō o kangaeru [Contemplating Security]*, September 1. http://anpokon.or.jp/pdf/kaishi_820.pdf.

Tangredi, Sam. 2013. *Anti-Access Warfare: Countering A2AD Strategies*. Naval Institute Press.

Taylor-Kendall, Andrea, and Michael Kofman. 2025. "Putin's Point of No Return How an Unchecked Russia Will Challenge the West." *Foreign Affairs* 104 (1). January/February. https://www.foreignaffairs.com/russia/putins-point-no-return.

Terada, Hiroyuki. 2019. "Kaijō Jieitai kanbu gakkō no kyōiku kaikaku [Education Reform at the JMSDF Command and Staff College]." *JMSDF Command and Staff College Review* 8 (2): 6–17.

Thornton, Rod. 2015. "The Changing Nature of Modern Warfare: Responding to Russian Information Warfare." *RUSI Journal* 160 (4): 40-48. DOI: 10.1080/03071847.2015.1079047.

Tokariuk, Olga. 2024. "Ukraine's Gamble in Kursk Restores Belief It Can Beat Russia – It Requires a Western Response." *Chatham House*. August 19. https://www.chathamhouse.org/2024/08/ukraines-gamble-kursk-restores-belief-it-can-beat-russia-it-requires-western-response.

Toki, Masato. 2009. "Japan's Evolving Security Policies: Along Came North Korea's Threats." *Nuclear Threat Initiative*. June 3. https://www.nti.org/analysis/articles/japans-evolving-security-policies/.

Tollefson, James. 2018. "Fixing Army Doctrine: A Network Approach." *Military Review* 98 (1): 71-79. https://www.armyupress.army.mil/Journals/Military-Review/English-Edition-Archives/January-February-2018/Fixing-Army-Doctrine-A-Network-Approach/.

Tournès, René. 1936. "The French Army, 1936," *Foreign Affairs*. April 1. https://www.foreignaffairs.com/articles/france/1936-04-01/french-army-1936.

TRADOC Military History & Heritage Office Staff. *Victory Starts Here: A Short 50-Year History of the US Training and Doctrine Command.* Fort Leavenworth: Combat Institute Press, 2023.

Tsvetkova, Maria, et al. 2025. "Russia Building Major New Explosives Facility as Ukraine War Drags On," *Reuters.* May 8. https://www.reuters.com/investigations/russia-building-major-new-explosives-facility-ukraine-war-drags-2025-05-08/.

Udagawa, Daizo, et al. 2025. "Kaijō Jieitai Reiwa 7 nendo gaisan yōkyū no jūten [Draft on the Budget of JMSDF for FY 2025]." *Ships of the World,* January, 156–163.

Vergun, David. 2024. "Air Force General Says Ukraine Needs Ammunition." *DOD News.* February 28. https://www.defense.gov/News/News-Stories/Article/Article/3689717/air-force-general-says-ukraine-needs-ammunition/.

Yamazaki, Koji. 2024. "Tōgō un'yō no genjō to kadai [Current Status and Issues of Joint Operations]." *National Defense Academy of Japan Alumni Association,* March 2, 2024. Last modified March 2, 2024. https://www.bodaidsk.com/news_topics/docs/20240302koen_tougou-unyou.pdf.

Yanagida, Osamu. 2018. "The U.S. Military's Air Tasking Cycle and Its Challenges: From the Viewpoint of Establishing an 'Operation Cycle.'" *Air Power Studies* 5: 138–158.

Vershinin, Alex. 2020. "The Challenge of Dis-Integrating A2AD Zone." *Joint Force Quarterly* 97 (2): 13-19. https://ndupress.ndu.edu/Portals/68/Documents/jfq/jfq-97/jfq-97_13-19_Vershinin.pdf?ver=2020-03-31-125227-110.

Vial, Philippe. 2019. "1932-1961: Unifying Defense." *Inflexions* 21 (3). June 6. https://doi.org/10.3917/infle.021.0011

Wald, Emmanuel. 1992. *The Wald Report: The Decline of Israeli National Security Since 1967.* Westview.

Walt, Stephen. 2005. "The Relationship between Theory and Policy in International Relations." *Annual Review of Political Science* 8: 23-48. https://www.annualreviews.org/content/journals/10.1146/annurev.polisci.7.012003.104904.

Watling, Jack, et al. 2024. "Russian Military Objectives and Capacity in Ukraine Through 2024." *RUSI.* February 13. https://www.rusi.org/explore-our-research/publications/commentary/russian-military-objectives-and-capacity-ukraine-through-2024.

Weisgerber, Marcus. 2016. "The US is Raiding its Global Bomb Stockpiles to Fight ISIS." *Defense One*. May 26. https://www.defenseone.com/threats/2016/05/us-raiding-its-global-bomb-stockpiles-fight-isis/128646/.

Weiss, Michael, et al. 2025. "Can Europe Back Ukraine's Fight Alone?" *New Line Magazine*. March 3. https://newlinesmag.com/argument/can-europe-back-ukraines-fight-alone/.

Wesolowski, Kathrin. 2025. "Fact Check: Russia's Influence on Germany's 2025 Election." *DW*. February 18. https://www.dw.com/en/russian-disinformation-aims-to-manipulate-german-2025-election/a-71664788.

Wiener, Norbert. 2013. *Cybernetics: Or, Control and Communication in the Animal and the Machine*. MIT Press.

Work, Robert. 2015. "Deputy Secretary of Defense Speech," *Army War College Strategy Conference*. April 8. https://www.defense.gov/News/Speeches/Speech-View/Article/606661/army-war-college-strategy-conference/.

Wylie, J.C. 2014. *Military Strategy: A General Theory of Power Control*. Naval Institute Press.

Zenko, Micah. 2018. "America's Military Is Nostalgic for World Wars." *Foreign Policy*, March 13. https://foreignpolicy.com/2018/03/13/americas-military-is-nostalgic-for-great-power-wars/.

Zweibelson, Ben. 2024. *Beyond the Pale: Designing Military Decision-Making Anew*. Air University Press.

CONTRIBUTING AUTHORS

Dr Andrew Carr is a Senior Lecturer in the Strategic and Defence Studies Centre at the Australian National University. His research focuses on Strategy and Australian Defence Policy. He has published in outlets such as *Survival, Parameters, Journal of Strategic Studies, Australian Foreign Affairs, International Theory, The Washington Quarterly*, and *Comparative Strategy*. He has a sole authored book with Melbourne University Press and has edited books with Oxford University Press and Georgetown University Press. He is a member of the ANU-Defence Strategic Policy History Project, writing a history of Australian Defense White Papers from 1976-2020.

Andre Carvalho is reading for a PhD in Military Sciences at the Brazilian Army Command and General Staff College (CGSC), where he is also an associate researcher at Minerva, a research group focused on conflict, strategy and intelligence. He is also the Assistant Editor of the Meira Mattos Journal of Military Sciences, the Brazilian Army CGSC's peer-reviewed academic publication. Andre was an invited lecturer at the Pontifical Catholic University of Minas Gerais to teach modules of Geopolitics, New Technologies in Warfare, and worked as a defense consultant at the Integrated Manufacturing and Technology Center of Brazil. Research interests include operational art, force design, military doctrine, land warfare, air-land integration, military innovation and adaptation, the future of warfare, and defense technology.

Dr J.P. Clark, Colonel, U.S. Army (retired) was the lead author of the *U.S. Army in Multi-Domain Operations, 2028*. In twenty-six years as an armor officer, he also served as an exchange officer in the British Army Headquarters, as the chief of the Strategy Division on the Army Staff, and as an instructor at both West Point and the U.S. Army War College. J.P. holds a Ph.D. and M.A. in history from Duke University, an M.S.S. from the Army War College, and a B.S. in Russian and German from West Point. He is the author of *Preparing for War: The Emergence of the Modern U.S. Army, 1815–1917* (2017).

Dr Thomas Crosbie is an associate professor at the Royal Danish Defence College. The author of *The Political Army: How the U.S. Military Learned to Manage the Media and Public Opinion*, his work explores the intersection of politics and military operations. He is co-editor of the Oxford Handbook of Professional Military Education (with Andrew Stewart), is the series editor of Military Politics with Berghahn Books, was a founding lead editor of the Scandinavian Journal of Military Studies, and has written or contributed to dozens of articles, book chapters, and edited volumes.

Davis Ellison is a strategic analyst at the Hague Centre for Strategic Studies (HCSS), specializing in security and defense affairs and chairing the HCSS Initiative on the Future of Transatlantic Relations. His work focuses on deterrence, arms control, and strategy. Previously, he served as a strategist at NATO Allied Command Transformation, where he co-authored the NATO Warfighting Capstone Concept, contributed to the Secretary General's NATO 2030 initiative, and worked across a range of military-strategic issues. Ellison is a PhD candidate in the Department of War Studies at King's College London, researching civil-military relations in NATO. He holds a master's degree in political science from the University of North Carolina – Chapel Hill, with studies at Humboldt and Free Universities in Berlin and the University of Bath, and bachelor's degrees in political science and international studies from Indiana University – Bloomington, where he also studied at the London School of Economics.

Dr Amos C. Fox, Lieutenant Colonel, U.S. Army (Retired) is a Professor of Practice with the Future Security Initiative at Arizona State University where he teaches about armed conflict, the future of war, and security studies. He is also the Managing Editor for Small Wars Journal, and he also hosts the Revolution in Military Affairs podcast. He received his PhD from the University of Reading, his masters degrees from Ball State University and the School of Advanced Military Studies, and he received his bachelors degree from Indiana University-Indianapolis. He is the author of *Conflict Realism: Understanding the Causal Logic of Modern War and Warfare* (Howgate 2024).

Franz-Stefan Gady is an Austrian defense analyst, writer, and advisor focusing on modern warfare, military reform, and future conflict. He serves as an Adjunct Senior Fellow in the Defense Program at the Center for a New American Security (CNAS) in Washington, D.C., and is affiliated with the International Institute for Strategic Studies (IISS) in London. Gady has conducted extensive field research embedded with military forces in Afghanistan, Iraq, and Ukraine, examining how armies adapt to the realities of high-intensity warfare. His writing has appeared in *Foreign Affairs*, *Foreign Policy*, *The International New York Times*, *Financial Times*, and *The Diplomat*, where he is a contributing editor. He is the author of *Die Rückkehr des Krieges* (2024) and *How the United States Would Fight China* (2025), and advises European and U.S. defense institutions on doctrine, capability development, and the enduring relevance of attrition and adaptation in modern war.

Rintaro Inoue is a Ph.D. student at the Graduate School of Law, Keio University, focusing his dissertation on the history of the U.S.-Australia defense relationship. His research areas include the U.S. alliance policy in Asia, Japanese defense and security policy, and Australian defense policy. Mr. Inoue is also a research assistant at the Institute of Geoeconomics (IOG), focusing on Japan's counterstrike capabilities and defense industry policies. He has worked as a consultant at the International Institute for Strategic Studies. Mr. Inoue has written articles in academic journals and media outlets, such as the *Journal of International Security* (Japanese), *Ships of the World* (Japanese), and *The Japan Times*.

Michael Kofman is a senior fellow in the Russia and Eurasia Program at the Carnegie Endowment for International Peace, where he focuses on the Russian military, Ukrainian armed forces, and Eurasian security issues. Prior to joining Carnegie in 2023, he served as director of the Russia Studies Program at the Center for Naval Analyses, where he led a team conducting research on the capabilities, strategy, and military thought of the Russian Armed Forces. Widely recognized as one of the leading authorities on the Russian military, and the Russo-Ukrainian war, Kofman has led foundational work in the field, and is routinely cited in major publications.

Yuka Koshino is a Tokyo-based Associate Fellow at the International Institute for Strategic Studies and a doctoral student at Keio University Graduate School of Law, focusing on U.S. alliances and security, defense, and advanced technology cooperation. Previously, she served as a Research Fellow for Security and Technology Policy (2022-2023) and the inaugural Research Fellow for Japanese Security and Defense Policy (2020-2021) at the IISS, London. She is the co-author of *Japan's effectiveness as a Geo-Economic*

Actor: Navigating Great-Power Competition (2023). She holds a masters in Asian Studies from the Edmund A. Walsh School of Foreign Service at Georgetown University and a BA in Law from Keio University.

Holger Lindhardtsen is a military analyst at the Institute for Military Operations, Royal Danish Defence College. He is a member of the "Educating Future Warfighters" project at the Royal Danish Defence College, which aims to comparatively research how officers can best build competencies for future conflicts in the Nordic states. As part of this project, he has co-authored several research articles. His additional research interests include the lessons learned from special operations in the ongoing Russo-Ukrainian War, and the integration of multidomain operations in professional military education.

Dr Jeffrey W. Meiser is an Associate Professor in the Political Science Department and Global Affairs at the University of Portland in Portland, Oregon, USA. He previously worked for the U.S. Department of Defense as an Associate Professor and Director of the South and Central Asia Program at the College of International Security Affairs, at the National Defense University in Washington, DC.

William (Bill) Murray, Lieutenant Colonel, U.S. Army, is the deputy engineer for U.S. European Command, in Stuttgart, Germany. Previously, he served as the squadron commander for the 2nd Regimental Engineer Squadron, 2nd Cavalry Regiment in Vilseck, Germany. He commanded the 42nd Clearance Company in Ghazni, Afghanistan in 2012–13 and he was an engineer platoon leader in 2–7IN, in Hit, Iraq in 2007–08. LTC William Murray is the recipient of the United States Army Europe MacArthur Award for Leadership 2012 and Commander for the Best Engineer Company in the Army "Itschner Award" for 2012. He graduated in 2005 from the U.S. Military Academy, West Point, New York and is a 2017 graduate of the School of Advanced Military Studies, Command and General Staff College in Fort Leavenworth, Kansas.

William F. (Wilf) Owen is a military theorist, writer, and editor whose career spans active service, defense consulting, and strategic scholarship. He served twelve years in the British Army in both regular and reserve infantry and intelligence roles, experience that informs his work on doctrine, capability, and land warfare. After leaving the Army, Owen worked on defense advisory and security projects in West Africa, the Middle East, and the Far East, focusing on command, organization, and operational readiness. His work emphasizes preparing land forces for realistic combat conditions through doctrine and structures grounded in operational experience rather than technological idealism. He is the co-founder and Editor-in-Chief of *Military Strategy Magazine* (formerly *Infinity Journal*), and his writing has appeared in the *RUSI Journal*, the *British Army Review*, and other defense publications challenging prevailing ideas on maneuver warfare and force design. Wilf is the author of *Euclid's Army: Preparing Land Forces for Warfare Today* (Howgate 2025).

Bryan J. Quinn, Major, U.S. Army, is a Strategic Planner at United States European Command. He holds a B.A. and M.S. from the University of Southern Mississippi and an M.S. from the U.S. Naval War College. His assignments include tours in Korea, Germany, and Afghanistan and deployments with Operation Enduring Freedom and Resolute Support.

Robert G. Rose, Major, U.S. Army, is a Lt. Gen. James M. Dubik writing fellow and serves as the commander for Alpine Troop, 3rd Squadron, 4th Security Force Assistance Brigade. He holds an undergraduate degree from the United States Military Academy and graduate degrees from Harvard University and, as a Gates Scholar, from Cambridge University.

Dr Tim Sweijs is the Director of Research at the Hague Centre for Strategic Studies (HCSS) and a leading expert on strategy, defense, and security. His work spans political science, technology, and war studies, with publications on deterrence, coercion, emerging technologies, and the future of war. For nearly two decades, Dr Sweijs has advised international organizations, governments, and defense departments worldwide, providing expert testimony to the UN Security Council, NATO's Parliamentary Assembly, and the European and Dutch Parliaments. He teaches on the future of war at the Netherlands Defence Academy, Leiden University, and King's College London. At HCSS, he oversees the institute's research portfolio and a global network of experts. Dr. Sweijs is also a Research Affiliate at Georgia Tech, a board member of the European Initiative for Security Studies, and Scientific Advisor to the Global Commission on Responsible AI in the Military Domain.

Dr Heather P. Venable is an associate professor of Military and Security Studies in the Department of Airpower at the Air Command and Staff College where she is the course director of the Air Strategy and Operations Course.

She also teaches an elective entitled "On Killing: The Historical Experience of Combat." As a visiting professor at the US Naval Academy, she taught naval and Marine Corps history. She received her PhD in military history from Duke University. She also has attended the Space Operations Course as well as the Joint Firepower Course. She has written *How the Few Became the Proud: The Making of the Marine Corps' Mythos, 1874-1918* (2019). She is currently editing a volume entitled *The Future of Air and Space Power: Intersections of Theory and Technology.*

INDEX

A2AD, 3, 12–13, 35, 63, 83–84, 102, 123, 132, 180–81, 214, 232

active defense, 63–64, 176, 211, 264–65, 267–68, 296

ADF. *See* Australian Defense Force

AFC. *See* Army Futures Command

air defenses, 34–35, 52, 109, 111, 125, 153, 162, 166

Air Force Future Operating Concept, 25, 284

air-ground littoral, 257

AirLand Battle, 39, 60, 63–64, 83, 85, 105, 148, 176–77, 210–11, 215, 263–64, 266–67, 269

airpower, 112, 114, 148, 153, 211, 213, 243, 256, 266-267, 269, 290, 315

Air-Sea Battle, 12, 24–26, 106, 171, 180, 284

air superiority, 3, 86, 93, 96, 102, 108–10, 137, 153–55, 157, 194–95, 198, 297

all-domain operations, 35, 254

ANZUS (Australia, New Zealand, and United States), 226, 236

APS (active protection systems), 47, 187–88

Arab-Israeli War, 3, 105, 148

ARCIC. *See* Army Capabilities Integration Center

Army Capabilities Integration Center (ARCIC), 2, 11, 13–14, 19, 82, 86

Army Futures Command (AFC), 11, 78, 82, 86, 98, 202, 290, 299

Army Modernization Strategy, 177–78, 180, 288

Army's Future Combat Systems Program, 67, 72, 304

Army Operating Concept (AOC), 74, 178, 305

Army Tactical Missile System (ATACMS), 161, 178

artillery, 43–44, 48–49, 51, 54, 58, 90–91, 112, 138, 153, 155, 187–88

ATACMS (Army Tactical Missile System), 161, 178

attrition, ix, 148–67, 196, 209, 262–63, 271, 275, 281, 313

AUKUS (Australia, United Kingdom, and United States), 236

Australian Army, 226, 228–31, 234

Australian Defense Force (ADF), 6, 225–31, 235, 238

Battle of France, 86–87

Battle of Kyiv, 46

Battle of Mosul, 121, 125, 129, 296, 301, 305

BCT. *See* Brigade Combat Team

Brigade Combat Team (BCT), 2, 40–41, 43–44, 46–49, 52–57, 59–60, 278–80

British Army, 267, 314

China, 6, 22, 24, 31–32, 34–35, 83, 106–7, 132–33, 137, 170, 172–76, 180–81, 183, 192–96, 207–8, 214, 232, 274–75

China's Anti-Access Area Denial, 170, 285, 292

Chinese People's Liberation Army, 190, 214, 294

close battle, 141, 150–51, 159–60, 164–67

close combat, 3, 40–41, 137, 166, 278–79

close combat zone, 42–43, 44–45, 50, 53–54, 56, 60

close fight, 159–60, 163–65, 167, 276

Cold War, 35, 39, 191–92, 227, 229, 233, 240

combined arms, 16, 26, 43, 102, 104, 108, 110, 143, 268, 288

command, joint all-domain, 107, 145, 178, 221

command and control, xii–xiii, 20, 22, 58, 62, 75, 155, 162, 166, 200–201, 248

Command and General Staff College (CGSC), 57, 62, 68, 105, 129, 297, 311, 314

concept development, 13, 25, 35, 83, 86, 149, 215, 218, 223–24

conflict continuum, 7, 113–14

convergence, 16–17, 21, 80–81, 117, 123, 125–27, 129, 142–44, 146–47, 255, 276–78

Crimea, 10–11, 15, 124–25, 159–60, 214, 251

cross-domain operations (CDO), 5, 110, 190–91, 197–200, 202–6

decision dominance, 149, 165

decision superiority, 75, 195–96, 198, 202

deep battle, 48, 152, 155, 159, 164–65, 211, 267

defeat, 35, 78, 83, 87, 89, 98, 130, 141–43, 152, 210–11, 222, 267–68, 277–78, 291–92

defeat mechanisms, ix, 210, 215–16, 222, 273

Democratic People's Republic of Korea's (DPRK), 170, 173, 174, 182

Distributed Maritime Operations (DMO), 26, 254

DPRK (Democratic People's Republic of Korea's), 170, 173, 174, 182

drones, 2, 55, 67, 147, 155, 159–60, 162, 194, 224, 275, 278

East Asia, 169–72, 178–81, 184, 188–89, 193, 291

East China Sea, 183, 191, 195

Effects-Based Operations (EBO), 5, 212–13, 261, 269–70

electronic warfare, 20, 55, 57–58, 74, 138, 177, 184, 187, 235, 269, 275

Expeditionary Advanced Base Operations (EABO), 26, 254, 293

fait accompli, 2, 15, 35–36, 87, 137–38, 140, 180, 232

FCC (Futures and Concept Center), 11, 86

FCS. See Future Combat System

first island chain, 172, 175, 182–83

force design, 26, 279, 282, 311, 315

force development, 11, 15, 17, 19,
    22–23
Forces Command (FORSCOM), 85,
    98
force structure, 2, 26, 166–67, 211,
    231, 279, 283
forms of warfare, 280–81
Full-Spectrum Operations (FSO),
    63, 176
Future Combat System (FCS), 14,
    67, 289
Future Vertical Lift (FVL), 178

German army, 92, 94–95
Germany, 86–88, 90–94, 99, 134, 136,
    161, 210, 217, 219–21, 265, 314–15
Golan Heights, 265–66
ground combat, 44, 59, 276, 279
ground forces, 20, 33–34, 51, 57,
    64–66, 73–74, 77, 147, 153, 169–73,
    175–79, 181–89

Hezbollah-Israeli War, 96–97, 302
High Mobility Artillery Rocket
    System (HIMARS), 45, 50–51,
    161, 231
Homeland Defense, 172, 184, 302

IDF. See Israel Defense Forces
indirect approach, 109, 157–59
Indo-pacific, 4–6, 169–71, 174–75,
    177–81, 184–89, 226, 232, 234–36,
    274, 283, 295, 299, 302, 304
infantry fighting vehicles, 47, 58,
    186, 231
Information Age, 27, 29
information operations, 17, 138–39
information warfare, 25, 113, 204,
    252
Integrated Air and Missile Defense
    (IAMD), 188, 195, 199

Intelligence Surveillance and
    Reconnaissance (ISR), 41, 52, 57,
    59, 145, 160, 162, 196, 198, 200
Iraqi security forces, 121, 301
Israel, 86, 94–95, 97, 105, 151,
    209–10, 213, 217, 219–20, 222, 274,
    279
Israeli Defense Forces (IDF), 94–97,
    146 213, 219, 267, 300, 307

JACD. See Joint and Army Concepts
    Division
JADO. See joint-all domain
    operations
Japan, 5, 171, 174, 181–85, 190–99,
    201–6, 233–34, 286–88, 298, 304,
    313
Japan Ground Self-Defense Force
    (JGSDF), 182, 198–200, 204–6, 306
Japan's Self-Defense Forces (JSDF),
    182, 190–207
joint all-domain operations (JADO),
    6, 107–8, 165, 239–41, 249, 254,
    279
Joint and Army Concepts Division
    (JACD), 13–15, 19, 32
joint operations, 3, 5–7, 56, 63, 177,
    200–201, 204, 224, 227, 229, 231,
    241–42, 244–50, 252, 255–59
Joint Warfighting Concept, x, 11, 254

Korean Peninsula, 170, 173, 174,
    178, 181–82, 186

land forces, 8, 50, 125, 137, 178–79,
    181, 188–89, 234, 243, 275, 281–82
land warfare, 59, 186, 206, 311, 314
large-scale combat operations
    (LSCO), 4, 33, 84, 122, 126, 176,
    239–40, 283, 289
Liddell Hart, 262, 265

littoral operations, 26, 230, 254
Littoral Operations in a Contested
Environment (LOCE), 26, 254
long-range fires, 3, 8, 70, 101, 110,
114–15, 132, 138, 158, 266
long-range precision fires (LRPFs),
5, 40, 123, 158, 161, 177–78,
184–85, 187–88, 222, 275
long-range precision strikes, 8, 152,
155, 166–67, 280

maneuver, 4-8, 20, 26–27, 33, 104,
114–15, 141–42, 148–67, 185–86,
262–64, 271, 280–81
Marine Corps, 13–15, 26, 83, 102,
107, 184, 219, 242, 254, 301, 316
maritime domains, 231, 242, 257
MBD. See Multidomain Battle
McMaster, H.R., 11–14, 82
MDO. See multidomain operations
MDTFs. See multidomain task
forces
military strategy, 7, 109, 117, 130–31,
194, 222, 310
military thinking, ix, 106, 210, 212
mission command, xii–xiii, 20, 112,
142, 148, 262, 271
mobility, 5, 47–48, 51, 66, 122, 155,
163, 171, 187–88, 231, 234
modern warfare, 40, 49, 55, 59–60,
145, 151, 196, 206, 226, 294, 312
multidomain, 39, 104, 107, 165, 171,
218, 236–37, 241, 249, 271, 296
Multidomain Battle (MDB), 13–16,
18–19, 21–22, 26, 30, 35–36, 83,
215, 302, 304
multidomain integration, 40, 174,
217, 221, 238, 254
multidomain operations (MDO),
ix–x, xii–xiii, 2–8, 10–14, 16–37,
39–40, 42–147, 149–50, 152–320

multidomain task forces (MDTFs),
2, 4, 8, 32, 69, 83, 180, 184–85, 188,
279–80, 282, 304
multiple dilemmas, 69, 73, 115, 128,
142–43, 177
Multiple Launch Rocket System
(MLRS), 50, 161

Nansei Islands, 193–95, 197, 199, 207
National Training Center (NTC),
80–81, 99
NATO, 6–7, 15, 85, 104–5, 114–15,
130, 134, 209–10, 216–17, 221–23,
250–51, 276
network-centric warfare (NCW),
195, 210, 212–13, 215

operational art, 201–2, 271, 311
operational level, 200, 247, 269, 271
Operation Desert Storm, 109, 291
Operation Eagle Claw, 245, 248

People's Liberation Army (PLA),
xi, 23, 92, 190–91, 193–94, 205,
233–35
People's Republic of China (PRC),
24, 34, 193, 208, 232, 289
Perkins, David G., 30, 83, 250–51
PPBE (Planning, Programming,
Budgeting, and Execution),
97–98, 297
precision fires, 8, 50, 162
precision-guided munitions
(PGMs), 3, 46, 50, 54–55, 123, 125,
129, 160–62, 164
precision munitions, 49, 118, 163,
274
precision strike, 4, 8, 40–41, 54–55,
58, 118, 151, 161, 220
precision strike missile (PrSM), 32,
178, 185, 187, 289, 304

professional military education
(PME), 3, 6–8, 240–41, 312, 314

reconnaissance strike complexes,
211, 214, 224
Revolution in Military Affairs
(RMA), 27, 209, 211–12, 270, 312
Russia, 11–12, 14–15, 22–25, 31–32,
34–35, 84, 118, 120–21, 125,
132–35, 146, 153–55, 162, 164–65,
213–14, 221, 251, 278
Russia-Ukraine War, 74–75, 84, 103,
150, 153–54, 162, 305, 313

South China Sea (SCS), 4, 24, 170,
172, 174, 213–14, 224, 297, 299
South Korea, 16, 171, 173, 174, 183
Soviet Army, 42–43, 48, 265
Soviet doctrine, 42, 96, 267, 269
strategy, offset, 65, 214–15, 306

theater high-altitude air defense
(THAAD), 32
theory of failure, 4, 145–46
theory of success, 4–5, 131, 138–39,
141–42, 145, 147, 209–10, 212, 216,
222, 273
theory of the challenge, 4, 131–32,
135, 137–39, 141, 147
theory of victory, 2–3, 62–66, 71, 81,
85, 234, 276–77, 279

UGVs (unmanned ground
vehicles), 47, 52, 55
Ukraine, 4, 35, 45–48, 50–52, 55–59,
84, 114–15, 118, 120–21, 123–24,
134, 136, 148, 152–67, 274–75, 278
Unified Land Operations (ULO),
62–63, 70, 73, 79, 176, 284, 291
United Kingdom, 212, 219–21, 227,
245

United States, 24, 133–34, 136,
169–70, 172, 180, 219–21, 227,
237–38, 244–45, 313
unmanned aerial vehicles (UAV),
44, 47, 50–55, 58, 125, 199
US Air Force, 10, 12, 25, 102, 109,
137, 201, 254, 284
US Army, xii–xiii, 2–3, 12–14, 39–43,
47–50, 52–53, 56–57, 59–61, 63–64,
68–70, 123, 126–27, 129–32,
141–42, 175–76, 180–81, 183–85,
271–73, 298–300, 304–5

warfare
attritional, 6, 8, 155, 262
maneuver. See Maneuver
network-centric, 195, 210, 212–13
positional, 8, 115, 281, 297
temporal, 5, 205–6
western military thinking, 102, 105,
164, 166
World War I, 48, 86–89, 92–93, 121,
150, 156, 229, 243
World War II, 3, 25, 27, 30, 75, 84, 86,
91, 108, 229, 244–45

zones of proximal dominance
(ZoPD), 4, 117, 122–23, 125, 129

www.ingramcontent.com/pod-product-compliance
Lightning Source LLC
Chambersburg PA
CBHW051255020426
42333CB00026B/3224